Composite and Reinforced Elements of Constructions

Composite and Reinforced Elements of Construction

Alexander L. Kalamkarov
Ecole Centrale de Paris
Laboratoire de Mécanique
des Sols, Structures et Matériaux
France

JOHN WILEY & SONS
Chichester · New York · Brisbane · Toronto · Singapore

Other Wiley Editorial Offices

John Wiley & Sons, Inc., 605 Third Avenue,
New York, NY 10158–0012, USA

Jacaranda Wiley Ltd, G.P.O. Box 859, Brisbane,
Queensland 4001, Australia

John Wiley & Sons (Canada) Ltd, 22 Worcester Road,
Rexdale, Ontario M9W 1L1, Canada

John Wiley & Sons (SEA) Pte Ltd, 37 Jalan Pemimpin 05-04,
Block B, Union Industrial Building, Singapore 2057

Library of Congress Cataloging-in-Publication Data

Kalamkarov, Alexander L.
Composite and reinforced elements of construction / Alexander L. Kalamkarov.
p. cm.
Includes bibliographical references (p.) and index.
ISBN 0 471 93593 X
1. Composite materials—Mechanical properties. 2. Composite materials—Thermomechanical properties. 3. Fibrous composites. 4. Composite construction. I. Title.
TA418.9.C6K34 1992
624.1'8—dc20 92–10150
CIP

British Library Cataloguing in Publication Data

A catalogue record for this book is available from the British Library

ISBN 0 471 93593 X

Typeset in 10/12 pt Times by Thomson Press (India) Ltd, New Delhi
Printed and bound in Great Britain by Bookcraft [Bath] Ltd.

Contents

Preface

Little excuse is necessary for a book on composite materials in general and on the mechanics of composite materials in particular. The dramatic increase in the use of composite materials and structural elements for almost all conceivable applications increasingly calls for the development of rigorous mathematical models capable of predicting their mechanical behaviour under any given set of conditions.

The composites with a regular structure that are the subject matter of this book are not only worth considering because of their numerous practical uses in a wide range of fields but also have an obvious advantage over many other composite material types in being more amenable to theoretical analysis.

A very effective tool that has been developed for the treatment of such systems is the asymptotic method of homogenization of differential equations with rapidly oscillating coefficients. The method has been given a strict mathematical justification and yields an asymptotically correct solution when used in its domain of applicability. It is by application of this method that I have discussed the elastic, heat conduction and thermoelastic properties of the composite materials and composite or structurally nonhomogeneous reinforced thin-walled elements of construction.

The book is largely based on the author's original work carried out over the last decade, and contains also the relevant results provided by the Russian researchers which, to my knowledge, have not been available in the West so far in a systematic way.

The results are presented either in numerical form or, when possible, in the form of formulas which are—or so I hope—easy to interpret and amenable to design evaluation and application.

The book is offered in the hope that it may prove useful to researchers, teachers and students in the field of composite materials as well as to practising engineers engaged in the design, fabrication and use of these materials in various fields.

It is a pleasure to acknowledge the valuable assistance of my friend, Professor Boris Kudryavtsev of Moscow Institute of Chemical Engineering.

I wish to thank Professor Dominique François, the Director of Laboratory of Materials at Ecole Centrale Paris, for his kind attention and support.

Finally, I wish to express my gratitude to the publishers for making the printing of this book possible.

ALEXANDER L. KALAMKAROV
Paris, February 1992

Introduction

If throughout history it has been that the evolution of technology has been controlled by the materials available, then it is increasingly so today, when progress in aviation, rocketry, shipbuilding, mechanical and civil engineering crucially depends on the reasonable compromise between high strength, stiffness, corrosion resistance and perhaps other material properties, on the one hand, and reduced weight and cost on the other. The use of composite materials, i.e. continuous or discontinuous objects embedded in suitable metallic or non-metallic matrices, is therefore receiving ever wider attention in these and other areas of technology owing to the inherent tailoring of the properties of these materials.

Composite materials have actually been known since well before the dawn of recorded history, for certainly it is to this class that we must relate mixtures of clay, sand, straw and other naturally occurring components, those early building materials which, however primitive, anticipated in many respects the basic composite material concepts developed on a more sophisticated level in this our technologically ambitious age.

In modern times it is the advent of composite materials such as concrete and then reinforced concrete that was destined to be a major breakthrough in construction. The development of new building materials is one of the most identifiable trends in today's technology. In a growing number of applications composite materials have come to take over the work that was previously done by metals and metal alloys. In the United States, for example, advanced non-metallic materials account for about 20% of all structural materials. In Japan, it is anticipated that the figure will rise to fully 50% somewhere around the year 2000.

For most practical purposes, a composite material may be defined to be any material that fulfils the following conditions:

(a) it is man-made;

(b) it contains two or more chemically dissimilar substances (constituents);

(c) it consists of a number of regions with distinct interfaces between them, each occupied by a different constituent;

(d) its properties are different from those of either (each) of its constituents.

Depending on the chemical nature of the matrix material, the composite materials currently being used in industry may be divided into three classes.

(1) Composite materials with a polymeric matrix (glass-, carbon- and basalt-reinforced plastics, organic plastics, filled and highly-filled plastics).

(2) Composite materials with a metal matrix. These include aluminium, magnesium and titanium with either low-plasticity reinforcements (silicon carbide, alumina, boron or carbon fibres or whiskers of high-melting-point materials) or high-plasticity reinforcements (beryllium, tungsten, molybdenum or steel wire).

(3) Composite materials with a ceramic matrix (oxides, carbides, borides, intermetallids or carbon). Reinforcements may include fibres of silicon carbide, alumina, boron, carbon, etc.).

The above composite materials have low density, high corrosion resistance and high mechanical strength.

Apart from the choice of constituents, the arrangement of the reinforcing elements is a factor affecting—indeed in some cases controlling—the properties of a particular composite material. According to the arrangement used, composite materials may be classified into two types, laminar and non-laminar (or essentially three-dimensional).

A laminar composite is a system of single plies stacked one on another, in each of which all fibres are impregnated with a binder and have the same orientation. In practice, such composites are manufactured either by the 'wet' winding of fibres (or ribbons), or by first impregnating the fibres with a binder and subsequently moulding and solidifying them.

The non-laminar composites differ in the types of three-dimensional mechanical bonds within the material and may be divided into four classes:

(1) two-filament systems (a straight warp filament and a curved weft filament);

(2) three filaments in either a rectangular or a cylindrical coordinate system;

(3) more than three filaments, some of which are mutually orthogonal in three directions, while the others lie in planes oriented to these three;

(4) three-dimensional bonds formed by whiskers or similar discrete elements resulting either from heat treatment or from other pre-manufacture treatments.

The large-scale introduction of composite materials has created a need for further progress in such classical areas of mechanics as the theory of anisotropic and non-homogeneous deformable solids, and the theory of optimization.

It is well known that, given the same external conditions, the response of a composite material differs from that of a homogeneous (or monolithic) material and so too, of course, do the strength and service life of the structural elements fabricated from these materials. It is also well known—and quite evident—that the experimental determination of the properties of composite materials for all possible reinforcement types is often impracticable because of the volume—and often the cost—of the measurements needed.

Hence the need for theoretical models capable of predicting both the average characteristics of the composite materials and the local structure of the processes occurring in them under various types of environment.

The mathematical framework from which to predict the material behaviour of a highly non-homogeneous medium is generally provided by equations, or sets of equations, with rapidly oscillating coefficients, these latter characterizing the properties of the individual phases of the composite material. Taken in this rigorous formulation, the corresponding boundary value problems are difficult to handle even with the help of a high-speed modern computer, and it is quite natural, therefore, to seek mathematical models that describe a highly non-homogeneous medium by means of equations with some averaged coefficients. Clearly, there is a necessary condition that must be fulfilled in constructing such averaged (or 'homogenized') equations, namely that the solution of the resulting boundary value problem be close enough to the solution of the original problem. To develop a rigorous mathematical model of a composite medium, it is necessary to specify—to some extent at least—the structural configuration of the problem, and since composites with a regular, or nearly regular, structure are in extremely wide use today (see, for example, Figs 0.1 and 0.2) we seem to be fully justified in choosing them as the subject matter of this book. It should be noted that although this point will be taken up later in the book, the scale of non-homogeneity in a particular problem is always assumed to be small compared with the dimensions of the solid as a whole.

As far as the mathematical description of composites with a regular structure is concerned, the essential point is that the rapidly oscillating coefficients involved in the pertinent equations are periodic functions of spatial coordinates and these equations can therefore be treated by means of the asymptotic method of homogenization, theoretically justified and well developed by a number of authors during the 1970s (Babuška 1976, 1977, Duvaut 1976, 1977, Bensoussan *et al.* 1978, Sanchez-Palencia 1980, Lions 1981; see also Bakhvalov and Panasenko 1984). The method makes it possible to predict

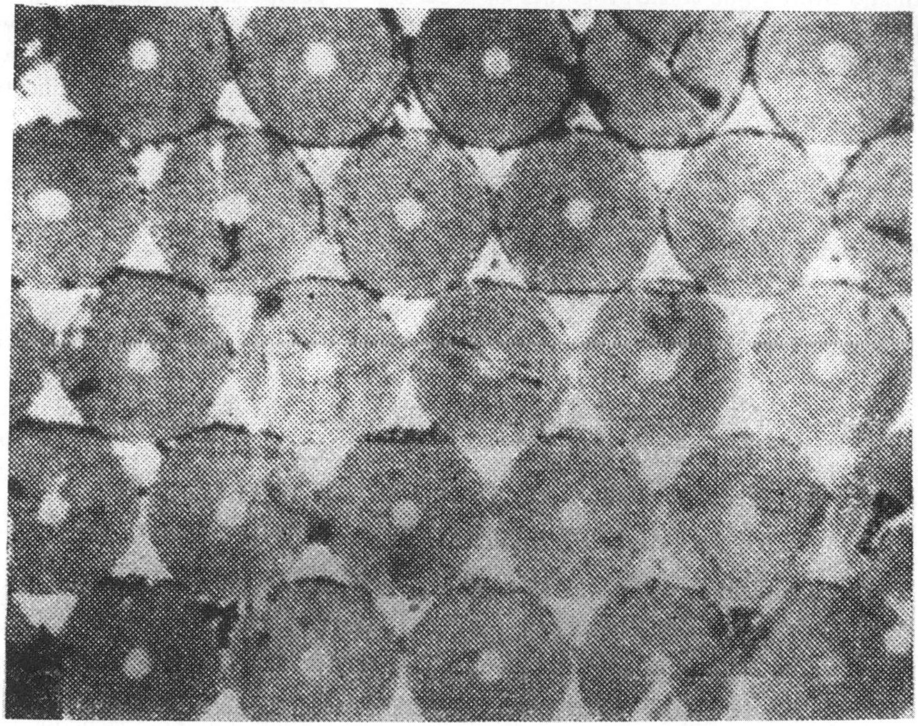

Fig. 0.1 *Structure of a fibre composite fabricated on the basis of a metal matrix by the method of continuous impregnation of the filament assembly.*

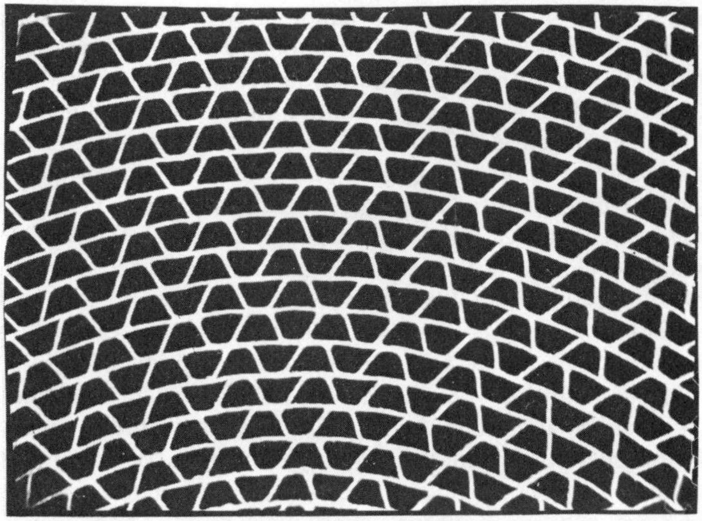

Fig. 0.2 *Honeycomb structure of a spirally-wound metallic catalytic converter* (*Ecole des Mines, Paris*).

both the overall and local properties of processes in composites by first solving the appropriate local problems set on the unit cell of the material, and subsequently solving a boundary value problem for a homogeneous (or quasi-homogeneous) material with effective material properties obtained at the first step. The mathematical justification for the homogenization method crucially depends on the proof that the solutions of the homogenized problem converge to those of the original problem in the limit as the dimensions of unit cell of the structure tend to zero; this having been proved, an error estimate can be given.

As pointed in Bakhvalov and Panasenko (1984), the method of homogenization yields an asymptotically correct representation of the solution, and the complexity of the method is fully compensated for by its accuracy which, in terms of the closeness of the approximation to the 'true' solution, cannot be achieved by any other method of solution within the accepted model.

Considering the amount of work that has been done on purely mathematical aspects of the homogenization method, the application of the method to practical situations—particularly to the mechanics of composites and of the structures of regular configurations—appears to be disproportionately limited. Most recently, however, quite a number of important results of various degrees of generality have been obtained which seem to be capable of raising the theoretical study of the mechanical behaviour of composite materials to a qualitatively higher level. Collecting these results together, with an understandable emphasis on the author's own work, was in fact the *raison d'être* of this book.

The book covers a wide range of subjects, as a view of the contents will reveal, including linear and geometrically non-linear elasticities, the heat conduction properties and thermoelasticity of composite materials and structurally non-homogeneous reinforced thin-walled elements of construction, all possessing a regular structure with a non-homogeneity scale much less than the dimensions of the solid. Considerable attention is given to the construction of the homogenized models of composite material plates and shells. We calculate both the overall (effective) properties and local properties of various types of composites and reinforced thin-walled structural members now widely used in many fields, such as laminated and fibrous composites; ribbed, wafer, honeycomb, network and angle-ply reinforced plates and shells; as well as plates and shells with corrugated surfaces.

It is not amiss to remark that the rigorous methods we present in this book provide corrections, occasionally appreciable ones, to effective moduli results that have been obtained earlier by other—approximate—methods.

Naturally, the material of this book is restricted, to some extent, by the interests of the author. There are, of course, many other topics in the field of composite materials, of much practical importance and in a very active state of development, which are not covered here but may be found discussed in detail in many other sources; see, for example, Sendeckyj (1974), Jones (1975), Christensen (1979), Cherepanov (1983), Hashin (1983), Vanin (1985), Vinson and Sierakowski (1986), Vasil'ev (1988), Vasil'ev and Tarnopol'skii (1990), François *et al.* (1991).

1 Mechanics of Inhomogeneous Deformable Solids: Basic Concepts and Problem Formulations

A major problem in continuum mechanics (or mechanics of deformable media) consists of predicting the deformations and internal forces arising in a material body subject to a given set of external forces. In what may be called the classical formulation of the problem, only forces of a purely mechanical nature are considered, but as far as composite materials and their behaviour in various environments are concerned—particularly when piezoelectric composites are to be dealt with—the effects of temperature and the presence of electromagnetic fields have to be introduced and—in mathematical terms—a coupled system of equations considered.

The mechanics of composites is a subject of considerable current interest, both theoretically and experimentally. A wide variety of heterogeneous media have been studied and the development of fundamental methods for reliably predicting their mechanical behaviour is the main objective of those involved in the field. The present chapter discusses some of the methods commonly employed in the analysis of the macroscopic behaviour of different types of heterogeneous media (or composites) and describes the relationships between the effective properties of such media and the characteristics of their individual constituent materials. We develop the subject primarily from the point of view of linear elasticity (thermoelasticity) theory, being perfectly aware, of course, that many aspects of mechanical behaviour are left unaccounted for in this approach. We refer the unsatisfied reader to monographs in which a level of higher generality is adopted.

As is customary in studies of continuum mechanics, we too express material properties in tensor form. Another feature of our discussion is the thermodynamic expressions for various combinations of the electric, magnetic, elastic and thermal energies, which enable us to determine all the equilibrium properties by simple tensor derivatives of a number of thermodynamic potentials.

1. BASIC RELATIONS OF CONTINUUM MECHANICS

1.1. Kinematics of a moving material particle. The theory of deformation

Suppose a body at time $t = 0$ occupies a region V^0 (Fig. 1.1). After a time t has elapsed, the body is deformed and occupies a region V. An arbitrary material point M^0 within

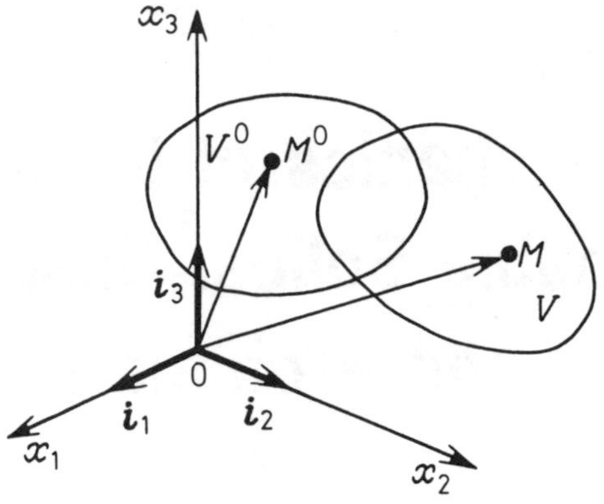

Fig. 1.1 *Deformation of a material body V.*

the body, located by the vector $\boldsymbol{x}^0(x_1^0, x_2^0, x_3^0)$, moves during the deformation to the spatial point, M, located by the vector $\boldsymbol{x}(x_1, x_2, x_3)$, both the vectors being referred to rectangular axes $x_i(i = 1, 2, 3)$ fixed in space. The coordinates $x_i^0(i = 1, 2, 3)$ specifying the point M^0 are called the material coordinates, the quantities $x_i(i = 1, 2, 3)$ are the spatial coordinates of the point M and the relations

$$\boldsymbol{x} = \boldsymbol{x}(\boldsymbol{x}^0, t) \tag{1.1}$$

or

$$x_i = x_i(x_1^0, x_2^0, x_3^0, t) \tag{1.2}$$

determine the motion of the material points of the body. The independent variables x_i^0 in (1.1) are usually referred to as the Lagrange variables.

We may now invert relations (1.1) assuming the coordinates x_i of point M in V to be unique functions of the time t and the coordinates x_i^0 of point M^0 in the reference state V^0. Clearly, the relations

$$\boldsymbol{x}^{(0)} = \boldsymbol{x}^{(0)}(\boldsymbol{x}, t) \tag{1.3}$$

or

$$x_i^{(0)} = x_i^{(0)}(x_1, x_2, x_3, t) \tag{1.4}$$

determine equally well the kinematics of the material body. The independent variables in this case are the spatial coordinates x_i (the Euler variables) corresponding to the coordinates of the particle that occupied point M^0 at the initial instant of time.

The strain tensor (describing the change in distance between various points in the deformed body) is conveniently defined by introducing the displacement vector

$$u_i = x_i - x_i^0. \tag{1.5}$$

If two material particles were located initially at points x_i^0 and $x_i^0 + \mathrm{d}x_i^0$, the squared distances between the particles will be determined by

$$\mathrm{d}s_0^2 = \mathrm{d}x_i^0\,\mathrm{d}x_i^0, \qquad \mathrm{d}s^2 = \mathrm{d}x_i\,\mathrm{d}x_i \tag{1.6}$$

for respective instants 0 and t, and the deformed state of the body is characterized by the difference

$$ds^2 - ds_0^2.$$

Working with the material variables x_i^0 and noting that

$$dx_i = \frac{\partial x_i}{\partial x_j^0} dx_j^0, \tag{1.7}$$

we find

$$\begin{aligned} ds^2 - ds_0^2 &= \frac{\partial x_i}{\partial x_j^0}\frac{\partial x_i}{\partial x_k^0} dx_j^0\, dx_k^0 - dx_j^0\, dx_j^0 \\ &= \left(\frac{\partial x_i}{\partial x_j^0}\frac{\partial x_i}{\partial x_k^0} - \delta_{jk}\right) dx_j^0\, dx_k^0 = 2\hat{\varepsilon}_{jk}\, dx_j^0\, dx_k^0, \end{aligned} \tag{1.8}$$

where δ_{jk} is the Kronecker delta and

$$\hat{\varepsilon}_{jk} = \frac{1}{2}\left(\frac{\partial x_i}{\partial x_j^0}\frac{\partial x_i}{\partial x_k^0} - \delta_{jk}\right)$$

is the strain tensor in Lagrange's representation.

Alternatively, we employ the spatial coordinates x_i as the independent variables and use the relation

$$dx_i^0 = \frac{\partial x_i^0}{\partial x_j} dx_j \tag{1.9}$$

to obtain:

$$\begin{aligned} ds^2 - ds_0^2 &= dx_j\, dx_j - \frac{\partial x_i^0}{\partial x_j}\frac{\partial x_i^0}{\partial x_k} dx_j\, dx_k \\ &= \left(\delta_{jk} - \frac{\partial x_i^0}{\partial x_j}\frac{\partial x_i^0}{\partial x_k}\right) dx_j\, dx_k = 2\varepsilon_{jk}\, dx_j\, dx_k, \end{aligned} \tag{1.10}$$

where

$$\varepsilon_{jk} = \frac{1}{2}\left(\delta_{jk} - \frac{\partial x_i^0}{\partial x_j}\frac{\partial x_i^0}{\partial x_k}\right)$$

is the strain tensor in Euler's representation.

Using (1.5) we now write

$$\hat{\varepsilon}_{jk} = \frac{1}{2}\left(\frac{\partial u_j}{\partial x_k^0} + \frac{\partial u_k}{\partial x_j^0} + \frac{\partial u_i}{\partial x_j^0}\frac{\partial u_i}{\partial x_k^0}\right) \tag{1.11}$$

and

$$\varepsilon_{jk} = \frac{1}{2}\left(\frac{\partial u_j}{\partial x_k} + \frac{\partial u_k}{\partial x_j} - \frac{\partial u_i}{\partial x_j}\frac{\partial u_i}{\partial x_k}\right) \tag{1.12}$$

for the finite strain tensor in, respectively, Lagrange's and Euler's representation, and it is worth repeating here that both expressions refer to the same Cartesian coordinate system.

Since

$$\frac{\partial x_i}{\partial x_k^0}\frac{\partial x_k^0}{\partial x_j} = \delta_{ij}, \tag{1.13}$$

the $\hat{\varepsilon}_{jk}$ and ε_{jk} tensors are readily found to be related by

$$\hat{\varepsilon}_{jk} = \varepsilon_{sn}\frac{\partial x_s}{\partial x_j^0}\frac{\partial x_n}{\partial x_k^0}, \qquad \varepsilon_{jk} = \hat{\varepsilon}_{sn}\frac{\partial x_s^0}{\partial x_j}\frac{\partial x_n^0}{\partial x_k}, \tag{1.14}$$

showing that, if one of the tensors has all its components zero, so too does the other. (The case in point is the situation in which the material medium moves as a rigid body; we then have $ds = ds_0$ and $\hat{\varepsilon}_{jk} = \varepsilon_{jk} = 0$.)

In many practical applications it proves possible to neglect the products (or squares) of derivatives in equations (1.11) and (1.12), and the following well-known infinitesimal strain expression is the result:

$$\varepsilon_{ij} = \frac{1}{2}\left(\frac{\partial u_i}{\partial x_j} + \frac{\partial u_j}{\partial x_i}\right). \tag{1.15}$$

It is often desirable that the components of the strain tensor be expressed in terms of some set of curvilinear coordinates $\theta^i (i = 1, 2, 3)$ defined by

$$x_i^0 = x_i^0(\theta^1, \theta^2, \theta^3). \tag{1.16}$$

Assuming that the $x_i^0(\theta^1, \theta^2, \theta^3)$ are single-valued continuous functions with continuous partial derivatives, the position of point M^0 in the non-deformed state is given by the vector

$$\boldsymbol{r}^0 = \boldsymbol{r}^0(\theta^1, \theta^2, \theta^3), \tag{1.17}$$

and the position this point assumes in the deformed state is determined by the vectors

$$\boldsymbol{r} = \boldsymbol{r}(\theta^1, \theta^2, \theta^3) \quad \text{or} \quad x_i = x_i(\theta^1, \theta^2, \theta^3). \tag{1.18}$$

In the initial non-deformed state

$$ds_0^2 = g_{jk}(\theta^1, \theta^2, \theta^3)\,d\theta^j\,d\theta^k, \tag{1.19}$$

where

$$g_{jk} = \boldsymbol{g}_j\boldsymbol{g}_k = \frac{\partial \boldsymbol{r}^0}{\partial \theta^j}\frac{\partial \boldsymbol{r}^0}{\partial \theta^k} = \frac{\partial x_i^0}{\partial \theta^j}\frac{\partial x_i^0}{\partial \theta^k}$$

is the metric tensor of the curvilinear coordinate system for the non-deformed state, $\boldsymbol{g}_j$ being the basis vectors of this state.

During the deformation the element $\mathrm{d}s_0$ transforms into the element $\mathrm{d}s$ and

$$\mathrm{d}s^2 = G_{jk}(\theta^1, \theta^2, \theta^3)\, \mathrm{d}\theta^j\, \mathrm{d}\theta^k, \tag{1.20}$$

where

$$G_{jk}(\theta^1, \theta^2, \theta^3) = \boldsymbol{G}_j \boldsymbol{G}_k = \frac{\partial \boldsymbol{r}}{\partial \theta^j} \frac{\partial \boldsymbol{r}}{\partial \theta^k} = \frac{\partial x_i}{\partial \theta^j} \frac{\partial x_i}{\partial \theta^k}$$

is the metric tensor and $\boldsymbol{G}_j$ are the basis vectors of the deformed state.

The Lagrange strain tensor is now defined by

$$\mathrm{d}s^2 - \mathrm{d}s_0^2 = 2\hat{\varepsilon}_{jk}\, \mathrm{d}\theta^j \mathrm{d}\theta^k$$

and, by means of (1.19) and (1.20),

$$\hat{\varepsilon}_{jk} = \tfrac{1}{2}(G_{jk} - g_{jk}). \tag{1.21}$$

Defining the displacement vector by

$$\boldsymbol{u} = u_n \boldsymbol{g}^n \tag{1.22}$$

and making use of the relations

$$\boldsymbol{r} = \boldsymbol{r}^0 + u_n \boldsymbol{g}^n$$

and

$$\boldsymbol{g}^n \boldsymbol{g}_k = \delta^n_k,$$

we get, in terms of $g_{jk} = \boldsymbol{g}_j \boldsymbol{g}_k$,

$$\begin{aligned} G_{jk} &= \frac{\partial \boldsymbol{r}}{\partial \theta^j} \frac{\partial \boldsymbol{r}}{\partial \theta^k} = \left[\frac{\partial \boldsymbol{r}^0}{\partial \theta^j} + (\nabla_j u_n)\boldsymbol{g}^n\right]\left[\frac{\partial \boldsymbol{r}^0}{\partial \theta^k} + (\nabla_k u_m)\boldsymbol{g}^m\right] \\ &= [\boldsymbol{g}_j + (\nabla_j u_n)\boldsymbol{g}^n][\boldsymbol{g}_k + (\nabla_k u_m)\boldsymbol{g}^m] \\ &= g_{jk} + \nabla_j u_k + \nabla_k u_j + (\nabla_j u_n)(\nabla_k u^n), \end{aligned} \tag{1.23}$$

where ∇_j denotes covariant differentiation involving the metric tensors g_{jk} and g^{jk}. From (1.21) and (1.23), the finite strain tensor is given by

$$\hat{\varepsilon}_{jk} = \tfrac{1}{2}(\nabla_j u_k + \nabla_k u_j + \nabla_j u_n \nabla_k u^n). \tag{1.24}$$

An alternative representation of this tensor may be obtained by writing

$$\boldsymbol{u} = U_n \boldsymbol{G}^n \tag{1.25}$$

for the displacement vector. Noting that

$$\boldsymbol{r}^0 = \boldsymbol{r} - U_n \boldsymbol{G}^n, \tag{1.26}$$

we then find

$$g_{jk} = \frac{\partial \boldsymbol{r}^0}{\partial \theta^j}\frac{\partial \boldsymbol{r}^0}{\partial \theta^k} = \left[\frac{\partial \boldsymbol{r}}{\partial \theta^j} - (\bar{\nabla}_j U_n)\boldsymbol{G}^n\right]\left[\frac{\partial \boldsymbol{r}}{\partial \theta^k} - (\bar{\nabla}_k U_m)\boldsymbol{G}^m\right]$$
$$= [\boldsymbol{G}_j - (\bar{\nabla}_j U_n)\boldsymbol{G}^n][\boldsymbol{G}_k - (\bar{\nabla}_k U_m)\boldsymbol{G}^m]$$
$$= G_{jk} - \bar{\nabla}_j U_k - \bar{\nabla}_k U_j + \bar{\nabla}_j U_n \bar{\nabla}_k U^n, \tag{1.27}$$

where $\bar{\nabla}_j$ denotes covariant differentiation involving the metric tensors G_{ij} and G^{ij}. We thus can write

$$\varepsilon_{jk} = \tfrac{1}{2}(G_{jk} - g_{jk}) = \tfrac{1}{2}(\bar{\nabla}_j U_k + \bar{\nabla}_k U_j - \bar{\nabla}_j U_n \bar{\nabla}_k U^n). \tag{1.28}$$

Employing the metric tensor g^{ij} of the non-deformed state or the metric tensor G^{ij} of the final, deformed state, it is an easy matter to deduce a mixed-type tensor from the covariant strain tensor. In particular,

$$\hat{\varepsilon}_i^k = g^{ks}\hat{\varepsilon}_{si} = \tfrac{1}{2}(g^{ks}G_{si} - \delta_i^k).$$

For small deformations we have

$$\bar{\nabla}_i(\cdots) \cong \nabla_i(\cdots), \qquad u_k \cong U_k.$$

and neglecting the terms quadratic in u reduces both (1.24) and (1.28) to

$$\varepsilon_{ik} = \tfrac{1}{2}(\nabla_i u_k + \nabla_k u_i), \tag{1.29}$$

where

$$\nabla_i u_k = \frac{\partial u_k}{\partial \theta^i} - \Gamma^s_{ki} u_s \tag{1.30}$$

and

$$\Gamma^s_{ki} = \tfrac{1}{2} g^{sp}\left(\frac{\partial g_{pk}}{\partial \theta^i} + \frac{\partial g_{pi}}{\partial \theta^k} - \frac{\partial g_{ik}}{\partial \theta^p}\right) \tag{1.31}$$

are the Christoffel symbols of the second kind.

Let us consider a system of orthogonal curvilinear coordinates with a fundamental quadratic form given by

$$ds^2 = H_i^2(d\theta^i)^2 = H_1^2(d\theta^1)^2 + H_2^2(d\theta^2)^2 + H_3^2(d\theta^3)^2. \tag{1.32}$$

The components of the metric tensor in this case are

$$\begin{aligned} &g_{11} = H_1^2, \quad g_{22} = H_2^2, \quad g_{33} = H_3^2, \quad g_{12} = g_{13} = g_{23} = 0, \\ &g^{11} = \frac{1}{H_1^2}, \quad g^{22} = \frac{1}{H_2^2}, \quad g^{33} = \frac{1}{H_3^2}, \quad g^{12} = g^{13} = g^{23} = 0, \end{aligned} \tag{1.33}$$

and for the Christoffel symbols we find

$$\Gamma^s_{ki} = 0, \quad k \neq s \neq i,$$
$$\Gamma^k_{ki} = \Gamma^k_{ik} = \frac{1}{H_k}\frac{\partial H_k}{\partial \theta^i}, \qquad \Gamma^k_{kk} = \frac{1}{H_k}\frac{\partial H_k}{\partial \theta^k}, \tag{1.34}$$
$$\Gamma^s_{kk} = -\frac{H_k}{H_s^2}\frac{\partial H_k}{\partial \theta^s}.$$

If we denote by $u_{\theta 1}$, $u_{\theta 2}$ and $u_{\theta 3}$ the physical components of the displacement vector, then

$$u_1 = H_1 u_{\theta 1}, \qquad u_2 = H_2 u_{\theta 2}, \qquad u_3 = H_3 u_{\theta 3} \tag{1.35}$$

and, in accordance with (1.29), the components of the strain tensor may be written (for small deformations) as

$$e_{11} = \frac{1}{H_1^2}\varepsilon_{11} = \frac{1}{H_1}\frac{\partial u_{\theta 1}}{\partial \theta^1} + \frac{1}{H_1 H_2}\frac{\partial H_1}{\partial \theta^2}u_{\theta 2} + \frac{1}{H_1 H_3}\frac{\partial H_1}{\partial \theta^3}u_{\theta 3},$$
$$e_{22} = \frac{1}{H_2^2}\varepsilon_{22} = \frac{1}{H_2}\frac{\partial u_{\theta 2}}{\partial \theta^2} + \frac{1}{H_2 H_1}\frac{\partial H_2}{\partial \theta^1}u_{\theta 1} + \frac{1}{H_2 H_3}\frac{\partial H_2}{\partial \theta^3}u_{\theta 3},$$
$$e_{33} = \frac{1}{H_3^2}\varepsilon_{33} = \frac{1}{H_3}\frac{\partial u_{\theta 3}}{\partial \theta^3} + \frac{1}{H_3 H_1}\frac{\partial H_3}{\partial \theta^1}u_{\theta 1} + \frac{1}{H_3 H_2}\frac{\partial H_3}{\partial \theta^2}u_{\theta 2},$$
$$e_{23} = \frac{1}{H_2 H_3}\varepsilon_{23} = \frac{1}{2}\left(\frac{1}{H_2}\frac{\partial u_{\theta 3}}{\partial \theta^2} + \frac{1}{H_3}\frac{\partial u_{\theta 2}}{\partial \theta^3} - \frac{1}{H_1 H_2}\frac{\partial H_3}{\partial \theta^2}u_{\theta 3} - \frac{1}{H_2 H_3}\frac{\partial H_2}{\partial \theta^3}u_{\theta 2}\right), \tag{1.36}$$
$$e_{13} = \frac{1}{H_1 H_3}\varepsilon_{13} = \frac{1}{2}\left(\frac{1}{H_3}\frac{\partial u_{\theta 1}}{\partial \theta^3} + \frac{1}{H_1}\frac{\partial u_{\theta 3}}{\partial \theta^1} - \frac{1}{H_1 H_3}\frac{\partial H_1}{\partial \theta^3}u_{\theta 1} - \frac{1}{H_3 H_1}\frac{\partial H_3}{\partial \theta^1}u_{\theta 3}\right),$$
$$e_{12} = \frac{1}{H_1 H_2}\varepsilon_{12} = \frac{1}{2}\left(\frac{1}{H_1}\frac{\partial u_{\theta 2}}{\partial \theta^1} + \frac{1}{H_2}\frac{\partial u_{\theta 1}}{\partial \theta^2} - \frac{1}{H_2 H_1}\frac{\partial H_2}{\partial \theta^1}u_{\theta 2} - \frac{1}{H_1 H_2}\frac{\partial H_1}{\partial \theta^2}u_{\theta 1}\right).$$

1.2. Stressed state at a point. Equilibrium equations

Select ΔS to be an elementary area within the deformed body V, and let $\boldsymbol{\Delta T}$ be the force which the part of the body on one side of ΔS exerts on the part on the other side. If ΔS is allowed to approach zero, the vector $\boldsymbol{\Delta T}/\Delta S$ becomes a vector $\boldsymbol{t}$ which describes the stresses arising in the area element ΔS and corresponds to a force acting on a unit area within the material.

Referring to Fig. 1.2, let the small tetrahedron $MABC$ have the representative point M as its apex and the edges MA, MB and MC coincide with coordinate lines $\theta^i = \text{const}$. If a linear element along a coordinate line is defined by

$$\mathrm{d}\boldsymbol{s}_i = \boldsymbol{G}_i \mathrm{d}\theta^i \qquad (i \text{ not summed}) \tag{1.37}$$

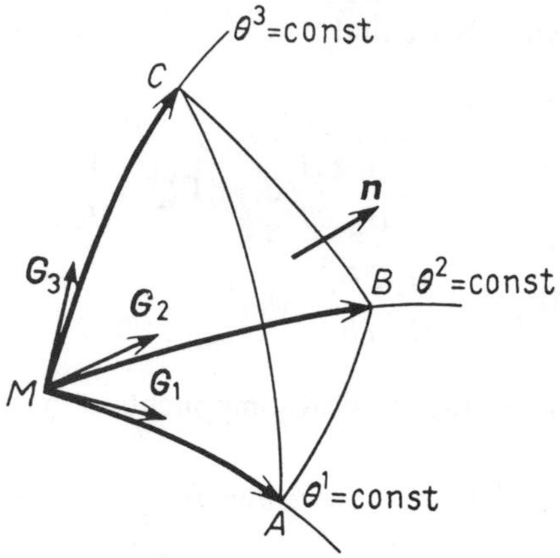

Fig. 1.2 *Elemental volume MABC in a deformed body V.*

then the elemental area on the surface $\theta^i = \text{const.}$ will be given by

$$\mathrm{d}S_i = \tfrac{1}{2}|\mathrm{d}\boldsymbol{s}_j \times \mathrm{d}\boldsymbol{s}_k| = \tfrac{1}{2}|\boldsymbol{G}_j \times \boldsymbol{G}_k|\mathrm{d}\theta^j\mathrm{d}\theta^k \quad (i \neq j \neq k)$$

which, by the definition of reciprocal coordinate vectors

$$\boldsymbol{G}_i \times \boldsymbol{G}_j = \sqrt{G}\boldsymbol{G}^k \quad (i \neq j \neq k),$$

$$G = \det G_{ij},$$

may be written as

$$\mathrm{d}S_i = \tfrac{1}{2}\sqrt{GG^{ii}}\mathrm{d}\theta^j\mathrm{d}\theta^k \quad (i \neq j \neq k, i \text{ not summed}) \tag{1.38}$$

or, in vector form,

$$\mathrm{d}\boldsymbol{S}_i = \frac{\boldsymbol{G}^i\mathrm{d}S_i}{\sqrt{G^{ii}}}. \tag{1.39}$$

Denoting by $\mathrm{d}S$ the area of the face ABC and by $\boldsymbol{n}$ the unit normal to the face, we may write

$$\boldsymbol{n}\mathrm{d}S = \sum_{i=1}^{3} \frac{\boldsymbol{G}^i\mathrm{d}S_i}{\sqrt{G^{ii}}}, \tag{1.40}$$

and setting

$$\boldsymbol{n} = n_i\boldsymbol{G}^i \tag{1.41}$$

we obtain:

$$n_i\sqrt{G^{ii}}\mathrm{d}S = \mathrm{d}S_i \quad (i \text{ not summed}). \tag{1.42}$$

Turning now to the derivation of equilibrium equations for the elemental tetrahedron $MABC$, it is expedient to introduce a stress vector $\boldsymbol{t}_i$ associated with elementary areas on the surfaces $\theta^i = \text{const}$. Clearly the vector $\boldsymbol{t}$ at point M of the deformed body is defined by

$$\boldsymbol{t}\mathrm{d}S = \boldsymbol{t}_i \,\mathrm{d}S_i \tag{1.43}$$

so that, by (1.42),

$$\boldsymbol{t} = \sum_{i=1}^{3} n_i \boldsymbol{t}_i \sqrt{G^{ii}}. \tag{1.44}$$

Since the vector $\boldsymbol{t}$ is invariant and the vector $\boldsymbol{n}$ covariant, it follows that the vector quantities $\boldsymbol{t}_i \sqrt{G^{ii}}$ (i not summed) may be represented in the form

$$\boldsymbol{t}_i \sqrt{G^{ii}} = \tau^{ij} \boldsymbol{G}_j, \tag{1.45}$$

where τ^{ij} is the contravariant stress tensor.

The mixed and covariant stress tensors will be defined by

$$\tau^i_j = G_{sj} \tau^{is}, \qquad \tau_{ij} = G_{is} \tau^s_j. \tag{1.46}$$

If the covariant and contravariant components of the stress vector $\boldsymbol{t}$ are defined by

$$\boldsymbol{t} = t_j \boldsymbol{G}^j, \qquad \boldsymbol{t} = t^j \boldsymbol{G}_j, \tag{1.47}$$

it follows from (1.44) and (1.45) that

$$t^j = \tau^{ij} n_i, \qquad t_j = n_i \tau^i_j. \tag{1.48}$$

Rewriting (1.45) in the form

$$\boldsymbol{t}_i = \sum_{j=1}^{3} \frac{\tau^{ij}}{\sqrt{G^{ii}}} \boldsymbol{G}_j = \sum_{j=1}^{3} \sqrt{\frac{G_{jj}}{G^{ii}}} \tau^{ij} \left(\frac{\boldsymbol{G}_j}{\sqrt{G_{jj}}} \right) = \sum_{j=1}^{3} \sigma_{ij} \left(\frac{\boldsymbol{G}_j}{\sqrt{G_{jj}}} \right), \tag{1.49}$$

and noting that the $\boldsymbol{G}_j / \sqrt{G_{jj}}$ are the unit vectors, we conclude that the quantities

$$\sigma_{ij} = \sqrt{\frac{G_{jj}}{G^{ii}}} \tau^{ij} \qquad (i, j \text{ not summed}) \tag{1.50}$$

are the physical components of the stress tensor at the elementary area the normal of which coincides with the vector $\boldsymbol{G}_i$.

In a body occupying a volume V, bounded by a surface S, the equations of equilibrium of the volume are

$$\iint_S \boldsymbol{t}\mathrm{d}S + \iiint_V \rho \boldsymbol{P} \mathrm{d}V = 0 \tag{1.51}$$

and

$$\iint_S (\boldsymbol{r} \times \boldsymbol{t}) \mathrm{d}S + \iiint_V (\boldsymbol{r} \times \rho \boldsymbol{P}) \mathrm{d}V = 0, \tag{1.52}$$

where $\rho \boldsymbol{P}$ is the body force vector and ρ the mass density in the deformed state. Equations (1.51) and (1.52) express the requirements that, respectively, the forces acting on V and the momenta of these forces should sum up to zero.

Applying the Green–Gauss theorem,

$$\iint_S F^i n_i \,\mathrm{d}S = \iiint_V \bar{\nabla}_i F^i \,\mathrm{d}V \quad (\boldsymbol{n} = n_i \boldsymbol{G}^i),$$

and using (1.47) and (1.48), the surface integrals in (1.51) and (1.52) are converted to

$$\iint_S \boldsymbol{t}\,\mathrm{d}S = \iint_S t^j \boldsymbol{G}_j \,\mathrm{d}S = \iint_S (\tau^{ij}\boldsymbol{G}_j) n_i \,\mathrm{d}S$$

$$= \iiint_V \bar{\nabla}_i(\tau^{ij}\boldsymbol{G}_j)\mathrm{d}V,$$

$$\iint_S (\boldsymbol{r} \times \boldsymbol{t})\mathrm{d}S = \iint_S (\boldsymbol{r} \times t^j \boldsymbol{G}_j)\,\mathrm{d}S = \iint_S (\boldsymbol{r} \times \tau^{ij}\boldsymbol{G}_j) n_i \mathrm{d}S$$

$$= \iiint_V \bar{\nabla}_i(\boldsymbol{r} \times \tau^{ij}\boldsymbol{G}_j)\,\mathrm{d}V,$$

giving

$$\bar{\nabla}_i(\tau^{ij}\boldsymbol{G}_j) + \rho \boldsymbol{P} = 0 \tag{1.53}$$

and

$$\bar{\nabla}_i(\boldsymbol{r} \times \tau^{ij}\boldsymbol{G}_j) + (\boldsymbol{r} \times \rho \boldsymbol{P}) = 0 \tag{1.54}$$

in view of the arbitrariness of the volume being considered. Combining the last two equations gives

$$\begin{aligned}(\boldsymbol{G}_i \times \boldsymbol{G}_j)\tau^{ij} &= \varepsilon_{ijs}\tau^{ij}\boldsymbol{G}^s \\ &= (\tau^{23} - \tau^{32})\boldsymbol{G}^1 + (\tau^{31} - \tau^{13})\boldsymbol{G}^2 + (\tau^{12} - \tau^{21})\boldsymbol{G}^3 = 0.\end{aligned} \tag{1.55}$$

showing the symmetry of the stress tensor: $\tau^{ij} = \tau^{ji}$.

Equation (1.53) represents the vector form of the differential equations of equilibrium and may, if desired, be rewritten as

$$\frac{1}{\sqrt{G}} \frac{\partial}{\partial \theta^i}(\sqrt{G}\tau^{ij}\boldsymbol{G}_j) + \rho \boldsymbol{P} = 0, \tag{1.56}$$

using the well-known result (see Sedov 1972)

$$\bar{\nabla}_i F^i = \frac{1}{\sqrt{G}} \frac{\partial}{\partial \theta^i}(\sqrt{G} F^i).$$

Instead of the deformed volume V, the non-deformed volume V^0 may be considered in (1.51). Using

$$dV^0 = \sqrt{g/G}\,dV \quad (g = \det g_{ij}),$$

the equilibrium equations may then be written in the form

$$\nabla_i(\tau^{ij}\boldsymbol{G}_j\sqrt{G/g}) + \rho^0\boldsymbol{P}^0 = 0, \tag{1.57}$$

where $\boldsymbol{P}^0$ is the body force vector referred to the non-deformed volume.

To an accuracy of the order of strain, we have

$$G \cong g, \qquad \nabla_i \cong \bar{\nabla}, \qquad \rho^0\boldsymbol{P}^0 \cong \rho\boldsymbol{P}$$

in (1.53) and (1.57), and the equilibrium equations may be written

$$\nabla_i(\tau^{ij}\boldsymbol{G}_j) + \rho\boldsymbol{P} = 0 \tag{1.58}$$

or

$$\nabla_i\tau^{ij} + \rho P^i = 0, \tag{1.59}$$

where $\rho\boldsymbol{P} = \rho P^i\boldsymbol{G}_i$. The physical components of the stress tensor are given by

$$\sigma_{ij'} = \tau^{ij}\sqrt{g_{jj}/g^{ii}}. \tag{1.60}$$

For an orthogonal system of curvilinear coordinates with quadratic form (1.32), the physical components of the stress tensor are defined as

$$\begin{aligned} &\sigma_{11} = \tau^{11}H_1^2, \qquad \sigma_{12} = \tau^{12}H_1H_2, \qquad \sigma_{13} = \tau^{13}H_1H_3, \\ &\sigma_{22} = \tau^{22}H_2^2, \qquad \sigma_{23} = \tau^{23}H_2H_3, \qquad \sigma_{33} = \tau^{33}H_3^2, \end{aligned} \tag{1.61}$$

and expanding (1.59) in terms of these we have

$$\begin{aligned} &\frac{1}{H_1H_2H_3}\left[\frac{\partial}{\partial\theta^1}(H_2H_3\sigma_{11}) + \frac{\partial}{\partial\theta^2}(H_3H_1\sigma_{12}) + \frac{\partial}{\partial\theta^3}(H_1H_2\sigma_{13})\right] \\ &\quad + \frac{1}{H_1H_2}\frac{\partial H_1}{\partial\theta^2}\sigma_{12} + \frac{1}{H_1H_3}\frac{\partial H_1}{\partial\theta^3}\sigma_{13} - \frac{1}{H_1H_2}\frac{\partial H_2}{\partial\theta^1}\sigma_{22} - \frac{1}{H_1H_3}\frac{\partial H_3}{\partial\theta^1}\sigma_{33} + \rho P_1 = 0, \\ &\frac{1}{H_1H_2H_3}\left[\frac{\partial}{\partial\theta^1}(H_2H_3\sigma_{12}) + \frac{\partial}{\partial\theta^2}(H_3H_1\sigma_{22}) + \frac{\partial}{\partial\theta^3}(H_1H_2\sigma_{23})\right] \\ &\quad + \frac{1}{H_2H_3}\frac{\partial H_2}{\partial\theta^3}\sigma_{23} + \frac{1}{H_2H_1}\frac{\partial H_2}{\partial\theta^1}\sigma_{12} - \frac{1}{H_3H_2}\frac{\partial H_3}{\partial\theta^2}\sigma_{33} - \frac{1}{H_1H_2}\frac{\partial H_1}{\partial\theta^2}\sigma_{11} + \rho P_2 = 0, \\ &\frac{1}{H_1H_2H_3}\left[\frac{\partial}{\partial\theta^1}(H_2H_3\sigma_{13}) + \frac{\partial}{\partial\theta^2}(H_3H_1\sigma_{23}) + \frac{\partial}{\partial\theta^3}(H_1H_2\sigma_{33})\right] \\ &\quad + \frac{1}{H_3H_1}\frac{\partial H_3}{\partial\theta^1}\sigma_{13} + \frac{1}{H_3H_2}\frac{\partial H_3}{\partial\theta^2}\sigma_{23} - \frac{1}{H_1H_3}\frac{\partial H_1}{\partial\theta^3}\sigma_{11} - \frac{1}{H_2H_3}\frac{\partial H_2}{\partial\theta^3}\sigma_{22} + \rho P_3 = 0, \end{aligned} \tag{1.62}$$

where $P_i = P^iH_i$ (i not summed) are the physical components of the body force vector.

1.3. Elastic potential. Stress–strain relations in an elastic body

Let $\delta\boldsymbol{u}$ be the variation of the displacement vector $\boldsymbol{u}$. Multiplying the equation of motion (1.56) by δu and integrating over the deformed volume, since $\mathrm{d}V=\sqrt{G}\,\mathrm{d}\theta^1\mathrm{d}\theta^2\mathrm{d}\theta^3$, (1.56) becomes

$$\iiint_V \left[\frac{\partial}{\partial\theta^i}(\sqrt{G}\tau^{ij}\boldsymbol{G}_j)\delta\boldsymbol{u}+\rho\sqrt{G}\boldsymbol{P}\delta\boldsymbol{u}\right]\mathrm{d}\theta^1\mathrm{d}\theta^2\mathrm{d}\theta^3=0 \tag{1.63}$$

or

$$\iiint_V \left[\frac{\partial}{\partial\theta^i}(\sqrt{G}\tau^{ij}\boldsymbol{G}_j\delta\boldsymbol{u})-\sqrt{G}\tau^{ij}\boldsymbol{G}_j\frac{\partial}{\partial\theta^i}(\delta\boldsymbol{u})+\rho\sqrt{G}\boldsymbol{P}\delta\boldsymbol{u}\right]\mathrm{d}\theta^1\mathrm{d}\theta^2\mathrm{d}\theta=0. \tag{1.64}$$

After using the Green–Gauss theorem on the left-hand side of (1.64) we obtain:

$$\iint_S \boldsymbol{t}\delta\boldsymbol{u}\mathrm{d}S+\iiint_V \rho\boldsymbol{P}\delta\boldsymbol{u}\mathrm{d}V=\iiint_V \tau^{ij}\boldsymbol{G}_j\frac{\partial}{\partial\theta^i}(\delta\boldsymbol{u})\mathrm{d}V, \tag{1.65}$$

where we have also used the relation $n_i\tau^{ij}\boldsymbol{G}_j=\boldsymbol{t}$ in which n_i is the unit vector normal to the bounding surface S. Since τ^{ij} is a symmetric tensor, as discussed in the previous section, the integrand of the right-hand side of (1.65) can be written as

$$\tau^{ij}\boldsymbol{G}_j\frac{\partial}{\partial\theta^i}(\delta\boldsymbol{u})=\tfrac{1}{2}\tau^{ij}\left[\boldsymbol{G}_j\frac{\partial}{\partial\theta^i}(\delta\boldsymbol{u})+\boldsymbol{G}_i\frac{\partial}{\partial\theta^j}(\delta\boldsymbol{u})\right]. \tag{1.66}$$

This can be transformed by noting that the variations of the basis vectors $\boldsymbol{G}_j$ are

$$\delta\boldsymbol{G}_j=\delta\left(\boldsymbol{g}_j+\frac{\partial\boldsymbol{u}}{\partial\theta^j}\right)=\delta\left(\frac{\partial\boldsymbol{u}}{\partial\theta^j}\right)=\frac{\partial}{\partial\theta^j}(\delta\boldsymbol{u}), \tag{1.67}$$

where $\delta\boldsymbol{g}_j=0$ because the vectors $\boldsymbol{g}_j$ are independent of $\boldsymbol{u}$. From (1.28), then,

$$\begin{aligned}2\delta\varepsilon_{ij}&=\delta G_{ij}=\delta(\boldsymbol{G}_i\boldsymbol{G}_j)\\&=\delta\boldsymbol{G}_i\boldsymbol{G}_j+\boldsymbol{G}_i\delta\boldsymbol{G}_j=\boldsymbol{G}_i\frac{\partial}{\partial\theta^j}(\delta\boldsymbol{u})+\boldsymbol{G}_j\frac{\partial}{\partial\theta^i}(\delta\boldsymbol{u})\end{aligned} \tag{1.68}$$

and (1.66) changes to

$$\tau^{ij}\boldsymbol{G}_j\frac{\partial}{\partial\theta^i}(\delta\boldsymbol{u})=\tau^{ij}\delta\varepsilon_{ij}. \tag{1.69}$$

Returning to (1.65), we are now in a position to write

$$\delta^*A=\delta U, \tag{1.70}$$

where

$$\delta^* A = \iint_S \boldsymbol{t}\delta\boldsymbol{u}\mathrm{d}S + \iiint_V \rho\boldsymbol{P}\delta\boldsymbol{u}\mathrm{d}V$$

is the virtual work done by the surface and body forces along the permissible displacement $\delta\boldsymbol{u}$ and

$$\delta U = \delta \iiint_V \Phi\rho\mathrm{d}V = \iiint_V \delta\Phi\rho\mathrm{d}V.$$
$$\delta\Phi = (\tau^{ij}/\rho)\delta\varepsilon_{ij}. \tag{1.71}$$

For an elastic body the function Φ depends on the strain,

$$\Phi = \Phi(\varepsilon_{ij}), \qquad \delta\Phi = \frac{\partial\Phi}{\partial\varepsilon_{ij}}\delta\varepsilon_{ij},$$

so that

$$\tau^{ij} = \tfrac{1}{2}\rho\left(\frac{\partial\Phi}{\partial\varepsilon_{ij}} + \frac{\partial\Phi}{\partial\varepsilon_{ji}}\right). \tag{1.72}$$

The function Φ is usually referred to as the elastic potential (per unit mass of the material) and is meaningful for reversible isothermal and adiabatic processes. In the former case Φ is in fact the free energy of the body and depends on strain and temperature; in the latter case this is the internal energy, a function of strain and (constant) entropy.

Introducing the elastic potential $W = \rho^0\Phi$ for a unit volume of a non-deformed body and using $\rho^0\sqrt{g} = \rho\sqrt{G}$, equation (1.72) becomes

$$\sqrt{G}\tau^{ij} = \tfrac{1}{2}\sqrt{g}\left(\frac{\partial W}{\partial\varepsilon_{ij}} + \frac{\partial W}{\partial\varepsilon_{ji}}\right). \tag{1.73}$$

For infinitesimal deformations, $G \cong g$ and (1.73) yields

$$\tau^{ij} = \frac{1}{2}\left(\frac{\partial W}{\partial\varepsilon_{ij}} + \frac{\partial W}{\partial\varepsilon_{ji}}\right). \tag{1.74}$$

If there are no stresses and no strains in the initial (virgin) state of the body, then assuming W to be invariant, we may define it by

$$W = \tfrac{1}{2}c^{ijkl}\varepsilon_{ij}\varepsilon_{kl}, \tag{1.75}$$

where c^{ijkl} is the tensor of elastic constants with the symmetry properties

$$c^{ijkl} = c^{jikl} = c^{klij} = c^{lkij}. \tag{1.76}$$

From (1.74) and (1.75) the small deformation linear elasticity relations take the form

$$\tau^{ij} = c^{ijkl}\varepsilon_{kl}. \tag{1.77}$$

Now if we introduce the mixed tensor of elastic constants,

$$c^{ij}_{kl} = c^{ijps} g_{pk} g_{ls} \tag{1.78}$$

with symmetry properties

$$c^{ij}_{kl} = c^{ji}_{kl} = c^{ij}_{lk} = c^{ji}_{lk},$$

we see from (1.77) that

$$\tau^{ij} = c^{ij}_{kl} \varepsilon^{kl}, \qquad \tau_{ij} = c^{kl}_{ij} \varepsilon_{kl}. \tag{1.79}$$

Since covariant and contravariant tensor components are the same in a rectangular coordinate system,

$$c_{ijkl} = c^{ijkl}$$

and the linear elasticity relations may be expressed as

$$\sigma_{ij} = c_{ijkl} e_{kl}. \tag{1.80}$$

It is sometimes convenient to introduce a contracted notation and to rewrite (1.80) in the compact matrix form

$$\sigma_\alpha = c_{\alpha\beta} e_\beta \quad (\alpha, \beta = 1,2,\cdots,6), \tag{1.81}$$

where

$$\sigma_1 = \sigma_{11}, \quad \sigma_2 = \sigma_{22}, \quad \sigma_3 = \sigma_{33}, \quad \sigma_4 = \sigma_{23}, \quad \sigma_5 = \sigma_{13}, \quad \sigma_6 = \sigma_{12},$$
$$e_1 = e_{11}, \quad e_2 = e_{22}, \quad e_3 = e_{33}, \quad e_4 = 2e_{23}, \quad e_5 = 2e_{13}, \quad e_6 = 2e_{12},$$

and the two-index components $c_{\alpha\beta}$ are obtained from the components of c_{ijkl} by the following algorithm:

$$11 \to 1, \quad 22 \to 2, \quad 33 \to 3, \quad 23 \to 4, \quad 13 \to 5, \quad 12 \to 6.$$

In its most general anisotropic form the (symmetric) matrix of elastic moduli, $c_{\alpha\beta}(\alpha, \beta = 1,2,3,4,5,6)$, has 21 independent components, but this number is reduced if there are some symmetry elements in the elastic properties of the material. Consider, for example, the case of symmetry with respect to three mutually orthogonal planes x_1x_2, x_1x_3 and x_2x_3 in a Cartesian coordinate system. This class is known as orthotropy, and in this case nine independent components of $c_{\alpha\beta}$ remain:

$$c_{\alpha\beta} = \begin{pmatrix} c_{11} & c_{12} & c_{13} & 0 & 0 & 0 \\ c_{12} & c_{22} & c_{23} & 0 & 0 & 0 \\ c_{13} & c_{23} & c_{33} & 0 & 0 & 0 \\ 0 & 0 & 0 & c_{44} & 0 & 0 \\ 0 & 0 & 0 & 0 & c_{55} & 0 \\ 0 & 0 & 0 & 0 & 0 & c_{66} \end{pmatrix}. \tag{1.82}$$

The hexagonal symmetry class occurs when the material is isotropic in one of the orthogonal planes. Letting x_3 be normal to the plane of isotropy we have five independent components:

$$c_{\alpha\beta} = \begin{pmatrix} c_{11} & c_{12} & c_{13} & 0 & 0 & 0 \\ c_{12} & c_{11} & c_{13} & 0 & 0 & 0 \\ c_{13} & c_{13} & c_{33} & 0 & 0 & 0 \\ 0 & 0 & 0 & c_{44} & 0 & 0 \\ 0 & 0 & 0 & 0 & c_{44} & 0 \\ 0 & 0 & 0 & 0 & 0 & c_{66} \end{pmatrix}. \tag{1.83}$$

with

$$c_{66} = \tfrac{1}{2}(c_{11} - c_{12}).$$

Finally, if the elastic medium exhibits hexagonal symmetry with respect to any two mutually orthogonal axes, we have complete isotropy and the matrix of elastic constants reduces to

$$c_{\alpha\beta} = \begin{pmatrix} c_{11} & c_{12} & c_{12} & 0 & 0 & 0 \\ c_{12} & c_{11} & c_{12} & 0 & 0 & 0 \\ c_{12} & c_{12} & c_{11} & 0 & 0 & 0 \\ 0 & 0 & 0 & c_{44} & 0 & 0 \\ 0 & 0 & 0 & 0 & c_{44} & 0 \\ 0 & 0 & 0 & 0 & 0 & c_{44} \end{pmatrix}, \tag{1.84}$$

where c_{11}, c_{12} and c_{44} are defined to be

$$c_{11} = \lambda + 2G, \qquad c_{12} = \lambda, \qquad c_{44} = G = \tfrac{1}{2}(c_{11} - c_{12}),$$

and λ and G are the Lamé constants related by

$$E = \frac{(2G + 3\lambda)G}{(\lambda + G)}, \qquad \nu = \frac{\lambda}{2(\lambda + G)} \tag{1.85}$$

to Young's modulus E and Poisson's ratio ν. The general form of the Cartesian components of the tensor of elastic constants is thus

$$c_{ijkl} = \lambda\delta_{ij}\delta_{kl} + G(\delta_{ik}\delta_{jl} + \delta_{il}\delta_{jk}) \tag{1.86}$$

for an isotropic medium.

2. BASIC EQUATIONS OF THERMOELASTICITY AND ELECTROELASTICITY

2.1. Generalized equation of heat conduction; thermoelastic constitutive equations

It is a well-known fact of thermodynamics that irreversible heat tranfer processes produce entropy in a solid body. On the basis of the energy conservation law, the equation of entropy transfer may be written in local form as (see Nowacki 1970)

$$\frac{ds}{dt}=-\frac{1}{T}\frac{\partial q_i}{\partial x_i}+\frac{w}{T}, \tag{2.1}$$

where q_i is the heat flow density, s is the entropy density, T is the absolute temperature and w is the heat source function.

Let us apply (2.1) to the determination of the rate of change of entropy during the process of heat conduction. Rewriting (2.1) in the form

$$\frac{ds}{dt}=-\frac{\partial}{\partial x_i}\left(\frac{q_i}{T}\right)-\frac{1}{T^2}\frac{\partial T}{\partial x_i}q_i+\frac{w}{T}, \tag{2.2}$$

integrating over a volume V and using the Green–Gauss theorem yields

$$\iiint_V \frac{ds}{dt}\,dV=-\iint_S \frac{1}{T}q_i n_i ds+\iiint_V\left(-\frac{1}{T^2}\frac{\partial T}{\partial x_i}q_i+\frac{w}{T}\right)dV, \tag{2.3}$$

showing that the time rate of change of entropy is associated with two factors: the heat flow across the bounding surface S (the first term on the right-hand side of (2.3)) and the production of entropy within the volume by heat sources and heat transfer processes (the second term).

By the second law of tehrmodynamics, the right-hand side of (2.3) is non-negative and

$$\sigma=\left(-q_iF_i+\frac{w}{T}\right)=\left(-q_i\frac{1}{T^2}\frac{\partial T}{\partial x_i}+\frac{w}{T}\right)>0. \tag{2.4}$$

The thermodynamic forces

$$F_i=-\frac{1}{T^2}\frac{\partial T}{\partial x_i},$$

just introduced, are thought of as causing irreversible thermal processes in a body and have to be related to the heat flow q_i in some way or another. In the simplest—linear—case,

$$q_i=\Lambda_{ij}F_j, \tag{2.5}$$

where $\Lambda_{ij}=\Lambda_{ji}$ (Onsager's reciprocity relations), and the vector q_i is related to the temperature through

$$q_i=-\frac{\Lambda_{ij}}{T^2}\frac{\partial T}{\partial x_j}=-\lambda_{ij}\frac{\partial T}{\partial x_j}, \tag{2.6}$$

which is Fourier's law for an anisotropic medium.

To derive the constitutive relations of thermoelasticity theory we employ the thermodynamic relation

$$dU=\sigma_{ij}de_{ij}+Tds, \tag{2.7}$$

which presents the total differential of the internal energy of the body as the sum of the increment of the work of deformation and the amount of heat introduced into the volume under study. Since the internal energy is a function of strain e_{ij} and entropy s,

$$dU = \left(\frac{\partial U}{\partial e_{ij}}\right)_s de_{ij} + \left(\frac{\partial U}{\partial s}\right)_e ds, \tag{2.8}$$

and hence, by comparison with (2.7),

$$\sigma_{ij} = \left(\frac{\partial U}{\partial e_{ij}}\right)_s, \qquad T = \left(\frac{\partial U}{\partial s}\right)_e. \tag{2.9}$$

The dependence of the stresses σ_{ij} on the strains e_{ij} and the temperature T is found by considering the free energy function

$$F = U - Ts, \tag{2.10}$$

which has the same strains and temperature as its independent variables. From

$$\begin{aligned} dF &= dU - Tds - sdT = \sigma_{ij}de_{ij} - sdT, \\ dF &= \left(\frac{\partial F}{\partial e_{ij}}\right)_T de_{ij} + \left(\frac{\partial F}{\partial T}\right)_e dT \end{aligned} \tag{2.11}$$

it is seen that

$$\sigma_{ij} = \left(\frac{\partial F}{\partial e_{ij}}\right)_T, \qquad s = -\left(\frac{\partial F}{\partial T}\right)_e. \tag{2.12}$$

For small deformations and small temperature changes $\theta = T - T_0 (\theta/T_0 \ll 1)$ we may expand $F(e_{ij}, T)$ in a power series of its arguments in the neighbourhood of the virgin state and may retain only linear and quadratic terms in the expansion. Introducing the notation

$$c_{ijkl} = \frac{\partial^2 F(0, T_0)}{\partial e_{ij} \partial e_{kl}}, \qquad \beta_{ij} = -\frac{\partial^2 F(0, T_0)}{\partial e_{ij} \partial T}, \qquad m = \frac{\partial^2 F(0, T_0)}{\partial T^2}$$

and remembering that $e_{ij} = 0$ and $T = T_0$ in the virgin state, we get

$$F(e_{ij}, T) = \tfrac{1}{2} c_{ijkl} e_{ij} e_{kl} - \beta_{ij} e_{ij} \theta + \frac{m}{2} \theta^2, \tag{2.13}$$

where use has been made of the relations

$$F(0, T_0) = 0, \qquad \frac{\partial F(0, T_0)}{\partial T} = 0, \qquad \frac{\partial F(0, T_0)}{\partial e_{ij}} = 0,$$

which follow from the fact that s and σ_{ij} are initially zero.

Using (2.13) in the first part of (2.12) now yields the equation

$$\sigma_{ij} = c_{ijkl}e_{ij} - \beta_{ij}\theta, \tag{2.14}$$

which expresses the Duhamel–Neumann law for an anisotropic body. when solved for strains, (2.14) gives

$$e_{ij} = s_{ijkl}\sigma_{kl} + \alpha_{ij}^{T}\theta, \tag{2.15}$$

where s_{ijkl} are the elastic compliances, and α_{ij}^{T} are the temperature coefficients of expansion and shear related to β_{ij} by

$$\beta_{ij} = \alpha_{kl}^{T}c_{ijkl}. \tag{2.16}$$

To determine the entropy as a function of strains e_{ij} and temperature T consider the total differential of the function $s(e_{ij}, T)$:

$$\mathrm{d}s = \left(\frac{\partial s}{\partial e_{ij}}\right)_T \mathrm{d}e_{ij} + \left(\frac{\partial s}{\partial T}\right)_e \mathrm{d}T. \tag{2.17}$$

Noting that

$$\left(\frac{\partial \sigma_{ij}}{\partial T}\right)_e = -\left(\frac{\partial s}{\partial e_{ij}}\right)_T,$$

$$\frac{\partial \sigma_{ij}}{\partial T} = -\beta_{ij}, \qquad T\left(\frac{\partial s}{\partial T}\right)_e = c_e$$

where c_e denotes the specific heat at constant strain, (2.17) becomes

$$\mathrm{d}s = \beta_{ij}\mathrm{d}e_{ij} + \frac{c_e}{T}\mathrm{d}T \tag{2.18}$$

which, when integrated under the virgin state conditions $e_{ij} = 0$, $T = T_0$, gives

$$s = \beta_{ij}e_{ij} + c_e \ln(1 + \theta/T_0) \tag{2.19}$$

or, for small temperature changes ($\theta/T \ll 1$),

$$s = \beta_{ij}e_{ij} + (c_e/T_0)\theta. \tag{2.20}$$

Inserting (2.20) and (2.6) into (2.1) and linearizing with respect to T yields the generalized heat conduction equation:

$$\frac{\partial}{\partial x_i}\left(\lambda_{ij}\frac{\partial \theta}{\partial x_j}\right) - c_e\frac{\partial \theta}{\partial t} - T_0\beta_{ij}\frac{\partial e_{ij}}{\partial t} = -w. \tag{2.21}$$

The third term on the left-hand side of (2.21) couples the temperature and strain fields of the problem and makes it necessary that, simultaneously with (2.21), the equations of motion (or equilibrium) be considered. In Cartesian coordinates these are

$$\frac{\partial \sigma_{ij}}{\partial x_i} + \rho P_i = \rho\frac{\partial^2 u_i}{\partial t^2}. \tag{2.22}$$

Substituting from (2.14) and noting that

$$e_{ij} = \frac{1}{2}\left(\frac{\partial u_i}{\partial x_j} + \frac{\partial u_j}{\partial x_i}\right),$$

the following system of equations is obtained:

$$c_{ijkl}\frac{\partial^2 u_k}{\partial x_l \partial x_j} + \rho P_i = \rho\frac{\partial^2 u_i}{\partial t^2} + \beta_{ij}\frac{\partial \theta}{\partial x_j}. \tag{2.23}$$

The differential equations (2.21) and (2.23) make a complete system of equations of the coupled dynamic thermoelastic problem for an anisotropic body and enable us, in principle, to describe the deformations due to non-stationary thermal and mechanical influences and the temperature changes associated with the deformations.

Of course we have to adjoin appropriate boundary and initial conditions to the above equations. At the initial instant of time it is necessary to specify temperature, displacements and velocities at all points within the volume V, that is, we must be able to write

$$\begin{gathered} \theta(M,0) = \theta_0(M), u_i(M,0) = u_i^0(M), \\ \frac{\partial u_i(M,0)}{\partial t} = v_i^0(M) \quad (M \in V, t=0), \end{gathered} \tag{2.24}$$

with θ_0, u_i^0 and v_i^0 as known functions of the point $M \in V$.

The most widely used boundary condition for the temperature function is the heat exchange condition of the third kind (known also as Newton's law), which states that the heat flow across the surface S of a body is proportional to the difference between the temperature of the surface, $\theta(P,t)$ $(P \in S)$, and the (known) temperature of the surrounding medium, θ_m. Mathematically this is expressed by writing

$$\lambda_{ij}\frac{\partial \theta(P,t)}{\partial x_j} n_i + \alpha_S[\theta(P,t) - \theta_m(P,t)] = 0 \quad (P \in S), \tag{2.25}$$

where α_S is the coefficient of heat exchange at S and n_i is the unit vector of the outward normal to S.

Alternatively, we may impose the boundary condition of the first kind,

$$\theta(P,t) = \theta_0(P,t), \tag{2.26}$$

or the second kind,

$$-\lambda_{ij}\frac{\partial \theta(P,t)}{\partial x_j} n_i = q_0(P,t), \tag{2.27}$$

with θ_0 and q_0 known functions of the point $P \in S$.

Turning to the mechanical boundary conditions for thermoelastic problems, we must require that external loadings p_i or displacements $\tilde{u}_i$ be specified on the surface S as functions of spatial coordinates and time:

$$\sigma_{ij}n_j = p_i(P,t), \qquad u_i = \tilde{u}_i(P,t) \quad (P \in S, t > 0). \tag{2.28}$$

If there are no mechanical influences in the problem and deformations are caused by a time-varying heating or cooling of the surface S of the body, we must set $P_i = 0$ in equations (2.23) and $p_i = 0$ in the boundary conditions (2.28).

The system (2.21)–(2.22) lends itself to considerable simplification. Dropping the term $T_0\beta_{ij}(\partial e_{ij}/\partial t)$ in the left-hand side of (2.21) we obtain the uncoupled equations of dynamic thermoelasticity:

$$\frac{\partial}{\partial x_i}\left(\lambda_{ij}\frac{\partial \theta}{\partial x_j}\right) - c_e\frac{\partial \theta}{\partial t} = -w, \tag{2.29}$$

$$c_{ijkl}\frac{\partial^2 u_k}{\partial x_l \partial x_j} + \rho P_i = \rho\frac{\partial^2 u_i}{\partial t^2} + \beta_{ij}\frac{\partial \theta}{\partial x_j}, \tag{2.30}$$

which can be solved by first finding the temperature field θ and then calculating the displacements u_i from (2.30). A further simplification consists of neglecting the inertial terms in (2.30) with the result that a system of equations of the quasi-static thermoelastic problem is obtained.

The basic relations and thermoelastic equations we have developed in this section are general enough to fit any type of symmetry, or the absence of symmetry for that matter. In an isotropic medium the equations are reduced to a simpler form by setting

$$\begin{gathered} \lambda_{ij} = \delta_{ij}\lambda, \qquad \alpha^{\mathrm{T}}_{ij} = \delta_{ij}\alpha^{\mathrm{T}}, \qquad \beta_{ij} = \delta_{ij}\beta, \\ c_{ijkl} = \lambda\delta_{ij}\delta_{kl} + G(\delta_{ik}\delta_{ij} + \delta_{il}\delta_{jk}), \end{gathered} \tag{2.31}$$

where $\beta = \alpha^{\mathrm{T}}(3\lambda + 2G)$ is the linear thermal-expansion coefficient of the material.

2.2. *Electroelastic equations for a piezoelectric medium*

The theory of thermoelasticity, i.e. the merger of classical elasticity theory and the theory of heat conduction, is a good illustrative example of a general theory of coupled fields which may be, and has been, developed in the framework of the mechanics of continuous media. Another very important aspect of the theory of coupled fields in such media is associated with electroelasticity and deals with phenomena caused by interactions between electric and mechanical fields. One of these phenomena is the so-called piezoelectric effect. Discovered experimentally in 1880, the direct piezoelectric effect is concerned with the appearance of electric charges on the surface of a body subjected to mechanical loading. The effect is conveniently described by the polarization vector P_i, which represents the electric moment per unit volume or the polarization charge per unit area, and is related to the stress tensor components σ_{kl} by the linear expression

$$P_i = d_{ikl}\sigma_{kl}, \tag{2.32}$$

where d_{ikl} is the third-rank tensor of piezoelectric moduli.

A simple thermodynamic argument shows that the inverse piezoelectric effect must also exist. In this effect the electric field components and (small) strains e_{ij} are related by the linear formula

$$e_{ij} = d_{kij}E_k, \tag{2.33}$$

where the tensor d_{kij} is symmetric with respect to interchanges of the indices i and j because of the symmetry of the e_{ij} tensor. It should be noted that both the direct and inverse modes of the piezoelectric effect are only possible if there is no centre of symmetry in the crystal; otherwise all components of the d_{kij} tensor tend to zero for reasons of symmetry.

In the pyroelectric effect, which also appears only for non-centrosymmetrical crystals, it is cooling or heating which changes the polarization of a material. For a small temperature change ΔT, the change ΔP_i in polarization can be written

$$\Delta P_i = P_i \Delta T, \tag{2.34}$$

where the p_i constants are known as pyroelectric coefficients.

The relations between the mechanical, electrical and thermal properties of real piezoelectric media can be demonstrated most easily from equilibrium thermodynamics and appear as the result of applying the so-called Maxwell relations to particular forms of thermodynamic potentials. If the strains e_{ij}, the electric field E_i and the temperature $T = T_0 + \theta$ are regarded as independent variables, the dependent variables will be the stresses σ_{ij}, the electric displacement $D_i = E_i + \varepsilon_0 P_i$ (ε_0 being the dielectric constant) and the entropy s. As is shown in books dealing with thermodynamics (e.g. Zheludev 1968), the thermodynamics potential corresponding to this choice of variables is the electric Gibbs function $G_{\ni}$:

$$G_{\ni} = \mathrm{U} - E_m D_m - sT, \tag{2.35}$$

where U denotes the internal energy, the differential form of which is

$$\mathrm{d}G_{\ni} = \sigma_{ij}\mathrm{d}e_{ij} - D_m \mathrm{d}E_m - s\mathrm{d}T. \tag{2.36}$$

From (2.36) we obtain the relations

$$\sigma_{ij} = \left(\frac{\partial G_{\ni}}{\partial e_{ij}}\right)_{E,T}, \qquad D_m = -\left(\frac{\partial G_{\ni}}{\partial E_m}\right)_{e,T}, \qquad s = -\left(\frac{\partial G_{\ni}}{\partial T}\right)_{e,E}, \tag{2.37}$$

in which the terms in parentheses indicate the variables that are held constant during differentiation. Differentiating (2.37) now gives

$$\begin{gathered} \left(\frac{\partial \sigma_{ij}}{\partial E_m}\right)_{e,T} = -\left(\frac{\partial D_m}{\partial e_{ij}}\right)_{E,T}, \\ \left(\frac{\partial \sigma_{ij}}{\partial T}\right)_{e,E} = -\left(\frac{\partial s}{\partial e_{ij}}\right)_{E,T}, \qquad \left(\frac{\partial D_m}{\partial T}\right)_{e,E} = \left(\frac{\partial s}{\partial E_m}\right)_{e,T}, \end{gathered} \tag{2.38}$$

and the perfect differentials of the dependent variables are found to be

$$\begin{aligned}
\mathrm{d}\sigma_{ij} &= \left(\frac{\partial \sigma_{ij}}{\partial e_{kl}}\right)_{E,T} \mathrm{d}e_{kl} + \left(\frac{\partial \sigma_{ij}}{\partial E_m}\right)_{E,T} \mathrm{d}E_m + \left(\frac{\partial \sigma_{ij}}{\partial T}\right)_{E,e} \mathrm{d}T, \\
\mathrm{d}D_m &= \left(\frac{\partial D_m}{\partial e_{ij}}\right)_{E,T} \mathrm{d}e_{ij} + \left(\frac{\partial D_m}{\partial E_k}\right)_{e,T} \mathrm{d}E_k + \left(\frac{\partial D_m}{\partial T}\right)_{e,E} dT, \\
\mathrm{d}s &= \left(\frac{\partial s}{\partial e_{ij}}\right)_{E,T} de_{ij} + \left(\frac{\partial s}{\partial E_k}\right)_{e,T} dE_k + \left(\frac{\partial s}{\partial T}\right)_{e,E} dT.
\end{aligned} \tag{2.39}$$

The partial derivatives may be regarded as being constant and equations (2.39) become integrable if the range of variables is assumed to be narrow enough. We then obtain the constitutive relations of a deformable piezoelectric medium:

$$\begin{aligned}
\sigma_{ij} &= c_{ijkl} e_{kl} - e_{ijm} E_m - \beta_{ij}\theta, \\
D_m &= \varepsilon_{mk} E_k + e_{mij} e_{ij} + p_m \theta, \\
\Delta s &= \beta_{ij} e_{ij} + p_m E_m + \frac{\rho c}{T_0}\theta,
\end{aligned} \tag{2.40}$$

where the partial derivatives have special names as follows:

$c_{ijkl} = (\partial\sigma_{ij}/\partial e_{kl})_{E,T}$ = elastic moduli;
$e_{ijm} = -(\partial\sigma_{ij}/\partial E_m)_{e,T}$ = piezoelectric constants;
$p_m = (\partial D_m/\partial T)_{e,T}$ = pyroelectric constants;
$\varepsilon_{mk} = (\partial D_m/\partial E_k)_{e,T}$ = dielectric constants;
$\beta_{ij} = -(\partial\sigma_{ij}/\partial T)_{e,E}$ = temperature-expansion coefficients;
$(\rho c/T_0) = (\partial s/\partial T)_{e,E}$ = specific heat per unit volume divided by T_0.

When dealing with specific electroelastic problems it is convenient to rewrite (2.40) in matrix form. Using the notation of (1.81) and the matrices

$$c_{\alpha\beta} = c_{ijkl}, \qquad e_{m\alpha} = e_{mij} \qquad (i, j, k, l = 1, 2, 3;\ \alpha, \beta = 1, 2, \ldots, 6),$$

we obtain:

$$\begin{aligned}
\sigma_\alpha &= c_{\alpha\beta} e_\beta - e_{\alpha m} E_m - \beta_\alpha \theta, \\
D_m &= \varepsilon_{mk} E_k + e_{m\alpha} e_\alpha + p_m \theta, \\
\Delta s &= \beta_\alpha e_\alpha + p_m E_m + \frac{\rho c}{T_0}\theta,
\end{aligned} \tag{2.41}$$

where the temperature-expansion coefficients are redefined by

$$\beta_{ij} = \beta_\alpha \quad (i, j = 1, 2, 3;\ \alpha = 1, 2, 6).$$

Situations in which the temperature effects may be neglected are of interest. In this case equations (2.41) reduce to

$$\begin{aligned} \sigma_\alpha &= c_{\alpha\beta} e_\beta - e_{\alpha m} E_m, \\ D_m &= \varepsilon_{mk} E_k + e_{m\alpha} e_\alpha. \end{aligned} \tag{2.42}$$

A brief discussion should be given here of the number of independent coefficients that represent the elastic and electric properties of the medium in (2.42). In the most general kind of anisotropic solid (triclinic system) equations (2.42) contain 21 elastic constants, 18 piezoelectric constants and 6 dielectric constants. Crystal symmetry eliminates some possible components of the material's tensors. In particular, for polarized ceramic (6 mm symmetry class), directing the x_3 axis along the polarization direction, we have

$$c_{\alpha\beta} = \begin{pmatrix} c_{11} & c_{12} & c_{13} & 0 & 0 & 0 \\ c_{12} & c_{11} & c_{13} & 0 & 0 & 0 \\ c_{13} & c_{13} & c_{33} & 0 & 0 & 0 \\ 0 & 0 & 0 & c_{44} & 0 & 0 \\ 0 & 0 & 0 & 0 & c_{44} & 0 \\ 0 & 0 & 0 & 0 & 0 & c_{68} \end{pmatrix}, \qquad c_{66} = \tfrac{1}{2}(c_{11} - c_{12}),$$

$$e_{\alpha m} = \begin{pmatrix} 0 & 0 & e_{13} \\ 0 & 0 & e_{13} \\ 0 & 0 & e_{33} \\ 0 & e_{15} & 0 \\ e_{15} & 0 & 0 \\ 0 & 0 & 0 \end{pmatrix}, \qquad \varepsilon_{mk} = \begin{pmatrix} \varepsilon_{11} & 0 & 0 \\ 0 & \varepsilon_{11} & 0 \\ 0 & 0 & \varepsilon_{33} \end{pmatrix}. \tag{2.43}$$

3. MECHANICAL MODELS OF COMPOSITE MATERIALS

The behaviour of composite materials (or of heterogeneous media, to take a broader view) is analysed in this book primarily from the standpoint of continuum mechanics, so when dealing with a particular problem, we usually invoke an idealized geometric model of the heterogeneous system under study and proceed to obtain the theoretical predictions of the macroscopic properties of the system in terms of the geometrical and physical properties of the individual constituent materials. The model must necessarily contain piecewise-continuous material dependences in its governing relations and the basic problem of the mechanics of composites is to determine the effective properties of some idealized homogeneous medium which, in so far as its response to external influences is concerned, is equivalent to the actual heterogeneous medium. In this book we limit ourselves to elastic and thermal influences (and properties), and therefore it is from the point of view of the theories of elasticity and thermoelasticity that the subject of the behaviour of composites is developed.

3.1. Effective mechanical characteristics of composite materials

We consider a composite system as consisting of a number of regions (or phases, or components), each described by governing elasticity relations of the form

$$\sigma_{ij} = c_{ijkl}(\boldsymbol{x})e_{kl}, \tag{3.1}$$

with the functions $c_{ijkl}(\boldsymbol{x})$ continuous in or as special case, independent of the spatial coordinate $\boldsymbol{x} = (x_1, x_2, x_3)$. We denote by L the characteristic dimension of the composite system. One of the components of the system is usually referred to as the 'matrix' while the others form what is called the 'reinforcement' and are usually termed 'inclusions'. Depending on the shape and distribution of the inclusions we may distinguish, for example, laminated or fibre (filamentary) composites, each of the terms being self-explanatory, or particular composites, if all three dimensions of the inclusions are of the same order and are much less than L. The choice—or design—of a composite material depends on its intended use, of course, and it must be kept in mind that combinations that enhance a particular property often involve the degradation of another property. Fibre materials, for example, provide good strength and stiffness properties; with rigid spherical inclusions, lower price and better dynamic response characteristics are achieved.

We follow the averaging procedure described in Sendeckyj 1974 (see also Christensen 1979) in defining the effective stiffness of a composite material. Assuming that the characteristic dimension of non-homogeneity, l, is much less than L, the averaged stress and averaged strain are defined by

$$\langle \sigma_{ij} \rangle = \frac{1}{V}\int_V \sigma_{ij}(\boldsymbol{x})\mathrm{d}V, \langle e_{ij} \rangle = \frac{1}{V}\int_V e_{ij}(\boldsymbol{x})\mathrm{d}V, \tag{3.2}$$

where V is the volume of the representative element of the heterogeneous medium. The effective stiffness properties $\tilde{c}_{ijkl}$ of the linearly elastic body may now be defined through their presence in the relation

$$\langle \sigma_{ij} \rangle = \tilde{c}_{ijkl}\langle e_{kl} \rangle. \tag{3.3}$$

To perform the above operations rigorously requires a knowledge of the stress and strain fields, $\sigma_{ij}(\boldsymbol{x})$ and $e_{ij}(\boldsymbol{x})$, in the heterogeneous medium, but in some special cases considerable simplifications can be made. Consider, for example, a two-material heterogeneous system, one material of which is continuous, the other is in the form of discrete inclusion, and both materials are isotropic. In this case (see Russel and Acrivos 1972)

$$\tilde{c}_{ijkl}\langle e_{kl} \rangle = \lambda_{\mathrm{M}}\delta_{ij}\langle e_{kk} \rangle + 2G_{\mathrm{M}}\langle e_{ij} \rangle + \frac{1}{V}\int_{V_I} (\sigma_{ij} - \lambda_{\mathrm{M}}\delta_{ij}e_{kk} - 2G_{\mathrm{M}}e_{ij})\,\mathrm{d}V, \tag{3.4}$$

where λ_{M} and G_{M} are the matrix Lamé constants and V_{I} is the volume of the inclusion. Note that only conditions within the inclusion are required for the evaluation of the $\tilde{c}_{ijkl}$ tensor.

By specializing (3.4) to the case of dilute suspension conditions, that is by assuming a small value of the volume fraction of the inclusion(s), rather simple expressions for the effective shear modulus $\tilde{G}$ (or for the bulk modulus $\tilde{k}$) can be derived (Christensen

1979) in terms of the homogeneous strain in the inclusion(s). In doing this, a result due to Eshelby (1957) may be employed, namely that a single ellipsoidal inclusion embedded in an infinite elastic medium undergoes a homogeneous deformation proportional to that imposed at large distances from the inclusion.

Alternatively, we may define the effective properties $\tilde{c}_{ijkl}$ through the equality of the strain energy stored in the heterogeneous medium to that stored in the equivalent homogeneous medium. We write, namely,

$$\frac{1}{2}\int_V \sigma_{ij}e_{ij}\mathrm{d}V = \tfrac{1}{2}\tilde{c}_{ijkl}\langle e_{ij}\rangle\langle e_{kl}\rangle. \tag{3.5}$$

This energy criterion may be used to determine the effective shear modulus in an elastic medium with a low concentration of spherical inclusions. Consider an isotropic medium in a state of simple shear deformation, with displacement components specified by

$$u_1 = \tau x_1, \quad u_2 = -\tau x_2, \quad u_3 = 0 \quad (\tau = \text{const.}) \tag{3.6}$$

at large distances from a single spherical inclusion of radius a. We now calculate—and subsequently equate to each other—the deformation energies of two spherical volumes, a composite one containing the inclusion, and an equivalent homogeneous one, of radius b. Assuming that $c = (a/b)^3 \ll 1$ (a small value of the volume fraction of the inclusions), this yields

$$\frac{\tilde{G}}{G_M} = 1 - \frac{15(1-\nu_M)(1-G_I/G_M)c}{7 - 5\,\nu_M + 2(4 - 5\,\nu_M)(G_I/G_M)}, \tag{3.7}$$

where ν_M is the matrix Poisson's ratio and G_I and G_M are the shear moduli of the inclusion and matrix materials, respectively.

One more method for determining the shear and bulk moduli of an elastic heterogeneous medium is due to Hashin (1962). In a composite spheres model he proposes that a system of spherical particles of different radii is embedded in a continuous unbounded matrix phase and associated with each particular particle (of radius a) is a spherical region of the matrix phase (of radius b). The composite particles fill up the entire volume being considered, and the ratio of radii a/b is taken to be a constant for each composite sphere, independent of its absolute size. To determine the effective bulk modulus within this model, we consider the deformation of the composite spherical particle of radius b and that of the equivalent homogeneous particle of the same radius, both particles undergoing a hydrostatic stress at $r = b$. Clearly, the strain states of these two particles will be identical in a volume-averaged sense if their radial displacements are equal at $r = b$. This condition yields for the effective bulk modulus

$$\tilde{k} = k_M + \frac{(k_I - k_M)c}{1 + (1-c)[(k_I - k_M)/(k_M + \frac{4}{3}G_M)]}, \tag{3.8}$$

where k_I and k_M denote the inclusion and matrix bulk moduli, respectively.

Turning now to the effective shear modulus, it should be admitted that its determination is a rather difficult task in the composite spheres model because when simple shear-type displacement components are prescribed on the surface of a composite sphere, the resulting boundary stresses are not those corresponding to a state of simple shear stress. To obviate this difficulty, different approaches are needed for the shear modulus problem. In the so-called three-phase model, for example, a single composite spherical particle is considered. The particle is placed within an equivalent homogeneous medium the effective characteristics $\tilde{G}$ and $\tilde{k}$ of which may be found, among other possibilities, from the condition that the energy stored in the composite equals that stored in an equivalent homogeneous medium.

A brief discussion of similar results for fibre composites is now in order. It will be observed first of all that a medium containing a system of parallel fibres has symmetry properties in the plane normal to the fibre direction. Choosing this latter to coincide with the x_1 axis, the elastic relations of this transversely isotropic medium may be written as

$$\begin{aligned}
\sigma_{11} &= c_{11}e_{11} + c_{12}e_{22} + c_{12}e_{33},\\
\sigma_{22} &= c_{12}e_{11} + c_{22}e_{22} + c_{23}e_{33},\\
\sigma_{33} &= c_{12}e_{11} + c_{23}e_{22} + c_{22}e_{33},\\
\sigma_{23} &= (c_{22} - c_{23})e_{23}, \qquad \sigma_{13} = 2c_{66}e_{23}, \qquad \sigma_{12} = 2c_{66}e_{12},
\end{aligned} \tag{3.9}$$

where we have introduced the conventional two-suffix notation for the five independent stiffness constants c_{ij} relevant to this particular symmetry. Note that in practical applications of (3.9), the experimentally measurable 'engineering properties' are more convinient-ly defined by

$$\begin{aligned}
E_{11} &= c_{11} - \frac{2c_{12}^2}{c_{22} + c_{23}}, \qquad \nu_{12} = \nu_{13} = \frac{c_{12}}{c_{22} + c_{23}},\\
k_{23} &= \tfrac{1}{2}(c_{22} + c_{23}),\\
G_{12} &= G_{31} = c_{66}, \qquad G_{23} = \tfrac{1}{2}(c_{22} - c_{23}).
\end{aligned} \tag{3.10}$$

Analogous to the case of spherical inclusions, we may employ the composite cylinders model to evaluate of the effective properties of fibre-reinforced composites (Hashin and Rosen 1964). That is to say, we associate a certain outer cylinder with each particular fibre and assume the ratio a/b to be constant for all the fibres, a and b denoting the respective radii of the fibre and the cylinder. The stress–strain analysis of the resulting composite cylinder then yields the following expressions for four of the five constants involved in (3.10):

$$E_{11} = cE_{\rm I} + (1 - c)E_{\rm M} + \frac{4c(1-c)(\nu_{\rm I} - \nu_{\rm M})^2 G_{\rm M}}{(1-c)G_{\rm M}/(k_{\rm I} + G_{\rm I}/3) + cG_{\rm M}/(k_{\rm M} + G_{\rm M}/3) + 1}, \tag{3.11}$$

$$\nu_{12} = (1-c)\nu_{\rm M} + c\nu_{\rm I} + \frac{c(1-c)(\nu_{\rm I} - \nu_{\rm M})[G_{\rm M}/(k_{\rm M} + G_{\rm M}/3) - G_{\rm M}(k_{\rm I} + G_{\rm I}/3)]}{(1-c)G_{\rm M}/(k_{\rm I} + G_{\rm I}/3) + cG_{\rm M}/(k_{\rm M} + G_{\rm M}/3) + 1}, \tag{3.12}$$

$$k_{23} = k_{\rm M} + \tfrac{1}{3}G_{\rm M} + \frac{c}{1/[k_{\rm I} - k_{\rm M} + \frac{1}{3}(G_{\rm I} - G_{\rm M})] + (1-c)/(k_{\rm M} + \frac{4}{3}G_{\rm M})}, \tag{3.13}$$

$$\frac{G_{12}}{G_{\rm M}} = \frac{G_{\rm I}(1+c) + G_{\rm M}(1-c)}{G_{\rm I}(1-c) + G_{\rm M}(1+c)}, \tag{3.14}$$

where $c=(a/b)^3$ is the volume fraction of fibres and the subscripts M and I refer to the matrix and fibre materials, respectively.

To determine the shear modulus G_{23} in the plane of isotropy, again the three-phase model should be preferred—this time with cylindrical inclusions—in which all the composite cylinders except one are replaced by an equivalent homogeneous medium. For low values of $c=(a/b)^3$ we then obtain:

$$\frac{G_{23}}{G_M}=1+\frac{c}{G_M/(G_I-G_M)+(k_M+\frac{7}{3}G_M)/(2k_M+\frac{8}{3}G_M)}. \tag{3.15}$$

Up to this point we have been concerned with the elastic characteristics of a composite material. The same methods may be applied to calculate other effective properties. In what follows the thermal conductivity of a composite material will be evaluated under the assumption of steady-state temperature fields, ignoring coupling between the mechanical and thermal variables. The effective thermal conductivities of the material will be defined by

$$\langle q_i\rangle=-\tilde{\lambda}_{ij}\frac{\partial\langle\theta\rangle}{\partial x_j}, \tag{3.16}$$

where $\langle q_i\rangle$ and $\langle\theta\rangle$ are the volume averaged values of the heat flow and temperature, respectively. We consider the spherical inclusion case to fix our ideas, and employ the three-phase model in our analysis. That is, we examine the temperature field in an unbounded region that contains a single composite spherical particle of radius b consisting of a spherical inclusion of radius a plus a spherical region of matrix material; the surrounding medium is thought of as an isotropic material with an unknown effective thermal conductivity $\tilde{\lambda}$. Specifying the temperature gradient by

$$\theta\to\beta x_3 \tag{3.17}$$

at large r and taking advantage of the symmetry with respect to the x_3 axis, the problem reduces to that of determining three harmonic functions of temperature that solve the equations

$$\begin{aligned} &\nabla^2\theta_I=0, \quad 0\leqslant r\leqslant a,\\ &\nabla^2\theta_M=0, \quad a\leqslant r\leqslant b,\\ &\nabla^2\theta=0, \quad b\leqslant r<\infty, \end{aligned} \tag{3.18}$$

subject to perfect thermal contact conditions of the form

$$\begin{aligned} &\theta_I=\theta_M, \quad \lambda_I\frac{\partial\theta_I}{\partial r}=\lambda_M\frac{\partial\theta_M}{\partial r}, \quad r=a,\\ &\theta_M=\theta, \quad \lambda_M\frac{\partial\theta_M}{\partial r}=\tilde{\lambda}\frac{\partial\theta}{\partial r}, \quad r=b, \end{aligned} \tag{3.19}$$

where λ_I and λ_M are the thermal conductivities of the inclusion and matrix materials, respectively. Note also that

$$\theta \to \beta r \cos\theta \quad \text{when } r \to \infty, \tag{3.20}$$

in accordance with (3.17). Changing now to spherical coordinates with axial symmetry relative to x_3 and noting that

$$\nabla^2 = \frac{1}{r}\frac{\partial}{\partial r}\left(r^2 \frac{\partial}{\partial r}\right) + \frac{1}{r^2 \sin\theta}\frac{\partial}{\partial \theta}\left(\sin\theta \frac{\partial}{\partial \theta}\right),$$

the solution to (3.18) is found to be

$$\begin{aligned}
\theta_{\rm I} &= A_1 r \cos\theta, && 0 \leqslant r \leqslant a,\\
\theta_{\rm M} &= \left(A_2 r + \frac{B_2}{r^2}\right)\cos\theta, && a \leqslant r \leqslant b,\\
\theta &= \left(Ar + \frac{B}{r^2}\right)\cos\theta, && b \leqslant r < \infty,
\end{aligned} \tag{3.21}$$

where $A = \beta$ and the constants A_1, A_2, B_2, and B are to be determined from conditions (3.19). We obtain, as a result:

$$\tilde{\lambda} = \lambda_{\rm M}\left[1 + \frac{c}{\lambda_{\rm M}/(\lambda_{\rm I} - \lambda_{\rm M}) + (1-c)/3}\right], \tag{3.22}$$

where, as always, c is the volume fraction of the (spherical) inclusions. It is of interest to point out that (3.22) is also applicable for evaluating the dielectric permittivity of the spherical inclusion model.

To summarize: any heterogeneous medium or composite described by the governing relations (3.1) may, for the purposes of analysis, be replaced by or associated with a certain idealized homogeneous medium, describable by the same relations but with different—effective—characteristics. An important point to be made about this homogeneous medium is that its symmetry properties differ from those of the individual constituent materials so that it may be anisotropic even when these latter are perfectly isotropic. The theory based on the use of effective characteristics is often spoken of as being an effective modulus theory having good predictive capability in many composite materials mechanics problems where the distribution of stress, strain, displacement and temperature fields in individual components is of no consequence. To take account of these micro characteristics, clearly more sophisticated approaches are needed.

3.2. Effective moduli from the Hashin–Shtrikman variational principle

Understandably, it is often desirable to have a lower and an upper bound for the effective characteristics of a composite material. One possible estimate for the effective bulk modulus k and the effective shear modulus G is given by the so-called Vougt–Reuss bounds which may be derived using two minimum energy theorems, namely the theorem of minimum potential energy and the theorem of minimum complementary energy. For a macroscopically isotropic N-phase inhomogeneous medium, denoting by a subscript 'i' the partial moduli and volume fractions, we have

$$\left(\sum_{i=1}^{N}\frac{c_i}{k_i}\right)^{-1}\leqslant\tilde{k}\leqslant\left(\sum_{i=1}^{N}c_ik_i\right),$$
$$\left(\sum_{i=1}^{N}\frac{c_i}{G_i}\right)^{-1}\leqslant\tilde{G}\leqslant\left(\sum_{i=1}^{N}c_iG_i\right),\tag{3.23}$$

which in the case $N=2$ reduce to

$$\left[\frac{\gamma}{k_1}+\frac{1-\gamma}{k_2}\right]^{-1}\leqslant\tilde{k}\leqslant\gamma k_1+(1-\gamma)k_2,$$
$$\left[\frac{\gamma}{G_1}+\frac{1-\gamma}{G_2}\right]^{-1}\leqslant\tilde{G}\leqslant\gamma G_1+(1-\gamma)G_2,\tag{3.24}$$

with the obvious notation $c_1=\gamma$ and $c_2=1-\gamma$. This estimate is, however, too rough to be of any practical significance.

We discuss here a more accurate approach, based on the variational principle due to Hashin and Shtrikman (1962). We follow the line of argument adopted by Pobedrya (1984), and limit ourselves to elastic composite materials.

Let the heterogeneous body occupy a volume V and be bounded by a surface S. We consider for this body the static elasticity problem

$$\frac{\partial\sigma_{ij}}{\partial x_j}=0,\qquad \sigma_{ij}=c_{ijkl}(\boldsymbol{x})e_{kl},\qquad e_{kl}=\frac{1}{2}\left(\frac{\partial u_k}{\partial x_l}+\frac{\partial u_l}{\partial x_k}\right),\tag{3.25}$$

$$u_k|_s=v_k^{(0)},\tag{3.26}$$

where

$$\sigma_{ij}=\frac{\partial U}{\partial e_{ij}},\qquad U(e)=\tfrac{1}{2}c_{ijkl}e_{ij}e_{kl},\tag{3.27}$$

and we consider also the elasticity problem

$$\frac{\partial\sigma_{ij}^{(c)}}{\partial x_j}=0,\qquad \sigma_{ij}^{(c)}=c_{ijkl}^{(c)}e_{kl}^{(c)},$$
$$e_{kl}^{(c)}=\frac{1}{2}\left(\frac{\partial u_k^{(c)}}{\partial x_l}+\frac{\partial u_l^{(c)}}{\partial x_k}\right),\tag{3.28}$$

$$u_k^{(c)}|_s=v_k^{(0)},\tag{3.29}$$

$$\sigma_{ij}^{(c)}=\frac{\partial U^{(c)}}{\partial e_{ij}^{(c)}},\qquad U^{(c)}(e^{(c)})=\tfrac{1}{2}c_{ijkl}^{(c)}e_{kl}^{(c)}e_{ij}^{(c)},\tag{3.30}$$

for the homogeneous body imagined to occupy the same volume and referred to as the 'comparison' body.

If the solution to (3.25) and (3.26) is taken in the form

$$u_k = u_k^{(c)} + u'_k, \tag{3.31}$$

then

$$e_{kl} = e_{kl}^{(c)} + e'_{kl}.$$

We introduce the notation

$$U^{(p)}(e) = U(e) - U^{(c)}(e) = (c_{ijkl}(\boldsymbol{x}) - c_{ijkl}^{(c)})e_{ij}e_{kl} \tag{3.32}$$

and define the symmetrical second-rank polarization tensor by

$$p_{ij} = \frac{\partial U^{(p)}}{\partial e_{ij}} = \sigma_{ij} - c_{ijkl}^{(c)}e_{kl} = (c_{ijkl}(\boldsymbol{x}) - c_{ijkl}^{(c)})e_{kl}. \tag{3.33}$$

Inverting (3.3) yields

$$e_{ij} = \frac{\partial W(p)}{\partial p_{ij}} = (c_{ijkl}(\boldsymbol{x}) - c_{ijkl}^{(c)})^{-1}p_{kl} = s_{ijkl}p_{kl}, \tag{3.34}$$

where

$$W(p) = \tfrac{1}{2}s_{ijkl}p_{kl}p_{ij} = \tfrac{1}{2}p_{ij}e_{ij}, \tag{3.35}$$

and combining (3.32) and (3.34) we find

$$U(e) - \tfrac{1}{2}c_{ijkl}^{(c)}e_{ij}e_{kl} + W(p) = p_{ij}e_{ij}. \tag{3.36}$$

Now, from (3.31) and (3.33).

$$\sigma_{ij} = p_{ij} + c_{ijkl}^{(c)}e_{kl} = p_{ij} + c_{ijkl}^{(c)}e_{kl}^{(c)} + c_{ijkl}^{(c)}e'_{kl} \tag{3.37}$$

which, when inserted into (3.25), (3.26), (3.28) and (3.29), yields the following problem for the displacement perturbation u'_k:

$$\frac{\partial p_{ij}}{\partial x_j} + c_{ijkl}^{(c)}\frac{\partial^2 u'_k}{\partial x_l \partial x_k} = 0, \tag{3.38}$$

$$u'_k|_S = 0. \tag{3.39}$$

Using the condition (3.39) it can be shown that

$$\int_V \sigma_{ij}e'_{ij}\,\mathrm{d}V = \int_V \left[\frac{\partial}{\partial x_j}(\sigma_{ij}u'_i) - \frac{\partial \sigma_{ij}}{\partial x_j}u'_i\right]\mathrm{d}V$$

$$= \int_S \sigma_{ij}u'_i n_i\,\mathrm{d}S = 0, \tag{3.40}$$

so that the Lagrangian of the problem (3.25)–(3.26) becomes

$$L=\int_V U(e)\,\mathrm{d}V=\int_V [U(e)-\tfrac{1}{2}\sigma_{ij}e_{ij}+\tfrac{1}{2}(\sigma_{ij}e_{ij}-\sigma_{ij}e'_{ij})]\,\mathrm{d}V. \tag{3.41}$$

The last term in the integrand of (3.41) is now rewritten as

$$\begin{aligned}\sigma_{ij}(e_{ij}-e'_{ij})&=\sigma_{ij}e^{(c)}_{ij}=p_{ij}e^{(c)}_{ij}+c^{(c)}_{ijkl}e_{kl}e^{(c)}_{ij}\\&=c^{(c)}_{ijkl}e^{(c)}_{ij}e_{kl}+2p_{ij}e^{(c)}_{ij}+p_{ij}e'_{ij}-p_{ij}e_{ij}\end{aligned} \tag{3.42}$$

to give

$$\begin{aligned}L=\int_V &[U(e)-\tfrac{1}{2}c^{(c)}_{ijkl}e_{ij}e_{kl}-\tfrac{1}{2}p_{ij}e_{ij}\\&+\tfrac{1}{2}(c^{(c)}_{ijkl}e_{kl}e^{(c)}_{ij}+2p_{ij}e^{(c)}_{ij}+p_{ij}e'_{ij}-p_{ij}e_{ij})]\,\mathrm{d}V\end{aligned} \tag{3.43}$$

or, substituting from (3.36),

$$\begin{aligned}L&=\tfrac{1}{2}\int_V [c^{(c)}_{ijkl}e^{(c)}_{ij}e_{kl}+2p_{ij}e^{(c)}_{ij}+p_{ij}e'_{ij}-2W(p)]\,\mathrm{d}V\\&=\tfrac{1}{2}\int_V [c^{(c)}_{ijkl}e^{(c)}_{ij}e_{kl}+p_{ij}e^{(c)}_{ij}]\,\mathrm{d}V,\end{aligned} \tag{3.44}$$

where use also has been made of the relation

$$2W(p)=p_{ij}e_{ij}=p_{ij}e^{(c)}_{ij}+p_{ij}e'_{ij}.$$

If we assume that the tensor p_{ij} is independent of σ_{ij} and $c^{(c)}_{ijkl}e_{kl}$, the term $\partial p_{ij}/\partial x_j$ may be interpreted as the body force in equations (3.38) and we are going to show now that these latter express the stationarity of the functional (3.44). Taking p_{ij} and e'_{ij} as variables in this functional we obtain, using (3.33):

$$\begin{aligned}\delta L&=\tfrac{1}{2}\int_V \left[2\delta p_{ij}e^{(c)}_{ij}+\delta p_{ij}e'_{ij}+p_{ij}\delta e'_{ij}-2\frac{\partial W}{\partial p_{ij}}\delta p_{ij}\right]\mathrm{d}V\\&=\tfrac{1}{2}\int_V (\delta e'_{ij}p_{ij}-\delta p_{ij}e'_{ij})\,\mathrm{d}V.\end{aligned} \tag{3.45}$$

We have to show now that the integral in (3.45) is zero. If we transform (3.38) by multiplying by $\delta u'_i$ ($\delta u'_i|_S=0$) and integrating over the volume V we find that

$$\begin{aligned}&\int_V \left(\frac{\partial p_{ij}}{\partial x_j}\delta u'_i+c^{(c)}_{ijkl}\frac{\partial e'_{kl}}{\partial x_j}\delta u'_i\right)\mathrm{d}V\\&=-\int_V (p_{ij}\delta e'_{ij}+c^{(c)}_{ijkl}e'_{kl}\delta e'_{ij})\,\mathrm{d}V=0.\end{aligned} \tag{3.46}$$

In a similar manner, integrating over V the product of

$$\frac{\partial}{\partial x_j}(\delta p_{ij} + c^{(c)}_{ijkl}\delta e'_{kl}) = 0$$

and u'_i $(u'_i|_S = 0)$ gives

$$\int_V \left(\frac{\partial \delta p_{ij}}{\partial x_j} u'_i + c^{(c)}_{ijkl}\frac{\partial \delta e'_{kl}}{\partial x_j} u'_i\right) \mathrm{d}V = -\int_V (\delta p_{ij} e'_{ij} + c^{(c)}_{ijkl} e'_{ij}\delta e'_{kl})\,\mathrm{d}V = 0. \tag{3.47}$$

Now using the symmetry $c^{(c)}_{ijkl} = c^{(c)}_{klij}$ in (3.47) we obtain the result

$$\delta L = \tfrac{1}{2}\int_V (\delta e'_{ij} p_{ij} - \delta p_{ij} e'_{ij})\,\mathrm{d}V = 0, \tag{3.48}$$

showing that the stationarity of the functional (3.44) leads to equations (3.38) and condition (3.39) provided (we recall) that the tensor p_{ij} is independent of σ_{ij} and $c^{(c)}_{ijkl}e_{kl}$.

Two statements concerning the functional (3.44) are important here: (1) the stationary point of the functional (3.44) is a maximum if $c^{(c)}_{ijkl} < c_{ijkl}(\boldsymbol{x})$, and (2) the stationary point of the functional (3.44) is a minimum if $c^{(c)}_{ijkl} > c_{ijkl}(\boldsymbol{x})$.

To prove statement (1), consider the second variation of the functional (3.44);

$$\begin{aligned}\delta^2 L &= \int_V \left[\delta p_{ij}\delta e'_{ij} - \frac{\partial^2 W(p)}{\partial p_{ij}\partial p_{kl}}\delta p_{kl}\delta p_{ij}\right]\mathrm{d}V \\ &= \int_V [\delta p_{ij}\delta e'_{ij} - s_{ijkl}\delta p_{kl}\delta p_{ij}]\,\mathrm{d}V.\end{aligned} \tag{3.49}$$

Noting that

$$\int_V [\delta p_{ij}\delta e'_{ij} + c^{(c)}_{ijkl}\delta e'_{kl}\delta e'_{ij}]\,\mathrm{d}V = 0 \tag{3.50}$$

(which follows from the equation $\partial p_{ij}/\partial x_j + c^{(c)}_{ijkl}\,\partial\delta e'_{kl}/\partial x_j = 0$) and using the condition $u'_i|_S = 0$, (3.49) can be expressed as

$$\delta^2 L = -\int_V [c^{(c)}_{ijkl}\delta e_{kl}\delta e_{ij} + s_{ijkl}\delta p_{kl}\delta p_{ij}]\,\mathrm{d}V. \tag{3.51}$$

Now since

$$s_{ijkl}\delta p_{kl} = (c_{ijkl} - c^{(c)}_{ijkl})^{-1}(c_{mnkl} - c^{(c)}_{mnkl})\delta e_{mn} = \delta e_{ij},$$

(3.51) may be written

$$\delta^2 L = -\int_V [c^{(c)}_{ijkl}\delta e_{kl}\delta e_{ij} + (c_{ijkl} - c^{(c)}_{ijkl})\delta e_{ij}\delta e_{kl}]\mathrm{d}V, \tag{3.52}$$

indicating that

$$\delta^2 L < 0 \quad \text{if } c^{(c)}_{ijkl} < c_{ijkl} \tag{3.53}$$

and so completing the proof of statement (1). The proof of statement (2) involves more general forms of the governing relations of continuum mechanics and is somewhat too cumbersome to be reproduced here. The interested reader may consult Pobedrya (1984) for details.

We now show that the above variational principle provides a more accurate estimate for composite material properties than does the Voigh–Reuss bounds of (3.23). To this end we rewrite the functional (3.44) in the form

$$A \equiv \frac{2L}{V} = A_1 + A_2 + A_3 + A_4, \tag{3.54}$$

where we have defined

$$\begin{aligned} A_1 &= \langle c^{(c)}_{ijkl} e^{(c)}_{ij} e_{kl} \rangle, \qquad & A_2 &= 2\langle p_{ij} e^{(c)}_{ij} \rangle, \\ A_3 &= -2\langle W(p) \rangle, \qquad & A_4 &= \langle p_{ij} e'_{ij} \rangle. \end{aligned} \tag{3.55}$$

The quantities A_1, A_2 and A_3 considered as functions of the tensor p_{ij} are determined from the problem (3.28)–(3.29) for a homogeneous medium, and from the problem (3.38)–(3.39) the quantities e'_{ij} as functions of the same argument are found. With these results we shall be able to calculate the extremum of the functional $A(p)$, which will be a maximum, $A_{\max}(p)$, or a minimum, $A_{\min}(p)$, depending, as discussed earlier, on whether $c^{(c)}_{ijkl}$ is greater or, respectively, smaller than c_{ijkl}. We may then write

$$A_{\min}(p) < A^*(p) < A_{\max}(p), \tag{3.56}$$

where $A^*(p)$ is the value of the functional as calculated from the effective modulus theory.

It will be convenient for later purposes to assume that the tensor p_{ij} is constant within each of the composite phases and that the boundary conditions (3.29) of the problem (3.28) are of the form

$$u^{(c)}_k|_S = e^{(0)}_{kj} x_j, \qquad e^{(0)}_{kj} = e^{(0)}_{jk} = \text{const.} \tag{3.57}$$

Then

$$\langle p_{ij} \rangle = \sum_{\alpha=1}^{N} c_\alpha p^{(\alpha)}_{ij}, \qquad p^{(\alpha)}_{ij} = \text{const.}, \tag{3.58}$$

so that the functional A will depend on the constant quantities $p^{(\alpha)}_{ij}$.

Omitting all the intermediate steps of derivation, the extremum value of A for an elastic isotropic composite is given by

$$A = \left[k^{(c)} + \frac{1}{9}\frac{m}{(1 + a_c m)} \right] (e^{(0)}_{kk})^2 + \left[2G^{(c)} + \frac{n}{1 + b_c n} \right] e^{(0)}_{ij} e^{(0)}_{ij}, \tag{3.59}$$

where

$$\begin{aligned} a_c &= -\tfrac{1}{3}(3k^{(c)} + 4G^{(c)})^{-1}, \\ b_c &= -\tfrac{3}{5}\frac{(k^{(c)} + 2G^{(c)})}{G^{(c)}(3k^{(c)} + 4G^{(c)})}, \\ m &= \sum_{\alpha=1}^{N} c_\alpha \left[\frac{1}{9(k_\alpha - k^{(c)})} - a_c \right]^{-1}, \\ n &= \sum_{\alpha=1}^{N} c_\alpha \left[\frac{1}{2(G_\alpha - G^{(c)})} - b_c \right]^{-1}. \end{aligned}$$

On the other hand, we have

$$A^* = k^*(e_{kk}^{(0)})^2 + 2G^* e_{ij}^{(0)} e_{ij}^{(0)} \tag{3.60}$$

for the same composite. Now if $k^{(c)} = k_{\min}$ and $G^{(c)} = G_{\min}$, where $k_{\min}$ and $G_{\min}$ are the least of all the partial moduli of the composite, then $c_{ijkl}^{(c)} < c_{ijkl}(\boldsymbol{x})$ and we have

$$A_{\max} = \left[k_{\min} + \frac{1}{9} \frac{m}{(1 + a_c m)} \right] (e_{kk}^{(0)})^2 + \left[2G_{\min} + \frac{n}{(1 + b_c n)} \right] e_{ij}^{(0)} e_{ij}^{(0)}. \tag{3.61}$$

By the same token, setting $k^{(c)} = k_{\max}$ and $G^{(c)} = G_{\max}$, where $k_{\max}$ and $G_{\max}$ are the greatest of all the partial moduli, we write

$$A_{\min} = \left[k_{\max} + \frac{1}{9} \frac{m}{1 + a_c m} \right] (e_{kk}^{(0)})^2 + \left[2G_{\max} + \frac{n}{1 + b_c n} \right] e_{ij}^{(0)} e_{ij}^{(0)}. \tag{3.62}$$

Turning now to inequalities (3.56) and comparing expressions (3.60), (3.61) and (3.62) yields the so-called Hashin–Shtrikman estimates:

$$k' < k^* < k'', \quad G' < G^* < G'', \tag{3.63}$$

where

$$\begin{aligned} k' &= k_{\max} + \frac{1}{9} \frac{m}{(1 + a_c m)}, \\ k'' &= k_{\min} + \frac{1}{9} \frac{m}{(1 + a_c m)}, \\ G' &= G_{\max} + \frac{1}{2} \frac{n}{(1 + b_c n)}, \\ G'' &= G_{\min} + \frac{1}{2} \frac{n}{(1 + b_c n)}. \end{aligned} \tag{3.64}$$

3.3. Effective properties of the periodic composite structures

A heterogeneous medium, or composite, is said to have a regular periodic structure if its mechanical behaviour is described by constitutional relations of the form

$$\sigma_{ij} = c_{ijkl}(\boldsymbol{x}) e_{kl}, \tag{3.65}$$

where $c_{ijkl}(\boldsymbol{x})$ is a periodic function of the spatial coordinates $\boldsymbol{x} = (x_1, x_2, x_3)$ in the same way that

$$c_{ijkl}(\boldsymbol{x} + n_p \boldsymbol{a}_p) = c_{ijkl}(\boldsymbol{x}), \tag{3.66}$$

where the n_p are arbitrary integer numbers and the constant vectors $\boldsymbol{a}_p$ determine the period of the structure.

We start our analysis by considering an unbounded isotropic medium (the matrix) containing periodically distributed inclusions of a dissimilar material. We consider the system to be composed of equal-size parallelepipedal cells Y and assume that there is only one inclusion in each cell, the region occupied by an inclusion being denoted by I.

If the (constant) stresses acting on the matrix are related to the strains by

$$\sigma_{ij}^{(0)} = c_{ijkl}^{(\mathrm{M})} e_{kl}^{(0)}, \tag{3.67}$$

where

$$c_{ijkl}^{(\mathrm{M})} = \lambda_{\mathrm{M}} \delta_{ij} \delta_{kl} + G_{\mathrm{M}} (\delta_{ik} \delta_{jl} + \delta_{il} \delta_{jk}), \tag{3.68}$$

then the periodically distributed inclusions with elastic moduli,

$$c_{ijkl}^{(\mathrm{I})} = \lambda_{\mathrm{I}} \delta_{ij} \delta_{kl} + G_{\mathrm{I}} (\delta_{ik} \delta_{jl} + \delta_{il} \delta_{jk}), \tag{3.69}$$

perturb the stress and strain fields in such a manner that, for given strains $e_{kl}^{(0)}$, the averaged cell stresses

$$\langle \sigma_{ij} \rangle = \frac{1}{V_Y} \int_Y \sigma_{ij} \, \mathrm{d}V$$

(V_Y being the cell volume) will be defined by the relation

$$\langle \sigma_{ij} \rangle = \tilde{c}_{ijkl} e_{kl}^{(0)}, \tag{3.70}$$

where $\tilde{c}_{ijkl}$ are the effective elastic moduli.

Following Nemat-Nasser *et al.* (1982) we designate by u_i and e_{ij} the displacement and strain perturbations due to the inclusion in the cell Y and we note that

$$e_{ij} = \frac{1}{2} \left(\frac{\partial u_i}{\partial x_j} + \frac{\partial u_j}{\partial x_i} \right)$$

and that $u_i = 0$ on ∂Y.

The full stress field in the cell Y will then be given by

$$\sigma_{ij}^{(\mathrm{M})} = c_{ijkl}^{(\mathrm{M})} (e_{kl}^{(0)} + e_{kl}) \quad \text{in } Y/I, \tag{3.71}$$

$$\sigma_{ij}^{(\mathrm{I})} = c_{ijkl}^{(\mathrm{I})} (e_{kl}^{(0)} + e_{kl}) \quad \text{in } I, \tag{3.72}$$

and it follows from the equation of equilibrium that

$$\frac{\partial \sigma_{ij}^{(\mathrm{M})}}{\partial x_j} = 0 \quad \text{in } Y/I, \qquad \frac{\partial \sigma_{ij}^{(\mathrm{I})}}{\partial x_j} = 0 \quad \text{in } I. \tag{3.73}$$

The boundary conditions to be satisfied at the surface of an inclusion are as follows:

$$[c_{ijkl}^{(\mathrm{M})} (e_{kl}^{(0)} + e_{kl})^{+} - c_{ijkl}^{(\mathrm{I})} (e_{kl}^{(0)} + e_{kl})^{-}] n_j = 0, \tag{3.74}$$

$$u_i^{+} - u_i^{-} = 0 \quad \text{on } \partial I, \tag{3.75}$$

where n_i is the unit vector in the outward normal direction to the surface ∂I and the minus and plus superscripts designate the value the quantity assumed in the immediate vicinity of ∂I inside and outside the inclusion, respectively.

Equation (3.73) is conveniently solved by introducting the so-called transformation strain e^*_{kl} such that

$$c^{(M)}_{ijkl}(e^{(0)}_{kl} + e_{kl} - e^*_{kl}) = c^{(I)}_{ijkl}(e_{kl} + e^{(0)}_{kl}) \quad \text{in } I \tag{3.76}$$

and

$$e^*_{kl} = 0 \quad \text{in } Y/I. \tag{3.77}$$

With this definition the equation

$$\frac{\partial \sigma^{(I)}_{ij}}{\partial x_j} = 0 \quad \text{in } I$$

becomes

$$\frac{\partial}{\partial x_j}[c^{(M)}_{ijkl}(e^{(0)}_{kl} + e_{kl} - e^*_{kl})] = 0 \quad \text{in } I \tag{3.78}$$

or, equivalently,

$$c^{(M)}_{sjkl}\frac{\partial^2 u_k}{\partial x_l \partial x_s} = c^{(M)}_{pjmn}\frac{\partial e^*_{mn}}{\partial x_p}. \tag{3.79}$$

In view of the spatial periodicity of the displacement components u_k it is expedient that both the solution of (3.79) and the quantities e^*_{mn} in (3.79) be expanded in the Fourier series:

$$u_k(\boldsymbol{x}) = \sum_{n_p=-\infty}^{\infty} \hat{u}_k(\boldsymbol{\xi})\,\mathrm{e}^{\mathrm{i}\boldsymbol{\xi x}}, \tag{3.80}$$

$$e^*_{mn} = \sum_{n_p=-\infty}^{\infty} \hat{e}^*_{mn}(\boldsymbol{\xi})\,\mathrm{e}^{\mathrm{i}\boldsymbol{\xi x}} \tag{3.81}$$

where

$$\boldsymbol{\xi} = (\xi_1, \xi_2, \xi_3), \quad \xi_p = \frac{2\pi n_p}{\Lambda_p} \quad (p = 1, 2, 3),$$

$$\boldsymbol{\xi x} = \xi_1 x_1 + \xi_2 x_2 + \xi_3 x_3,$$

$$\sum_{n_p=-\infty}^{\infty} = \sum_{n_1=-\infty}^{\infty}\sum_{n_2=-\infty}^{\infty}\sum_{n_3=-\infty}^{\infty}, \quad \mathrm{i} = \sqrt{-1}.$$

Upon substituting into (3.79) we then obtain:

$$\hat{u}_k(\boldsymbol{\xi}) = -\mathrm{i}N_{kj}(\boldsymbol{\xi})c^{(M)}_{pjmn}\hat{e}^*_{mn}(\boldsymbol{\xi})\xi_p, \tag{3.82}$$

$$u_k(\boldsymbol{x}) = -\mathrm{i}\sum_{n_p=-\infty}^{\infty} N_{kj}(\boldsymbol{\xi})c^{(M)}_{pjmn}\hat{e}^*_{mn}(\boldsymbol{\xi})\xi_p\,\mathrm{e}^{\mathrm{i}\boldsymbol{\xi x}}, \tag{3.83}$$

where

$$N_{kj}(\boldsymbol{\xi}) = (c_{sjkl}\xi_l\xi_s)^{-1}. \tag{3.84}$$

From (3.83) the perturbed strain components can be shown to be given by

$$e_{jk} = \sum_{n_p=-\infty}^{\infty}{}' g_{jkmn}(\boldsymbol{\xi})\, \hat{e}^*_{mn}(\boldsymbol{\xi})\, \mathrm{e}^{\mathrm{i}\boldsymbol{\xi}\boldsymbol{x}}, \tag{3.85}$$

where the prime indicates that the term with $n = \sqrt{n_p n_p} = 0$ is excluded from the sum, and, in terms of $\boldsymbol{\xi}^2 = \xi_k \xi_k$,

$$\begin{aligned} g_{jkmn}(\boldsymbol{\xi}) &= \tfrac{1}{2}(N_{ks}(\boldsymbol{\xi})\xi_j + N_{js}(\boldsymbol{\xi})\xi_k) c^{(\mathrm{M})}_{psmn} \xi_p \\ &= \frac{1}{2\boldsymbol{\xi}^2}\{\xi_k(\delta_{jn}\xi_m + \delta_{jm}\xi_n) + \xi_j(\delta_{kn}\xi_m + \delta_{km}\xi_n)\} \\ &\quad - \frac{1}{(1-\nu_{\mathrm{M}})} \frac{\xi_j \xi_k \xi_m \xi_n}{\boldsymbol{\xi}^4} + \frac{\nu_{\mathrm{M}}}{1-\nu_{\mathrm{M}}} \frac{\xi_j \xi_k}{\boldsymbol{\xi}^2} \delta_{mn}, \end{aligned} \tag{3.86}$$

where ν_{M} is the Poisson's ratio of the matrix material.

Now the Fourier coefficients in (3.81) are given by

$$\hat{e}^*_{mn}(\boldsymbol{\xi}) = \frac{1}{V_Y} \int_Y e^*_{mn}(\boldsymbol{x}')\, e^{-\mathrm{i}\boldsymbol{\xi}\boldsymbol{x}'}\, \mathrm{d}\boldsymbol{x}', \tag{3.87}$$

which, when substituted into (3.85), yields

$$e_{jk}(\boldsymbol{x}) = \frac{1}{V_Y} \sum_{n_p=-\infty}^{\infty}{}' g_{jkmn}(\boldsymbol{\xi}) \int_I e^*_{mn}(\boldsymbol{x}')\, \mathrm{e}^{\mathrm{i}\boldsymbol{\xi}(\boldsymbol{x}-\boldsymbol{x}')}\, \mathrm{d}\boldsymbol{x}'. \tag{3.88}$$

At this point use is made of the relation

$$e^{(0)}_{ij} = A_{ijkl}\, e^*_{kl}(\boldsymbol{x}) - e_{ij}(\boldsymbol{x}), \tag{3.89}$$

which follows from (3.76) and in which

$$\begin{aligned} A_{ijkl} &= (c^{(\mathrm{M})}_{ijmn} - c^{(\mathrm{I})}_{ijmn})^{-1} c^{(\mathrm{M})}_{mnkl} \\ &= \frac{G_{\mathrm{M}}}{2(G_{\mathrm{M}} - G_{\mathrm{I}})} (\delta_{ik}\delta_{jl} + \delta_{il}\delta_{jk}) \\ &\quad + \frac{(G_{\mathrm{M}}\lambda_{\mathrm{I}} - G_{\mathrm{I}}\lambda_{\mathrm{M}})\delta_{ij}\delta_{kl}}{(G_{\mathrm{M}} - G_{\mathrm{I}})[3(\lambda_{\mathrm{M}} - \lambda_{\mathrm{I}}) - 2(G_{\mathrm{M}} - G_{\mathrm{I}})]}. \end{aligned} \tag{3.90}$$

Substituting (3.88) into (3.89) yields the following integral equation for the quantities $e^*_{mn}(\boldsymbol{x})$:

$$\begin{aligned} e^{(0)}_{jk} &= A_{jksl}\, e^*_{sl}(\boldsymbol{x}) \\ &\quad - \frac{1}{V_Y} \sum_{n_p=-\infty}^{\infty}{}' g_{jkmn}(\boldsymbol{\xi}) \int_I e^*_{mn}(\boldsymbol{x}')\, \mathrm{e}^{\mathrm{i}\boldsymbol{\xi}(\boldsymbol{x}-\boldsymbol{x}')} \mathrm{d}\boldsymbol{x}'. \end{aligned} \tag{3.91}$$

Deferring until later the analysis of methods for treating (3.91), we now show that to determine the effective properties of a periodic composite, all that is needed is to calculate the volume average of the transformation strain over the inclusion

$$\langle e_{ij}^{*} \rangle = \frac{1}{V_I} \int_I e_{ij}^{*}(\boldsymbol{x}) \, \mathrm{d}\boldsymbol{x}. \tag{3.92}$$

To this end we note that for the periodicity cell subjected to a uniform strain $e_{ij}^{(0)}$ and thereby to a uniform stress $\sigma_{ij}^{(0)} = c_{ijkl}^{\mathrm{M}} e_{kl}^{(0)}$, the full strain energy is given by

$$W_0 = \tfrac{1}{2} \int_Y \sigma_{ij}^{(0)} e_{ij}^{(0)} \mathrm{d}\boldsymbol{x} = \tfrac{1}{2} V_Y \delta_{ij}^{(0)} e_{ij}^{(0)} \tag{3.93}$$

in the absence of an inclusion, and by

$$W_{\mathrm{e}} = \tfrac{1}{2} \int_Y c_{ijkl}^{(\mathrm{M})} (e_{kl}^{(0)} + e_{kl} - e_{kl}^{*})(e_{ij}^{(0)} + e_{ij}) \, \mathrm{d}\boldsymbol{x} \tag{3.94}$$

in the presence of an inclusion. Recalling that $e_{kl}^{*} = 0$ in Y/I and that $u_i = 0$ on ∂Y, equation (3.94) can be rewritten as

$$W_e = W_0 - \tfrac{1}{2} \gamma \sigma_{ij}^{(0)} \langle e_{ij}^{*} \rangle \, V_Y, \tag{3.95}$$

where $\gamma = V_I / V_Y$ is evidently the fraction volume of the inclusion. On the other hand, for a homogeneous cell with effective properties $\tilde{c}_{ijkl}$,

$$W_{\mathrm{e}} = \tfrac{1}{2} V_Y \tilde{c}_{ijkl} e_{kl}^{(0)} e_{ij}^{(0)}. \tag{3.96}$$

By comparing (3.95) and (3.96), the equation for determining $\tilde{c}_{ijkl}$ is found to be

$$\tilde{c}_{ijkl} e_{kl}^{(0)} e_{ij}^{(0)} = c_{ijkl}^{(\mathrm{M})} e_{kl}^{(0)} e_{ij}^{(0)} - \gamma c_{ijkl}^{(\mathrm{M})} e_{kl}^{(0)} \langle e_{ij}^{*} \rangle, \tag{3.97}$$

where only the (average) quantity (3.92) is involved.

Returning now to (3.91), we easily reduce this integral equation to an infinite system of linear algebraic equations for the variable

$$E_{jk}^{*}(\boldsymbol{\xi}) = \frac{1}{V_I} \int_I e_{jk}^{*}(\boldsymbol{x}') e^{-\mathrm{i}\boldsymbol{\xi}\boldsymbol{x}'} \mathrm{d}\boldsymbol{x}' \tag{3.98}$$

if we multiply (3.91) by $\exp(-\mathrm{i}\boldsymbol{\xi}\boldsymbol{x})$ and integrate over the volume I. We obtain:

$$e_{ij}^{(0)} Q(\boldsymbol{\eta}) = A_{ijkl} E_{kl}^{*}(\boldsymbol{\eta}) - \gamma \sum_{n_p = 0}^{\infty} g_{ijkl}(\boldsymbol{\xi}) Q(\boldsymbol{\eta} - \boldsymbol{\xi}) E_{kl}^{*}(\boldsymbol{\xi}), \tag{3.99}$$

where

$$Q(\boldsymbol{\eta}) = \frac{g_0(-\boldsymbol{\eta})}{V}, \quad g_0(-\boldsymbol{\eta}) = \int_I \mathrm{e}^{-\mathrm{i}\boldsymbol{\eta}\boldsymbol{x}} \, \mathrm{d}\boldsymbol{x},$$

$$\eta_i = 2\pi m_i / \Lambda_i \quad (i = 1, 2, 3).$$

The solution to (3.99) having been found, (3.98) gives

$$\langle e_{ij}^{*} \rangle = E_{ij}^{*}(0). \tag{3.100}$$

An alternative method for estimating the quantity $\langle e_{ij} \rangle$ is based on the replacement of the quantity $e^*_{ij}(\boldsymbol{x}')$ in (3.91) by its average value (3.92) (Nemat-Nasser *et al.* 1982). Integrating (3.91) over I then yields

$$e^{(0)}_{ij} = (A_{ijkl} - S_{ijkl})e^*_{kl}, \tag{3.101}$$

where

$$S_{ijkl} = \sum_{n_p=-\infty}^{\infty}{}' \; P(\xi)\, g_{ijkl}(\xi),$$
$$P(\xi) = \frac{1}{V_I V_Y} g_0(\xi) g_0(-\xi). \tag{3.102}$$

Equation (3.101) is now inverted, and e^*_{kl}, expressed in terms of $e^{(0)}_{ij}$, is substituted into (3.97) to give

$$[\tilde{c}_{ijkl} - c^{(M)}_{ijkl} + \gamma c^{(M)}_{ijmn}(A_{mnkl} - S_{mnkl})^{-1}]e^{(0)}_{kl} e^{(0)}_{ij} = 0. \tag{3.103}$$

By making use of the arbitrariness of $e^{(0)}_{ij}$ we conclude that

$$\tilde{c}_{ijkl} = c^{(M)}_{ijkl} - \gamma c^{(M)}_{ijmn}(A_{mnkl} - S_{mnkl})^{-1}. \tag{3.104}$$

Nemat-Nasser *et al.* (1982) used this formula to estimate the effective bulk modulus $\tilde{k}$ and the effective shear modulus $\tilde{G}$ of an isotropic elastic medium with a periodic array of spherical cavities. Figs 3.1 and 3.2, reproduced from this work, show the relations $\tilde{k}/k_M$ and $\tilde{G}/G_M$ as functions of the parameter γ for a cubic periodicity cell, the quantities $\tilde{k}$ and $\tilde{G}$ being defined by the relations $\tilde{k} = (1/9)\tilde{c}_{iijj}$ and $\tilde{G} = \tilde{c}_{2323} = \tilde{c}_{3131}$, respectively.

Nunan and Keller (1984) discussed methods for calculating the effective elastic moduli in an isotropic medium with a periodic array of rigid spherical inclusions. Using (3.4) and the fundamental periodic solution of the equations of the theory of elasticity, Nunan and Keller obtained the following expression for the effective elastic constants:

$$\tilde{c}_{ijkl} = (\lambda_M + G_M f)\delta_{ij}\delta_{kl} + G_M(1+\beta)(\delta_{ik}\delta_{jl} + \delta_{il}\delta_{jk}) + G_M(\alpha - \beta)\delta_{ijkl},$$

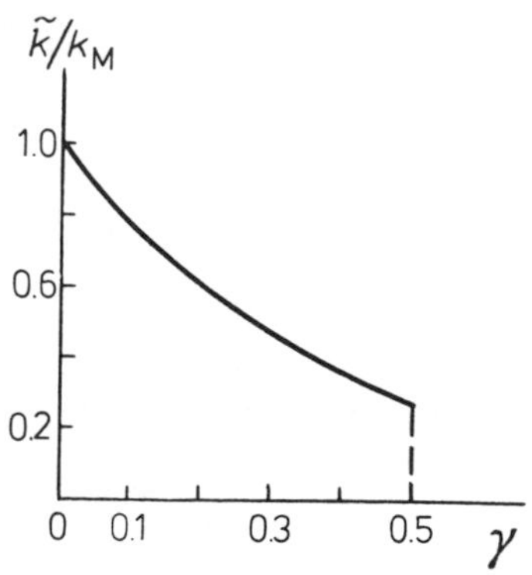

Fig. 3.1 *Effective bulk modulus versus the volume fraction of a spherical cavity in a composite unit cell.*

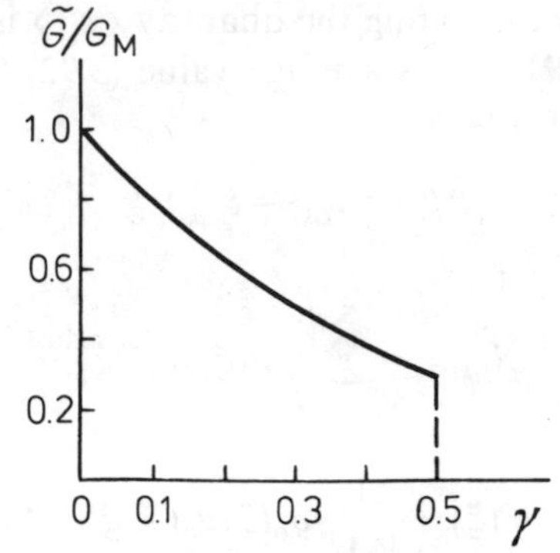

Fig. 3.2 *Effective shear modulus versus the volume fraction of a spherical cavity in a composite unit cell.*

where λ_M and G_M are the Lamé constants of the matrix material; $\delta_{ijkl} = 1$ if $i = j = k = l$ and $\delta_{ijkl} = 0$ if $i \neq j \neq k \neq l$; and α, β and f are functions of the inclusion concentration, the matrix Poisson's ratio and the parameters of the (periodic) arrangement, respectively.

2 Asymptotic Homogenization of Regular Structures

We have to admit the methods discussed in Section 3 of the preceding chapter may be regarded as of academic rather than practical interest. The assumptions and hypotheses which the methods have as their basis fail to give the boundaries of applicability of the results obtained; nor, as a rule, are stress susceptible later to any improvement in accuracy within the same mathematical framework. To obtain more useful results, it is general practice in the field of composite materials to develop rigorous mathematical models utilizing as much as possible the specific geometry of a particular problem. This is especially the case with composites that have a regular structure—the class of composites most widely used at the present time. The periodicity property often makes a problem readily amenable to theoretical treatment, and the physical processes occurring in such composites therefore have become the subject of extensive study in the last two decades or so.

Processes occurring in composite materials can be discussed by means of differential equations with rapidly oscillating coefficients. These latter are periodic functions of spatial coordinates in composites with a regular structure, and since the period is usually much smaller than the characteristic dimensions of the problem, it proves possible to apply to such composites the asymptotic method developed for the homogenization of equations with rapidly oscillating periodic coefficients (Babuška 1976, 1977, Bensoussan *et al.* 1978, Tartar 1978, Sanchez-Palencia 1980, Lions 1981, Bakhvalov and Panasenko 1984). In this method, the first step is to obtain average (or effective) characteristics of the material from a local problem formulated for the unit cell; in the second step, a boundary value problem for a (quasi-) homogeneous material described by these characteristics is considered.

4. HOMOGENIZATION TECHNIQUES FOR PERIODIC STRUCTURES

The physical behaviour of a heterogeneous medium (or composite) with a regular structure is governed by differential equations with rapidly oscillating coefficients dependent on the material properties of the individual components. Because of these coefficients, the relevant boundary value problem is extremely difficult to solve as it stands, even if a high-speed computer is available. To get round this difficulty it is necessary to model the inhomogeneous medium by a set of simpler equations in which some averaged coefficients would replace the exact ones; it is implicit, of course, that the boundary problem based on these averaged equations should give predictions differing as little as possible from those of the original problem. A mathematical framework from which to predict the mechanical behaviour of regularly inhomogeneous media has been developed under the assumption that there is an ordered micro structure in such media, describable by a characteristic inhomogeneity dimension (Bensoussan *et al.* 1978, Sanchez-Palencia 1980).

If we denote this (relative) dimension by ε, the partial differential equations of the problem have (rapidly oscillating) coefficients of the form $a(\boldsymbol{x}/\varepsilon)$, $a(\boldsymbol{y})$ being a periodic function of its argument, and the corresponding boundary value problem may be treated by asymptotically expanding the solution in powers of the small parameter ε with the help of the so-called two-scale expansion method known from the theory of ordinary differential equations. According to this method, two scales of spatial variables are considered, one for the microscopic description and the other for the macroscopic description of the process under study, and correspondingly the expansion coefficients are made to depend both on the (slow) macroscopic variable $\boldsymbol{x}$ and (rapid) microscopic variable $\boldsymbol{y} = \boldsymbol{x}/\varepsilon$. The former enable us to describe the global changes in the physical fields of interest, whereas the latter refer to the changes that take place at the local level.

The two-scale technique has already proved useful in the analysis of weakly perturbed periodic processes in the theory of vibrations. If we consider the simplest example of this kind, namely the motion of a pendulum in a weakly resisting medium, we see—referring to Fig. 4.1—that although two neighbouring 'periods' are almost equal in length, the difference gradually increases in time as the attenuation effect builds up. The usual way of treating this type of problem is by introducing two time scales, the rapid time $t^* = t$ and the slow time $\tau = \varepsilon t$ (ε being the small parameter), and by applying to the solution the asymptotic process

$$u_\varepsilon(t) = u_0(t, \tau) + \varepsilon u_1(t, \tau) + \cdots,$$

with $\varepsilon \to 0$.

The same principle can be extended to processes occurring in composite materials with a regular structure. Assigning to a composite material a coordinate system

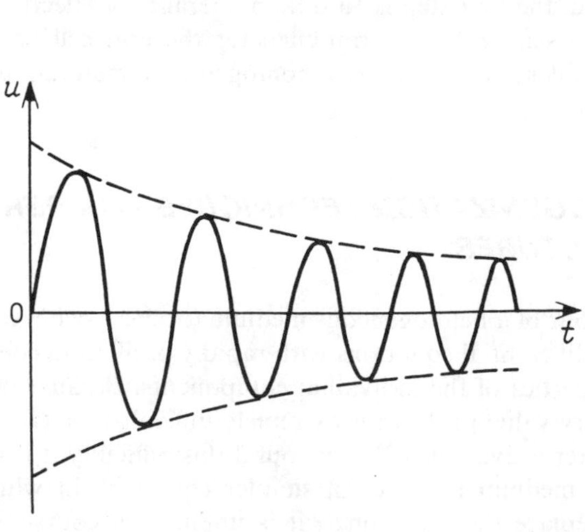

Fig. 4.1 *Motion of a pendulum in a medium with little resistance.*

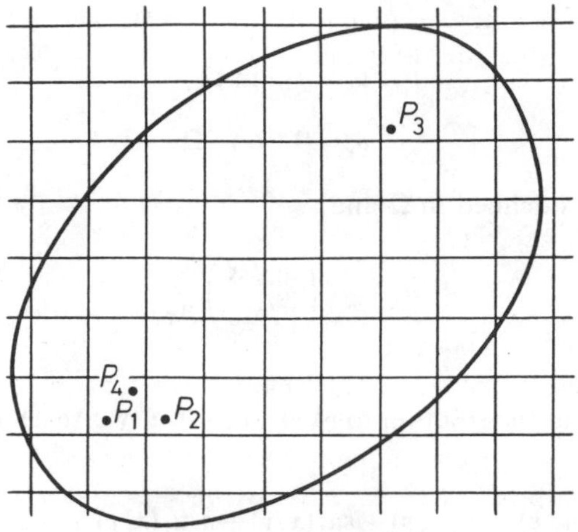

Fig. 4.2 *Characteristic points in a composite material of periodic structure.*

$\boldsymbol{x} = (x_1, x_2, x_3)$ in R^3 space and assuming periodicity in mechanical, thermal and other physical properties, let us identify in this composite a region Ω which will be regarded as a collection of parallelepiped unit cells of identical dimensions $\varepsilon Y_1, \varepsilon Y_2, \varepsilon Y_3$. In a local coordinate system $y = x/\varepsilon, y = (y_1, y_2, y_3)$, the sides of the parallelepiped will be y_1, y_2 and y_3. It seems more or less self-evident that the functions determining the material behaviour of the composite should be expanded as

$$u_\varepsilon(\boldsymbol{x}) = u_0(\boldsymbol{x}, \boldsymbol{y}) + \varepsilon u_1(\boldsymbol{x}, \boldsymbol{y}) + \cdots,$$

where $\boldsymbol{y} = \boldsymbol{x}/\varepsilon$, and the functions $u_0(\boldsymbol{x}, \boldsymbol{y})$, $u_1(\boldsymbol{x}, \boldsymbol{y})$, $\cdots$ are smooth with respect to $\boldsymbol{x}$ and $\boldsymbol{y}$-periodic in $\boldsymbol{y}$.

To gain some insight into the nature of the local periodicity of the functions $u_1(\boldsymbol{x}, \boldsymbol{y})$, let us compare the values of $u_1(\boldsymbol{x}, \boldsymbol{y})$ at congruent points P_1 and P_2 in two neighbouring cells, as shown in Fig. 4.2 (see Sanchez-Palencia 1987). Because of the assumed periodicity in $\boldsymbol{y}$ and the short distance between P_1 and P_2, the function $u_1(\boldsymbol{x}, \boldsymbol{y})$ has nearly equal values at these points. On the other hand, while $\boldsymbol{y}$-dependences are the same at points P_3 and P_1, $\boldsymbol{x}$-dependences are different because of the two points being rather far apart. Finally, if we compare the values of $u_1(\boldsymbol{x}, \boldsymbol{y})$ at points P_1 and P_2 within the same cell, it is easy to see that the $\boldsymbol{y}$-dependences are different in this case, whereas the $\boldsymbol{x}$-dependences are virtually the same.

A very simple problem involving the solution of the divergent elliptic equation will serve to illustrate some general features of the two-scale homogenization procedure (Bensoussan *et al.* 1978, Sanchez-Palencia 1980). This type of equation provides a basis for studying many physical processes, of which the (steady-state) heat conduction in a non-homogeneous solid with a regular structure may be cited as an example. We thus consider an unbounded medium with a periodic structures in a region Ω in R^3 space and note that the material properties in the parallelepiped unit cell Y are determined by a symmetrical matrix $a_{ij}(\boldsymbol{x}) = a_{ij}(\boldsymbol{y})$, where $\boldsymbol{y} = \boldsymbol{x}/\varepsilon$, $\boldsymbol{x} = (x_1, x_2, x_3)$ and the functions a_{ij} are periodic in the spatial variables $\boldsymbol{y} = (y_1, y_2, y_3)$. The boundary value problem to be dealt with is

$$A^{\varepsilon}u_{\varepsilon} = f \quad \text{in } \Omega, \tag{4.1}$$

$$u_{\varepsilon} = 0 \quad \text{on } \partial\Omega, \tag{4.2}$$

where the function f is defined in Ω and

$$A^{\varepsilon} = -\frac{\partial}{\partial x_i}\left(a_{ij}\left(\frac{\boldsymbol{x}}{\varepsilon}\right)\frac{\partial}{\partial x_j}\right) \tag{4.3}$$

is the elliptical operator.

To begin, we represent the solution to (4.1)–(4.2) as a two-scale asymptotic expansion of the form

$$u_{\varepsilon}(\boldsymbol{x}) = u_0(\boldsymbol{x},\boldsymbol{y}) + \varepsilon u_1(\boldsymbol{x},\boldsymbol{y}) + \varepsilon^2 u_2(\boldsymbol{x},\boldsymbol{y}) + \cdots, \tag{4.4}$$

where the functions $u_j(\boldsymbol{x},\boldsymbol{y})$ are Y-periodic in $\boldsymbol{y}$ ($\forall\ \boldsymbol{x}\in\Omega$), and we recall the rule of indirect differentiation to write

$$A^{\varepsilon} = \varepsilon^{-2}A_1 + \varepsilon^{-1}A_2 + \varepsilon^{0}A_3, \tag{4.5}$$

where

$$A_1 = -\frac{\partial}{\partial y_i}\left(a_{ij}(\boldsymbol{y})\frac{\partial}{\partial y_j}\right), \quad A_3 = -\frac{\partial}{\partial x_i}\left(a_{ij}(\boldsymbol{y})\frac{\partial}{\partial x_j}\right),$$

$$A_2 = -\frac{\partial}{\partial y_i}\left(a_{ij}(\boldsymbol{y})\frac{\partial}{\partial x_j}\right) - \frac{\partial}{\partial x_i}\left(a_{ij}(\boldsymbol{y})\frac{\partial}{\partial y_j}\right).$$

Using (4.4) and (4.5) the left-hand side of (4.1) can also be expanded in powers of ε and collecting terms of equal powers we find in the usual manner:

$$A_1u_0 = 0, \tag{4.6}$$

$$A_1u_1 + A_2u_0 = 0, \tag{4.7}$$

$$A_1u_2 + A_2u_1 + A_3u_0 = f, \ldots. \tag{4.8}$$

If $\boldsymbol{x}$ and $\boldsymbol{y}$ are considered as independent variables, equations (4.6)–(4.8) form a recurrent system of differential equations for the functions $u_0, u_1, \ldots$ parameterized by $\boldsymbol{x}$. Before proceeding to the analysis of this system, it will be useful to note that the equation

$$A_1u = F \quad \text{in } Y \tag{4.9}$$

for a Y-periodic function u has a unique solution if

$$\langle F\rangle = \frac{1}{|Y|}\int_Y F\,\mathrm{d}\boldsymbol{y} = 0, \tag{4.10}$$

where $|Y|$ denotes the volume of the unit cell.

It immediately follows from (4.6) that

$$u_0 = u(\boldsymbol{x}), \tag{4.11}$$

and upon substituting into (4.7) we find

$$A_1 u_1 = \frac{\partial a_{ij}(\boldsymbol{y})}{\partial y_i}\frac{\partial u(\boldsymbol{x})}{\partial x_j}. \tag{4.12}$$

Noting the separation of variables on the right-hand side of (4.12), the solution of this equation may be represented in the form

$$u_1(\boldsymbol{x},\boldsymbol{y}) = U_j(\boldsymbol{y})\frac{\partial u(\boldsymbol{x})}{\partial x_j} + \bar{u}_1(\boldsymbol{x}), \tag{4.13}$$

where $U_j(\boldsymbol{y})$ is the Y-periodic solution of the local equation

$$A_1 U_j(\boldsymbol{y}) = \partial a_{ij}(\boldsymbol{y})/\partial y_i \quad \text{in } Y. \tag{4.14}$$

We next turn to the problem of solving (4.8) for u_2 taking $\boldsymbol{x}$ as a parameter. It follows from condition (4.10) that (4.8) will have a unique solution if

$$\frac{1}{|Y|}\int_Y (A_2 u_1 + A_3 u_0)\,\mathrm{d}\boldsymbol{y} = f, \tag{4.15}$$

which when combined with (4.13) yields the following homogenized (macroscopic) equation for $u(\boldsymbol{x})$:

$$-\tilde{a}_{ij}\frac{\partial^2 u}{\partial x_i \partial x_j} = f, \tag{4.16}$$

where the quantities

$$\tilde{a}_{ij} = \frac{1}{|Y|}\int_Y \left(a_{ij}(\boldsymbol{y}) + a_{ik}(\boldsymbol{y})\frac{\partial U_j}{\partial y_k}\right)\mathrm{d}\boldsymbol{y} \tag{4.17}$$

are the effective coefficients of the homogenized operator

$$A = -\tilde{a}_{ij}\frac{\partial^2}{\partial x_i \partial x_j}.$$

We have thus demonstrated that if we retain only two terms in the asymptotic expansion (4.4), the homogenization of (4.1) splits into two problems: the determination of the local functions $U_j(\boldsymbol{y})$ from (4.14) is solved on the unit cell, while the other is concerned with one solution of the homogenized equation (4.16) set in Ω with the boundary condition $u = 0$ on $\partial\Omega$ (the function $u_1(\boldsymbol{x})$ in (4.13) may be considered arbitrary and set equal to zero). If we go to higher terms in the asymptotic series (4.4), we can follow the above procedure in deriving a homogenized equation of the type (4.16) for $u_1(\boldsymbol{x})$; in this case, one more local problem will have to be solved on the unit cell.

There is, however, a very wide class of problems in mechanics in which very realistic predictions can be made even within the framework of the two-term technique, that is, by the use of equations (4.14) and (4.16), and relations (4.17). This approach is sometimes referred to as the zeroth-order approximation theory.

It should be noted that the asymptotic process for equation (4.1) is amenable to the important extension to the case of piecewise-smooth (periodic) coefficients $a_{ij}(y)$. We may demand, for example, that the function $a_{ij}(y)$ assume constant values $a_{ij}^{(1)}$ and $a_{ij}^{(2)}$ in two regions $Y^{(1)}$ and $Y^{(2)}$ comprising the unit cell and having a smooth boundary Γ between them. It turns out that both the expansion (4.4) and expressions (4.17) for the averaged coefficients retain their forms in this case, but the local problem (4.14) for the Y-periodic functions $U_j(y)$ is replaced by the system

$$-\frac{\partial}{\partial y_i}\left(a_{ij}(y)+a_{ik}(y)\frac{\partial U_j(y)}{\partial y_k}\right)=0 \quad \text{in } Y^{(1)}, Y^{(2)}, \tag{4.18}$$

subject to the additional matching conditions

$$[U_j]=0, \qquad \left[\left(a_{ij}+a_{ik}\frac{\partial U_j}{\partial y_k}\right)n_i\right]=0 \tag{4.19}$$

across the boundary Γ, to which n_i is the unit normal.

The topic we consider next is the application of the above homogenization scheme to so-called perforated media. (We refer the reader to Lions (1987) for a comprehensive account of the work on the subject.)

Let a region Ω in R^n $(n=2,3)$ contain in it a system $\mathcal{O}_\varepsilon$ of periodically arranged holes and let Ω_ε designate the region excluding the holes,

$$\Omega_\varepsilon=\Omega\backslash\mathcal{O}_\varepsilon,$$

the boundary of which, $\partial\Omega_\varepsilon$, is specified to include the hole boundaries $\partial\mathcal{O}_\varepsilon$. A unit side cube is taken as the unit cell of the problem, and within the cube a region $\mathcal{O}$ bounded by a surface S with an outward unit normal ν_i is chosen, as shown in Fig. 4.3. The

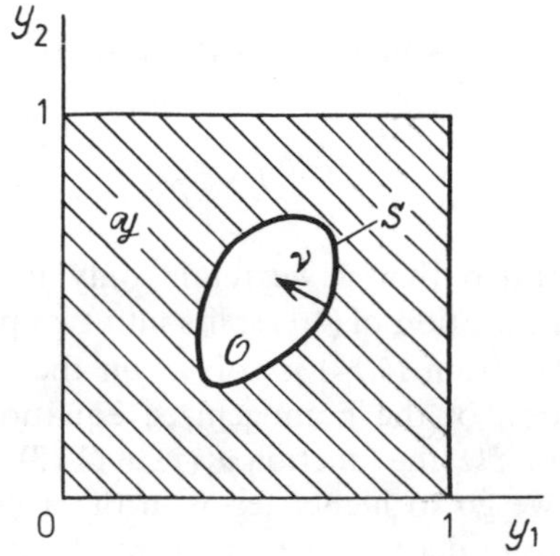

Fig. 4.3 *Unit cell with a hole.*

elementary boundary value problem we consider is

$$-\frac{\partial^2 u_\varepsilon}{\partial x_i \partial x_i} = f \quad \text{in } \Omega_\varepsilon, \tag{4.20}$$

$$u_\varepsilon = 0 \quad \text{on } \partial\Omega, \tag{4.21}$$

$$\frac{\partial u_\varepsilon}{\partial \nu} = 0 \quad \text{on } S, \tag{4.22}$$

where $f = f(\boldsymbol{x})$.

If we take the solution of this problem in the form

$$u_\varepsilon = u_0(\boldsymbol{x}, \boldsymbol{y}) + \varepsilon u_1(\boldsymbol{x}, \boldsymbol{y}) + \cdots, \tag{4.23}$$

where $u_k(\boldsymbol{x}, \boldsymbol{y})$ is Y-periodic in $\boldsymbol{y} = \boldsymbol{x}/\varepsilon$, we find

$$\varepsilon^{-2}\frac{\partial^2 u_0}{\partial y_i \partial y_i} + \varepsilon^{-1}\left(\frac{\partial^2 u_1}{\partial y_i \partial y_i} + 2\frac{\partial^2 u_0}{\partial x_i \partial y_i}\right) + \varepsilon^0\left(\frac{\partial^2 u_2}{\partial y_i \partial y_i} + 2\frac{\partial^2 u_1}{\partial x_i \partial y_i} + \frac{\partial^2 u_0}{\partial x_i \partial x_i}\right) + \cdots = f \quad \text{in } \Omega_\varepsilon, \tag{4.24}$$

$$\varepsilon^{-1}\nu_i\frac{\partial u_0}{\partial y_i} + \varepsilon^0\left(\frac{\partial u_0}{\partial x_i}\nu_i + \frac{\partial u_1}{\partial y_i}\nu_i\right) + \varepsilon\left(\frac{\partial u_2}{\partial y_i}\nu_i + \frac{\partial u_1}{\partial x_i}\nu_i\right) + \cdots = 0 \quad \text{on } S, \tag{4.25}$$

from (4.20) and (4.22), and comparing terms in equal powers in ε the following two problems result:

$$-\frac{\partial^2 u_0}{\partial y_i \partial y_i} = 0, \qquad \frac{\partial u_0}{\partial y_i}\nu_i = 0. \tag{4.26}$$

This gives us $u_0 = u_0(\boldsymbol{x})$ and hence the problem we are concerned with can be rewritten as

$$-\frac{\partial^2 u_1}{\partial y_i \partial y_i} = 0 \quad \text{in } \mathscr{Y}, \qquad \mathscr{Y} = Y/0, \tag{4.27}$$

$$\frac{\partial u_1}{\partial y_i}\nu_i + \frac{\partial u_0}{\partial x_i}\nu_i = 0 \quad \text{on } S, \tag{4.28}$$

$$\frac{\partial^2 u_2}{\partial y_i \partial y_i} + 2\frac{\partial^2 u_1}{\partial x_i \partial y_i} + \frac{\partial^2 u_0}{\partial x_i \partial x_i} = -f \quad \text{in } \mathscr{Y}, \tag{4.29}$$

$$\frac{\partial u_2}{\partial y_i}\nu_i + \frac{\partial u_1}{\partial x_i}\nu_i = 0 \quad \text{on } S. \tag{4.30}$$

Setting

$$u_1(\boldsymbol{x}, \boldsymbol{y}) = -\frac{\partial u_0(\boldsymbol{x})}{\partial x_k}U_k(\boldsymbol{y}), \tag{4.31}$$

with functions $U_k(\boldsymbol{y})$ 1-periodic ($\boldsymbol{x} \in \Omega$), equations (4.27) and (4.28) now lead to the local problem:

$$\frac{\partial^2 U_k(y)}{\partial y_i \partial y_i} = 0 \quad \text{in } \mathscr{Y}, \tag{4.32}$$

$$\frac{\partial U_k(y)}{\partial y_i} \nu_i = \nu_k \quad \text{on } S, \tag{4.33}$$

and (4.29) and (4.30) assume the respective forms

$$\frac{\partial^2 u_2}{\partial y_i \partial y_i} + \frac{\partial^2 u_0(x)}{\partial x_k \partial x_i}\left(\delta_{ik} - 2\frac{\partial U_k(y)}{\partial y_i}\right) = -f(x) \quad \text{in } \mathscr{Y}, \tag{4.34}$$

$$\frac{\partial u_2}{\partial y_i} \nu_i - \frac{\partial^2 u_0}{\partial x_i \partial x_k} - U_k(y)\nu_i = 0 \quad \text{on } S, \tag{4.35}$$

where (4.21) has been used.

The condition for the problem (4.34)–(4.35) to be solvable

$$\frac{1}{|\mathscr{Y}|}\int_S \frac{\partial u_2}{\partial y_i} \nu_i \, dy = \frac{1}{|\mathscr{Y}|}\int_{\mathscr{Y}} \left(\delta_{ik} - 2\frac{\partial U_k(y)}{\partial y_i}\right) dy \frac{\partial^2 u_0}{\partial x_k \partial x_i} - f(x), \tag{4.36}$$

is now combined with (4.35) to give the homogenized equation

$$-q_{ik}\frac{\partial^2 u_0}{\partial x_k \partial x_i} = f \quad \text{in } \Omega, \tag{4.37}$$

where the effective coefficients q_{ik} are defined by

$$q_{ik} = \frac{1}{|\mathscr{Y}|}\int_{\mathscr{Y}} \left(\delta_{ik} - \frac{\partial U_k(y)}{\partial y_i}\right) dy, \tag{4.38}$$

where $|\mathscr{Y}|$ denotes the volume of the region $\mathscr{Y}$. We thus conclude that the homogenized problem for a perforated medium reduces to the solution of equation (4.37) under the condition that u_0 should vanish at the boundary $\partial\Omega$ of the region Ω.

It can be shown (Lions 1987) that the above approach applies with equal force to the more general boundary value problem

$$\frac{\partial}{\partial x_i}\left(a_{ij}\left(\frac{x}{\varepsilon}\right)\frac{\partial u_\varepsilon}{\partial x_j}\right) = -f \quad \text{in } \Omega_\varepsilon, \tag{4.39}$$

$$u_\varepsilon = 0 \quad \text{on } \partial\Omega, \tag{4.40}$$

$$\nu_i a_{ij}\left(\frac{x}{\varepsilon}\right)\frac{\partial u_\varepsilon}{\partial x_j} = 0 \quad \text{on } S, \tag{4.41}$$

where the quantities a_{ij} may or may not be symmetric in their indices and are assumed to satisfy the condition

$$a_{ij}(y)\xi_i\xi_j \geqslant \alpha\xi_i\xi_j, \quad \alpha > 0.$$

If only two terms are retained in the asymptotic expansion

$$u_\varepsilon = u_0(\boldsymbol{x}) + \varepsilon u_1(\boldsymbol{x}, \boldsymbol{y}) + \cdots, \tag{4.42}$$

the function $u_0(\boldsymbol{x})$ turns out to solve the homogenized problem:

$$-q_{ik}\frac{\partial^2 u_0(\boldsymbol{x})}{\partial x_i \partial x_k} = f \quad \text{in } \Omega, \tag{4.43}$$

$$u_0(\boldsymbol{x}) = 0 \quad \text{on } \partial\Omega, \tag{4.44}$$

with

$$q_{ik} = \frac{1}{|\mathscr{Y}|}\int_{\mathscr{Y}}\left(a_{ik}(\boldsymbol{y}) - a_{ij}\frac{\partial U_k(\boldsymbol{y})}{\partial y_j}\right)\mathrm{d}\boldsymbol{y}, \tag{4.45}$$

and the 1-periodic functions $U_k(\boldsymbol{y})$ must be determined from the local problem

$$\frac{\partial}{\partial y_i}\left(a_{ij}(\boldsymbol{y})\frac{\partial U_k(\boldsymbol{y})}{\partial y_j}\right) = \frac{\partial a_{ik}(\boldsymbol{y})}{\partial y_i} \quad \text{in } \mathscr{Y}, \tag{4.46}$$

$$\nu_i a_{ij}(\boldsymbol{y})\frac{\partial U_k(\boldsymbol{y})}{\partial y_j} = \nu_i a_{ik}(\boldsymbol{y}) \quad \text{on } S, \tag{4.47}$$

and are defined by (cf. (4.31))

$$u_1(\boldsymbol{x}, \boldsymbol{y}) = -\frac{\partial u_0(\boldsymbol{x})}{\partial x_k}U_k(\boldsymbol{y}). \tag{4.48}$$

An important result that seems worthwhile mentioning here concerns the spectrum analysis of the perforated medium problem (Cioranescu and Paulin 1979, Kesavan 1979a, 1979b; see also Lions 1985):

$$-\frac{\partial}{\partial x_i}\left(a_{ij}\left(\frac{\boldsymbol{x}}{\varepsilon}\right)\frac{\partial u_\varepsilon}{\partial x_j}\right) = \lambda(\varepsilon)u_\varepsilon \quad \text{in } \Omega_\varepsilon, \tag{4.49}$$

$$u_\varepsilon = 0 \quad \text{on } \partial\Omega, \tag{4.50}$$

$$\nu_i a_{ij}\left(\frac{\boldsymbol{x}}{\varepsilon}\right)\frac{\partial u_\varepsilon}{\partial x_j} = 0 \quad \text{on } S, \tag{4.51}$$

and states that the eigenvalues $\lambda_m(\varepsilon)$ of this problem ($m = 1, 2, \ldots; 0 \leqslant \lambda_1(\varepsilon) \leqslant \lambda_2(\varepsilon) \leqslant \cdots$) have the property

$$\lambda_m(\varepsilon) \to \lambda_m \quad \text{as } \varepsilon \to 0, \tag{4.52}$$

where $\{\lambda_m\}$ is the spectrum of the Dirichlet problem:

$$-q_{ij}\frac{\partial u}{\partial x_i \partial x_j} = \lambda u \quad \text{in } \Omega, \tag{4.53}$$

$$u = 0 \quad \text{on } \partial\Omega, \tag{4.54}$$

with the coefficients q_{ij} defined by (4.45).

It should be remarked at this point that application of the two-scale asymptotic process in the version outlined above may present difficulties in satisfying the boundary condition of a homogenized problem. We can illustrate this by considering the problem of homogenization for a perforated region Ω_ε with Dirichlet conditions imposed at the boundaries of the holes (or cavities). That is, we write

$$-\varepsilon^2\frac{\partial^3 u_\varepsilon}{\partial x_i \partial x_i} = f \quad \text{in } \Omega_\varepsilon, \tag{4.55}$$

$$u_\varepsilon = 0 \quad \text{on } \partial\Omega \text{ and on } S. \tag{4.56}$$

As a solution of this problem we assume

$$u_\varepsilon = u_0(x, y) + \varepsilon u_1(x, y) + \cdots, \tag{4.57}$$

where the functions u_0 and u_1 are 1-periodic in y and

$$u_0(x, y) = u_1(x, y) = 0 \quad (y \in S,\ x \in \Omega).$$

When (4.57) is substituted into (4.55) the result is

$$-\frac{\partial^2 u_0(x, y)}{\partial y_i \partial y_i} = f(x), \tag{4.58}$$

$$-\frac{\partial^2 u_1(x, y)}{\partial y_i \partial y_i} - 2\frac{\partial^2 u_0(x, y)}{\partial x_i \partial y_i} = 0, \tag{4.59}$$

$$-\frac{\partial^2 u_2(x, y)}{\partial y_i \partial y_i} - 2\frac{\partial^2 u_1(x, y)}{\partial x_i \partial y_i} = 0, \ldots. \tag{4.60}$$

Now if we write

$$u_0(x, y) = W_0(y) f(x), \tag{4.61}$$

$$u_1(x, y) = W_k(y)\frac{\partial f(x)}{\partial x_k}, \ldots, \tag{4.62}$$

then the functions W_0 and W_k will have to be determined from the following local problems:

$$-\frac{\partial^2 W_0(y)}{\partial y_i \partial y_i} = 1 \quad \text{in } \mathscr{Y}, \tag{4.63}$$

$$W_0 = 0 \quad \text{on } S, \tag{4.64}$$

$$-\frac{\partial^2 W_k(y)}{\partial y_i \partial y_i} = 2\frac{\partial W_0(y)}{\partial y_k} \quad \text{in } \mathscr{Y}, \tag{4.65}$$

$$W_k = 0 \quad \text{on } S. \tag{4.66}$$

It is seen that if we take the solution of (4.55) in the form

$$u_\varepsilon = W_0(y) f(x) + \varepsilon W_k(y)\frac{\partial f(x)}{\partial x_k} + \cdots, \tag{4.67}$$

the boundary condition (4.56) on $\partial\Omega$ will be satisfied only if

$$f(x) = \frac{\partial f(x)}{\partial x_k} = 0 \quad \text{on } \partial\Omega. \tag{4.68}$$

Otherwise boundary layer type solutions will have to be introduced into the asymptotic process, the construction of which may be found discussed in the works of Panasenko (1979), Bakhvalov and Panasenko (1984), and Lions (1985); the paper by Sanchez-Palencia (1987) is also relevant.

Clearly, the solutions provided by locally periodic expansions like (6.4) are only adequate at points remote both from the boundary of the region Ω and from those regions within Ω where non-periodic effects due to cracks (discontinuities), inclusions and similar factors may be important. In the immediate vicinity of $\partial\Omega$, boundary layer solutions must be considered. If the boundary is taken to coincide with the plane $x_3 = 0$, as in Fig. 4.4, these solutions will have the form

$$u_\varepsilon(\boldsymbol{x}) = u_0^{(n)}(\boldsymbol{x}, \boldsymbol{y}) + \varepsilon u_1^{(n)}(\boldsymbol{x}, \boldsymbol{y}) + \cdots, \tag{4.69}$$

where the functions $u_i^{(n)}(\boldsymbol{x}, \boldsymbol{y})$ are periodic in y_1 and y_2 (with periods Y_1 and Y_2, respectively) but non-periodic in x_1, x_2 and y_3. Note that the boundary layer solution (4.69) is assumed to satisfy the true boundary conditions at $y_3 = 0$ and must be made consistent with solutions (4.4) in the limit as $y_3 \to \infty$, $x_3 \to 0$. Since the tangential variables x_1, x_2, y_1 and y_2 come simply as parameters in this case, one can write (Sanchez-Palencia 1987)

$$u_\varepsilon(\boldsymbol{x}) = u_0(x_3) + \varepsilon u_1(x_3) + \cdots \tag{4.70}$$

for the external (bulk) expansion, and

$$u_\varepsilon(\boldsymbol{x}) = u_0^{(n)}(y_3) + \varepsilon u_1^{(n)}(y_3) + \cdots \tag{4.71}$$

for the internal (boundary layer) expansion. It should be emphasized that the external form depends only on the external variables x_3 and the internal form depends only on $y_3 = x_3/\varepsilon$.

It is common practice in performing the matching procedure for an asymptotic expansion to define the external (internal) limit of the function $u_\varepsilon(\boldsymbol{x})$ as the value it attains in the limit as ε tends to zero at a fixed external (internal) variable x_3 (y_3). Referring to

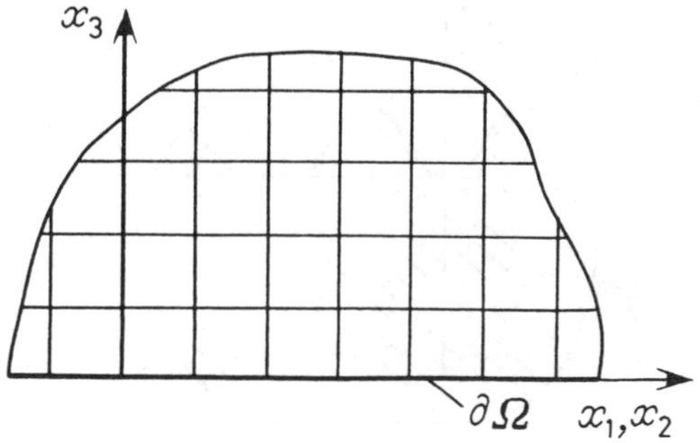

Fig. 4.4 *The boundary of a body made of a composite material of periodic structure.*

Nayfeh (1973) and Sanchez-Palencia (1987), the following simple rule expresses the required consistency between the external and internal solutions in the transition region:

the internal limit of the external expansion is equal to the external limit of the internal expansion.

According to this rule,

$$u_0(0) = u_0^{(n)}(\infty). \tag{4.72}$$

A general discussion of the matching procedure for an n-term asymptotic expansions is given in Nayfeh (1973).

5. HOMOGENIZATION METHOD FOR REGIONS WITH A WAVY BOUNDARY

Boundary value problems with rapidly oscillating coefficients are by no means the only area of application for the two-scale expansion method. The method has been successfully employed, for example, in the treatment of partial differential equations in regions bounded by small period wavy surfaces. We consider as an illustration a two-dimensional heat conduction problem for the half-plane Ω_ε with a wavy boundary $\partial\Omega_\varepsilon$, as shown in Fig. 5.1, taking the ambient temperature to be zero and assuming that a heat exchange condition is specified at $\partial\Omega_\varepsilon$. A general outline of the homogenization analysis of this problem has been given in Sanchez-Palencia (1980). Here a more detailed discussion seems to be appropriate.

Let the boundary $\partial\Omega_\varepsilon$ of the region $\Omega_\varepsilon = \{x_1, x_2\colon x_2 < \varepsilon F(x_1/\varepsilon), |x_1| < \infty\}$ be specified by the equation $x_2 = \varepsilon F(x_1/\varepsilon)$, where ε is a small positive parameter and $F(y_1)$ a smooth 1-periodic function. Assuming the thermal and physical characteristics of the medium to be constant and denoting $\boldsymbol{x} = (x_1, x_2)$, we write Fourier's law:

$$q_\alpha^{(\varepsilon)}(\boldsymbol{x}, t) = -\lambda_{\alpha\beta}\frac{\partial T^{(\varepsilon)}(\boldsymbol{x}, t)}{\partial x_\beta}, \tag{5.1}$$

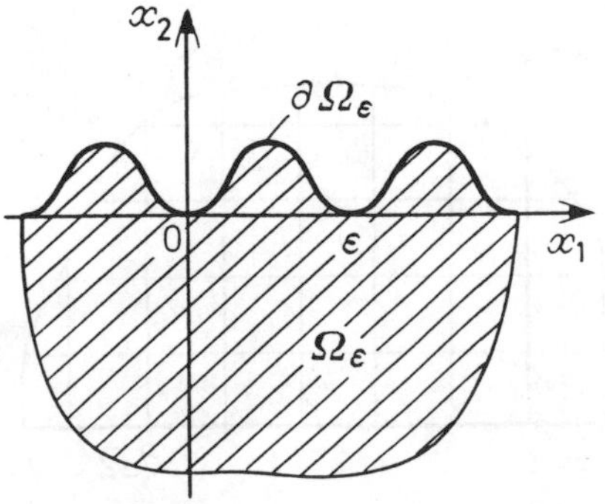

Fig. 5.1 *A region with a wavy boundary of periodic structure.*

and the heat balance condition:

$$-f(\boldsymbol{x})+c\rho\frac{\partial T^{(\varepsilon)}(\boldsymbol{x},t)}{\partial t}=-\frac{\partial q_\alpha^{(\varepsilon)}(\boldsymbol{x},t)}{\partial x_\alpha}\quad \text{in } \Omega_\varepsilon, \tag{5.2}$$

where the temperature $T^{(\varepsilon)}(\boldsymbol{x},t)$ is the unknown quantity of the problem, $q_\alpha^{(\varepsilon)}$ is the heat flow vector, $\lambda_{\alpha\beta}$ the (constant) thermal conductivities, c the specific heat, ρ the mass density and $f(\boldsymbol{x})$ a (known) heat source density function; the subscripts α and β assume values 1 and 2.

Let the solution of (5.1)–(5.2) satisfy the initial condition

$$T^{(\varepsilon)}(\boldsymbol{x},0)=0, \tag{5.3}$$

and the boundary conditions of the form

$$\begin{aligned} q_\alpha^{(\varepsilon)}(\boldsymbol{x},t)n_\alpha&=\alpha T^{(\varepsilon)}(\boldsymbol{x},t)\quad \text{on } \partial\Omega_\varepsilon,\\ T^{(\varepsilon)}(\boldsymbol{x},t)&\to 0\quad \text{as } x_2\to-\infty, \end{aligned} \tag{5.4}$$

where

$$n_1=-\frac{\mathrm{d}F}{\mathrm{d}y_1}\left[1+\left(\frac{\mathrm{d}F}{\mathrm{d}y_1}\right)^2\right]^{-1/2},\quad n_2=\left[1+\left(\frac{\mathrm{d}F}{\mathrm{d}y_1}\right)^2\right]^{-1/2}$$

and α is the heat transfer coefficient.

The asymptotic expansion postulated for the problem by Sanchez-Palencia (1980) reads:

$$T^{(\varepsilon)}(\boldsymbol{x},t)=T^{(0)}(\boldsymbol{x},t)+\varepsilon T^{(1)}(\boldsymbol{x},\boldsymbol{y},t)+\cdots, \tag{5.5}$$

where $\boldsymbol{y}=\boldsymbol{x}/\varepsilon$, $T^{(1)}(\boldsymbol{x},\boldsymbol{y},t)$ is 1-periodic in $\boldsymbol{y}$ and

$$\frac{\partial T^{(1)}}{\partial y_\alpha}\to 0\quad \text{as } y_2\to-\infty. \tag{5.6}$$

Substituting (5.5) into (5.1) and (5.2) we find

$$\frac{\partial q_\alpha^{(0)}(\boldsymbol{x},\boldsymbol{y},t)}{\partial y_\alpha}=0, \tag{5.7}$$

$$-f(\boldsymbol{x})+c\rho\frac{\partial T^{(0)}(\boldsymbol{x},t)}{\partial t}=-\frac{\partial q_\alpha^{(0)}(\boldsymbol{x},\boldsymbol{y},t)}{\partial x_\alpha}-\frac{\partial q_\alpha^{(1)}(\boldsymbol{x},\boldsymbol{y},t),}{\partial y_\alpha}, \tag{5.8}$$

where

$$q_\alpha^{(n)}(\boldsymbol{x},\boldsymbol{y},t)=-\lambda_{\alpha\beta}\left(\frac{\partial T^{(n)}(\boldsymbol{x},\boldsymbol{y},t)}{\partial x_\beta}+\frac{\partial T^{(n+1)}(\boldsymbol{x},\boldsymbol{y},t)}{\partial y_\beta}\right)\quad (n=0,1). \tag{5.9}$$

Now equation (5.7), corresponding to the local problem set on the unit cell

$B = \{y_1, y_2 : y_1 \in [0, 1],\ y_2 < F(y_1)\}$, takes the form

$$\lambda_{\alpha\beta} \frac{\partial^2 T^{(1)}(\boldsymbol{x}, \boldsymbol{y}, t)}{\partial y_\alpha \partial y_\beta} = 0 \quad \text{in } B, \tag{5.10}$$

and from the boundary condition (5.4) we have

$$-\lambda_{\alpha\beta} \frac{\partial T^{(1)}(\boldsymbol{x}, \boldsymbol{y}, t)}{\partial y_\beta}\bigg|_{y_2 = F(y_1)} n_\alpha(y_1) = \lambda_{\alpha\beta} \frac{\partial T^{(0)}(\boldsymbol{x}, \boldsymbol{y}, t)}{\partial x_\beta}\bigg|_{x_2 = \varepsilon F(x_1/\varepsilon)} n_\alpha(y_1) + \alpha T^{(0)}(\boldsymbol{x}, t)\bigg|_{x_2 = \varepsilon F(x_1/\varepsilon)} \tag{5.11}$$

Note that $T^{(1)}(\boldsymbol{x}, \boldsymbol{y}, t)$, viewed as a function of y_1, must satisfy condition (5.6) and be 1-periodic in y_1. It is also important to recall here the uniqueness condition for the (periodic) $T^{(1)}$ solving the problems (5.10), (5.11) and (5.6). This condition is obtained by integrating (5.10) over the region B, and in the limiting case of $\varepsilon \to 0$ it becomes

$$\lambda_{\alpha\beta} \frac{\partial T^{(0)}(\boldsymbol{x}, t)}{\partial x_\beta} + \alpha l_F T^{(0)}(\boldsymbol{x}, t) = 0 \quad \text{as } x_2 = \varepsilon F\left(\frac{x_1}{\varepsilon}\right) \to 0, \tag{5.12}$$

where l_F is the length of the arc of the curve $y_2 = F(y_1)$ $(0 \leqslant y_1 \leqslant 1)$.

With (5.12), the boundary condition (5.11) of the local problem (5.7) can be expressed as

$$-\lambda_{\alpha\beta} \frac{\partial T^{(1)}(\boldsymbol{x}, \boldsymbol{y}, t)}{\partial y_\beta}\bigg|_{y_2 = F(y_1)} n_\alpha(y_1) = \left(\lambda_{\alpha\beta} n_\alpha(y_1) - \frac{\lambda_{2\beta}}{l_F}\right) \frac{\partial T^{(0)}(\boldsymbol{x}, t)}{\partial x_\beta}\bigg|_{x_2 \to 0}, \tag{5.13}$$

and if we write

$$T^{(1)}(\boldsymbol{x}, \boldsymbol{y}, t) = U_\gamma(\boldsymbol{y}) \frac{\partial T^{(0)}(\boldsymbol{x}, t)}{\partial x_\gamma}, \tag{5.14}$$

then the (periodic in y_1) function $U_\gamma(\boldsymbol{y})$ will have to be found from the problem

$$\lambda_{\alpha\beta} \frac{\partial^2 U_\gamma}{\partial y_\alpha \partial y_\beta} = 0 \quad \text{in } B, \tag{5.15}$$

$$-\lambda_{\alpha\beta} \frac{\partial U_\gamma(\boldsymbol{y})}{\partial y_\beta}\bigg|_{y_2 = F(y_1)} n_\alpha(y_1) = \lambda_{\alpha\gamma} n_\alpha(y_1) - \frac{\lambda_{2\gamma}}{l_F}, \tag{5.16}$$

$$\frac{\partial U_\gamma(\boldsymbol{y})}{\partial y_\beta} \to 0 \quad \text{as } y_2 \to -\infty \quad (\gamma = 1, 2). \tag{5.17}$$

The macroscopic heat conduction equation is readily obtained by taking the volume average of (5.8) over the periodicity cell B. Since the region B is unbounded in the y_2 direction $(-\infty < y_2 < F(y_1))$, this is effected by first averaging over the bounded region $B_0 = \{y_1, y_2 : -h < y_2 < F(y_1), 0 < y_1 < 1\}$ and then passing to the limit as $h \to \infty$. Using condition (5.17) and recalling the periodicity of $U_\gamma(\boldsymbol{y})$ in y_1 then gives the result

$$-f(\boldsymbol{x}) + c\rho \frac{\partial T^{(0)}(\boldsymbol{x}, t)}{\partial t} = -\lambda_{\alpha\gamma} \frac{\partial^2 T^{(0)}(\boldsymbol{x}, t)}{\partial x_\alpha \partial x_\gamma} \tag{5.18}$$

in the region $\Omega_0 = \{x_1, x_2 : x_2 < 0, |x_1| < \infty\}$.

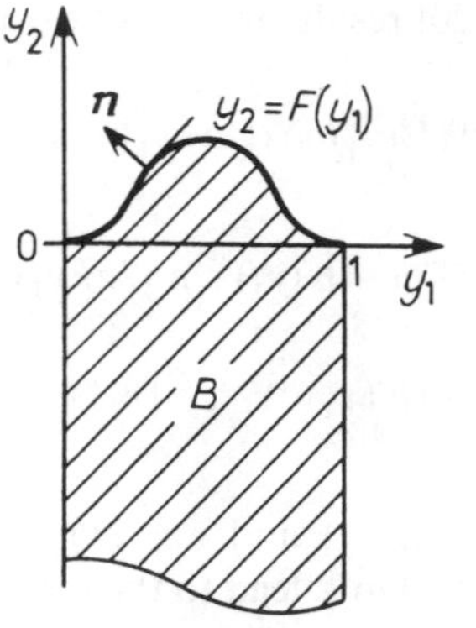

Fig. 5.2 *Unit cell of a region with a wavy boundary.*

With the expansion (5.5) taken to represent the solution, we thus see that we must start by determining its first term from the limiting problem for (5.18) in the half-plane $x_2 < 0$ under the boundary condition (5.12). This done, the function $T^{(1)}$ can be found up to a constant. A proof of the uniqueness of $T^{(1)}$ is given in Sanchez-Palencia (1980).

It is of interest to examine the solution of the local problem (5.15)–(5.17) in the special case where $\lambda_{\alpha\beta} = \lambda\delta_{\alpha\beta}$, and where the function $F(y_1)$, the shape of the cell boundary, is symmetric on the interval $(-1/2 < y_1 < 1/2)$; see Fig. 5.2. The problem (5.15)–(5.17) then reduces to that of finding the functions $U_\gamma(y_1, y_2)$ $(\gamma = 1, 2)$ periodic in y_1 and satisfying Laplace's equation,

$$\frac{\partial^2 U_\gamma}{\partial y_1} + \frac{\partial^2 U_\gamma}{\partial y_2} = 0 \quad \text{in B,} \tag{5.19}$$

subject to the conditions

$$\left(\frac{\partial U_\gamma}{\partial y_1} n_1 + \frac{\partial U_\gamma}{\partial y_2} n_2\right)\bigg|_{y_2 = F(y_1)} = \begin{cases} -n_1, & \gamma = 1, \\ -n_2 + (1/l_F), & \gamma = 2, \end{cases} \tag{5.20}$$

$$\frac{\partial U_\gamma}{\partial y_2} \to 0 \quad \text{as } y_2 \to -\infty. \tag{5.21}$$

We satisfy (5.21) and take account of the periodicity of $U_\gamma(y_1, y_2)$ in y_1 if we assume the solution to be of the form

$$U_1(y_1, y_2) = \sum_{k=1}^{\infty} \frac{1}{k} A_k e^{2\pi k y_2} \sin(2\pi k y_1), \tag{5.22}$$

$$U_2(y_1, y_2) = \sum_{k=1}^{\infty} \frac{1}{k} B_k e^{2\pi k y_2} \cos(2\pi k y_1), \tag{5.23}$$

with the constants A_k and B_k to be determined from the boundary conditions (5.20).

Using (5.22) and (5.23) in (5.20) results in the following pair of equations:

$$2\pi \sum_{k=1}^{\infty} A_k e^{2\pi k F(y_1)}(F'(y_1)\cos(2\pi k y_1) - \sin(2\pi k y_1)) = F'(y_1), \tag{5.24}$$

$$2\pi \sum_{k=1}^{\infty} B_k e^{2\pi k F(y_1)}(-F'(y_1)\sin(2\pi k y_1) - \cos(2\pi k y_1))$$
$$= 1 - \frac{1}{l_F}(1 + (F'(y_1))^2)^{1/2}, \quad |y_1| < \tfrac{1}{2}, \tag{5.25}$$

which when multiplied by $\sin(2\pi p y_1)$ and $\cos(2\pi p y_1)$, $p = 1, 2, \ldots$, respectively, and integrated with respect to y_1 from 0 to $\frac{1}{2}$, lead to the following infinite system of algebraic equations for determining the constants A_k and B_k:

$$\sum_{k=1}^{\infty} A_k \alpha_{kp} = a_p, \sum_{k=1}^{\infty} B_k \beta_{kp} = -b_p \quad (p = 1, 2, \ldots), \tag{5.26}$$

where

$$\alpha_{kp} = 2\pi \int_0^{1/2} e^{2\pi k F(y_1)}(F'(y_1)\cos(2\pi k y_1) - \sin(2\pi k y_1))\sin(2\pi p y_1)\,dy_1, \tag{5.27}$$

$$\beta_{kp} = 2\pi \int_0^{1/2} e^{2\pi k F(y_1)}(F'(y_1)\sin(2\pi k y_1) + \cos(2\pi k y_1))\cos(2\pi p y_1)\,dy_1, \tag{5.28}$$

$$a_p = \int_0^{1/2} F'(y_1)\sin(2\pi p y_1)\,dy_1, \tag{5.29}$$

$$b_p = \int_0^{1/2} \left[1 - \frac{1}{l_F}(1 + (F'(y_1))^2)^{1/2}\right]\cos(2\pi p y_1)\,dy_1. \tag{5.30}$$

To illustrate the use of the above result we consider the temperature distribution in an infinite cylinder with a regular wavy surface Σ (Fig. 5.3), making the following assump-

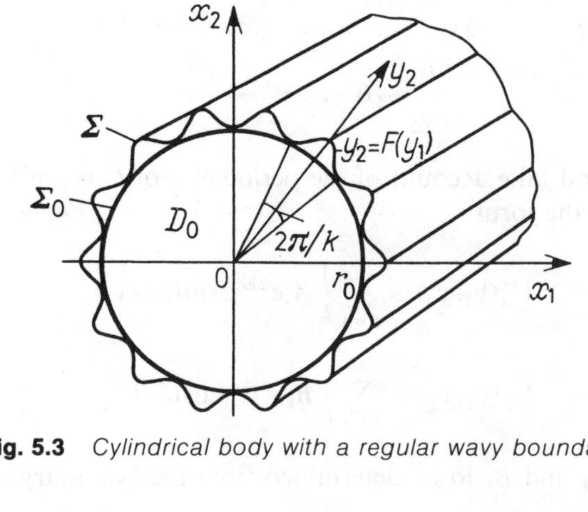

Fig. 5.3 *Cylindrical body with a regular wavy boundary.*

tions (see Kalamkarov 1986):

(a) T_0, the initial temperature of the cylinder, is constant (over the cylinder);
(b) T_c, the ambient temperature, is constant (both in space and time); and
(c) the heat exchange across Σ is governed by

$$-\lambda \frac{\partial T}{\partial n}\bigg|_{\Sigma} = \alpha(T_c - T|_{\Sigma}), \tag{5.31}$$

where $\boldsymbol{n}$ is outward normal to Σ, and $\lambda_{\alpha\beta} = \lambda\delta_{\alpha\beta}$. To describe the wavy shape of the surface Σ it is convenient to introduce the local coordinates

$$y_1 = \frac{r_0\varphi}{\varepsilon}, \qquad y_2 = \frac{r - r_0}{\varepsilon} = F(y_1), \tag{5.32}$$

where $\varepsilon = 2\pi r_0/k$, k being the number of sectors that comprise the circle D_0: $r < r_0$ in Fig. 5.3.

Using the asymptotic expansion

$$T(r, \varphi, t) = T^{(0)}(r, t) + \varepsilon T^{(1)}(r, y_1, y_2, t) + \cdots, \tag{5.33}$$

we obtain the following limiting problem for $T^{(0)}(r, t)$:

$$\frac{\partial^2 T^{(0)}}{\partial r^2} + \frac{1}{r}\frac{\partial T^{(0)}}{\partial r} - \frac{1}{\kappa^2}\frac{\partial T^{(0)}}{\partial t} = 0, \quad 0 < r < r_0,\ t > 0, \tag{5.34}$$

$$T^{(0)}(r, 0) = T_0, \tag{5.35}$$

$$-\lambda \frac{\partial T^{(0)}}{\partial r}\bigg|_{r=r_0} = l_F\alpha(T_c - T^{(0)}|_{r=r_0}), \tag{5.36}$$

where

$$l_F = \frac{k}{2\pi r_0}\int_0^1 \sqrt{1 + (F'(y_1))^2}\,\mathrm{d}y_1, \qquad \kappa^2 = \frac{\lambda}{c\rho}.$$

The function $T^{(1)}$ is written in the form

$$T^{(1)}(r, y_1, y_2, t) = U_\gamma(y)\frac{\partial T^{(0)}(r, t)}{\partial x_\gamma} = [U_1(y_1, y_2)\cos\varphi + U_2(y_1, y_2)\sin\varphi]\frac{\partial T^{(0)}(r, t)}{\partial r}, \tag{5.37}$$

and the functions $U_\gamma(y_1, y_2)$ are periodic in y_1 and solve the local problem (5.19)–(5.21).

For the region D_0: $r < r_0$, the solution of (5.34)–(5.36) is given by

$$T^{(0)}(r, t) = T_c + 2(T_c - T_0)\frac{l_F\alpha}{\lambda} r_0 \sum_{n=1}^{\infty} A_n \exp\left(-\frac{\kappa^2\mu_n^2 t}{r_0^2}\right) J_0\left(\frac{\mu_n r}{r_0}\right), \tag{5.38}$$

where

$$A_n = \left[J_0(\mu_n)(\mu_n^2 + l_F^2 \frac{\alpha^2}{\lambda^2} r_0^2) \right]^{-1}, \tag{5.39}$$

in which μ_n are the roots of the equation

$$\mu J'_0(\mu) + l_F \frac{\alpha}{\lambda} r_0 J_0(\mu) = 0.$$

6. LOCAL PROBLEMS AND EFFECTIVE COEFFICIENTS

It seems important to give some illustrations of how local problems arising from the use of asymptotic expansions of the form (4.4) can be treated. Consider, to start with, the two-dimensional, steady-state heat conduction problem for a rectangular region of a laminated material with lamina thickness ε. We assume, referring to Fig. 6.1, that the boundaries $x_2 = 0$ and $x_2 = l_2$ of the region D are thermally insulated and that at $x_1 = 0$ and $x_1 = l_1$ heat flows are prescribed. Under these circumstances the thermal and physical characteristics depend on x_1 alone (with period ε) and the problem of finding the temperature field $T^\varepsilon(x_1, x_2)$ reduces to the solution of the equation

$$\frac{\partial}{\partial x_\alpha}\left(\lambda_{\alpha\beta}\left(\frac{x_1}{\varepsilon}\right)\frac{\partial T^\varepsilon}{\partial x_\beta}\right) = 0 \quad \text{in } D, \tag{6.1}$$

subject to the conditions

$$\lambda_{2\beta}\frac{\partial T^\varepsilon}{\partial x_\beta} = 0 \quad \text{when } x_2 = 0,\ x_2 = l_2, \tag{6.2}$$

$$\mp \lambda_{1\beta}\frac{\partial T^\varepsilon}{\partial x_\beta} = \pm q(x_2) \quad \text{when } x_1 = 0,\ x_1 = l_1. \tag{6.3}$$

At large distances from the region D, the solution (6.1) takes the form

$$T^\varepsilon(\boldsymbol{x}) = T_0(\boldsymbol{x}) + \varepsilon T_1(\boldsymbol{x}, y_1) + \cdots \quad (y_1 = x_1/\varepsilon), \tag{6.4}$$

where, in accordance with (4.13),

$$T_1(\boldsymbol{x}, \boldsymbol{y}) = U_\gamma(y_1)\frac{\partial T_0(\boldsymbol{x})}{\partial x_\gamma} \quad (\gamma = 1, 2), \tag{6.5}$$

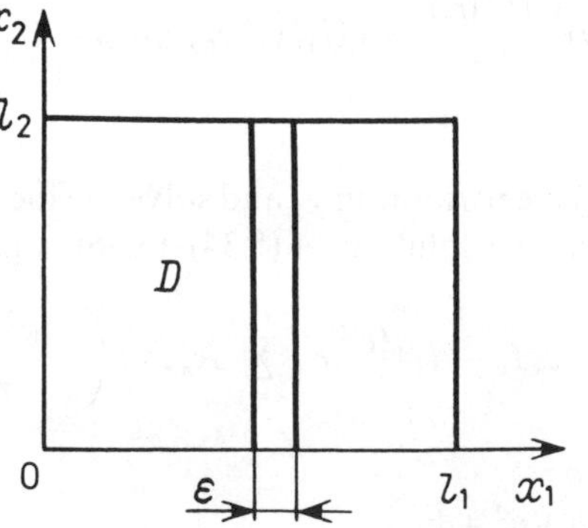

Fig. 6.1 *Laminated composite.*

and the functions $U_\gamma(y_1)$ are 1-periodic solutions of the one-dimensional local problem (cf. (4.14)):

$$-\frac{d}{dy_1}\left(\lambda_{11}(y_1)\frac{dU_\gamma(y_1)}{dy_1}\right)=\frac{d\lambda_{1\gamma}}{dy_1}, \quad 0<y_1<1. \tag{6.6}$$

Integrating (6.6) we get

$$U_\gamma(y_1)=-\int_0^{y_1}\lambda_{1\gamma}(\xi)\lambda_{11}^{-1}(\xi)\,d\xi+c_\gamma\int_0^{y_1}\lambda_{11}^{-1}(\xi)\,d\xi, \tag{6.7}$$

where the constants c_γ are to be determined from

$$U_\gamma(0)=U_\gamma(1). \tag{6.8}$$

It is easily verified that

$$c_\gamma=\int_0^1\lambda_{1\gamma}(\xi)\lambda_{11}^{-1}(\xi)\,d\xi\left(\int_0^1\lambda_{11}^{-1}(\xi)\,d\xi\right)^{-1}, \tag{6.9}$$

giving the solution

$$U_1(y_1)=\int_0^{y_1}\lambda_{11}^{-1}(\xi)\,d\xi\left(\int_0^1\lambda_{11}^{-1}(\xi)\,d\xi\right)^{-1}-y_1, \tag{6.10}$$

$$U_2(y_1)=\left(\int_0^1\lambda_{12}(\xi)\lambda_{11}^{-1}(\xi)\,d\xi\right)\left(\int_0^1\lambda_{11}^{-1}(\xi)\,d\xi\right)^{-1}\int_0^{y_1}\lambda_{11}^{-1}(\xi)\,d\xi - \int_0^{y_1}\lambda_{12}(\xi)\lambda_{11}^{-1}(\xi)\,d\xi, \tag{6.11}$$

so that the effective heat conductivities are found to be

$$\begin{aligned}
\tilde{\lambda}_{11}&=\left(\int_0^1\lambda_{11}^{-1}(\xi)\,d\xi\right)^{-1},\\
\tilde{\lambda}_{12}&=\left(\int_0^1\lambda_{12}(\xi)\lambda_{11}^{-1}(\xi)\,d\xi\right)\left(\int_0^1\lambda_{11}^{-1}(\xi)\,d\xi\right)^{-1},\\
\tilde{\lambda}_{22}&=\int_0^1\lambda_{22}(\xi)\,d\xi+\left(\int_0^1\lambda_{12}(\xi)\lambda_{11}^{-1}(\xi)\,d\xi\right)\left(\int_0^1\lambda_{11}^{-1}(\xi)\,d\xi\right)^{-1}\\
&\quad-\int_0^1\lambda_{12}^2(\xi)\lambda_{11}^{-1}(\xi)\,d\xi,
\end{aligned} \tag{6.12}$$

using (4.17). Averaging equation (6.1) for the homogenized medium now yields

$$\tilde{\lambda}_{11}\frac{\partial^2 T_0}{\partial x_1^2}+2\tilde{\lambda}_{12}\frac{\partial^2 T_0}{\partial x_1\partial x_2}+\tilde{\lambda}_{22}\frac{\partial^2 T_0}{\partial x_2^2}=0, \tag{6.13}$$

and by the use of (6.4) and (6.5) the boundary equations can be written as

$$\left(\lambda_{2\beta}(y_1)+\lambda_{21}(y_1)\frac{\mathrm{d}U_\beta}{\mathrm{d}y_1}\right)\frac{\partial T_0}{\partial x_\beta}=0 \quad \text{when } x_2=0,\ x_2=l_2, \tag{6.14}$$

$$\mp\left(\lambda_{1\beta}(y_1)+\lambda_{11}(y_1)\frac{\mathrm{d}U_\beta}{\mathrm{d}y_1}\right)\frac{\partial T_0}{\partial x_\beta}=\pm q(x_2) \quad \text{when } x_1=0,\ x_1=l_1. \tag{6.15}$$

giving

$$\tilde{\lambda}_{2\beta}\frac{\partial T_0}{\partial x_\beta}=0 \quad \text{when } x_2=0,\ x_2=l_2, \tag{6.16}$$

$$\mp\tilde{\lambda}_{1\beta}\frac{\partial T_0}{\partial x_\beta}=\pm q(x_2) \quad \text{when } x_1=0,\ x_1=l_1, \tag{6.17}$$

after the application of the averaging procedure.

It will be understood that the employment of averaged conditions at the boundaries $x_2=0$ and $x_2=l_2$ in the homogenized problems (6.13), (6.16) and (6.17) makes it necessary to invoke boundary layer type solutions in these vicinities. In what follows we consider the case $x_2=0$.

We introduce an additional term into (6.4) by writing

$$T^\varepsilon(\boldsymbol{x})=T_0(\boldsymbol{x})+\varepsilon(T_1(\boldsymbol{x},y_1)+T_1^{(n)}(\boldsymbol{x},\boldsymbol{y}))+\cdots \quad (\boldsymbol{y}=\boldsymbol{x}/\varepsilon,\ \boldsymbol{y}=(y_1,y_2)), \tag{6.18}$$

where $T_1(\boldsymbol{x},y_1)$ is determined from (6.5) and (6.6), and $T_1^{(n)}(\boldsymbol{x},\boldsymbol{y})$ may be represented as

$$T_1^{(n)}(\boldsymbol{x},\boldsymbol{y})=U_\beta^{(n)}(\boldsymbol{y})\frac{\partial T_0(\boldsymbol{x})}{\partial x_\beta}. \tag{6.19}$$

It can be shown that the function $U_\beta^{(n)}(\boldsymbol{y})$ solves the equation

$$\frac{\partial}{\partial y_\gamma}\left(\lambda_{\gamma\alpha}(y_1)\frac{\partial U_\beta^{(n)}(\boldsymbol{y})}{\partial y_\alpha}\right)=0 \tag{6.20}$$

in the strip $S=\{(y_1,y_2)\colon y_1\in[0,1],\ y_2\in[0,\infty)\}$ and is 1-periodic in y_1. Note also that

$$\frac{\partial U_\beta^{(n)}(\boldsymbol{y})}{\partial y_\alpha}\to 0 \quad \text{as } y_2\to\infty, \tag{6.21}$$

which, when combined with (6.20), gives

$$\int_0^1 \lambda_{2\alpha}(y_1)\frac{\partial U_\beta^{(n)}(\boldsymbol{y})}{\partial y_\alpha}\bigg|_{y_2=0}\mathrm{d}y_1=0, \tag{6.22}$$

using the periodicity property just mentioned. Now if the expansion (6.18) is introduced into (6.21) for $x_2=0$ and condition (6.16) is used, the condition to be satisfied by the boundary layer type solution may be expressed as

$$\lambda_{2\gamma}(y_1)\frac{\partial U_\beta^{(n)}(\boldsymbol{y})}{\partial y_\gamma}\bigg|_{y_2=0}=\tilde{\lambda}_{2\beta}-\lambda_{2\beta}(y_1)-\lambda_{21}(y_1)\frac{\mathrm{d}U_\beta(\boldsymbol{y})}{\mathrm{d}y_1}. \tag{6.23}$$

The problem of finding $U_\beta^{(n)}(y)$ thus reduces to that of solving (6.20) for a function satisfying conditions (6.21) and (6.23) which is 1-periodic in y_1.

In particular, if $\lambda_{\alpha\beta}(y_1) = \lambda(y_1)\delta_{\alpha\beta}$, the above problem is

$$\frac{\partial}{\partial y_2}\left(\lambda(y_1)\frac{\partial U_2^{(n)}(y)}{\partial y_\alpha}\right) = 0 \quad \text{in } S, \tag{6.24}$$

$$\lambda(y_1)\left.\frac{\partial U_2^{(n)}}{\partial y_2}\right|_{y_2=0} = \tilde{\lambda}_{22} - \lambda(y_1), \tag{6.25}$$

$$\frac{\partial U_2^{(n)}}{\partial y_\alpha} \to 0 \quad \text{as } y_2 \to \infty. \tag{6.26}$$

We note that in this case $U_1^{(n)} = 0$ and the set of non-zero effective coefficients is

$$\tilde{\lambda}_{11} = \left(\int_0^1 \lambda^{-1}(\xi)\,d\xi\right)^{-1}, \qquad \tilde{\lambda}_{22} = \int_0^1 \lambda(\xi)\,d\xi. \tag{6.27}$$

We further particularize by assuming the unit cell to be composed of two laminae with constant but different conductivities, $\lambda(y_1) = \lambda_1$ for $|y_1| < \gamma$ and $\lambda(y_1) = \lambda_2$ for $\gamma < |y_1| < \frac{1}{2}$ (Fig. 6.2). The function $U_2^{(n)}(y)$ will then satisfy Laplace's equation,

$$\frac{\partial^2 U_2^{(n)}}{\partial y_1^2} + \frac{\partial^2 U_2^{(n)}}{\partial y_2^2} = 0, \tag{6.28}$$

in regions

$$S_1 = \{(y_1, y_2): -\gamma < y_1 < \gamma, 0 < y_2 < \infty\}$$

and

$$S_2 = \{(y_1, y_2): \gamma < |y_1| < \tfrac{1}{2}, 0 < y_2 < \infty\},$$

and the interfacial contact conditions

$$\begin{aligned} U_2^{(n)}(-\gamma + 0, y_2) &= U_2^{(n)}(-\gamma - 0, y_2), \\ U_2^{(n)}(\gamma - 0, y_2) &= U_2^{(n)}(\gamma + 0, y_2), \end{aligned} \tag{6.29}$$

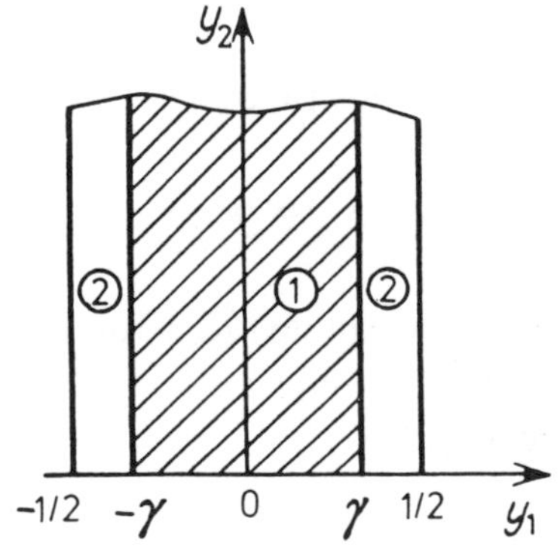

Fig. 6.2 *Unit cell of a two-component laminated composite.*

$$\lambda_1 \frac{\partial U_2^{(n)}}{\partial y_1}\bigg|_{y_1=-\gamma+0} = \lambda_2 \frac{\partial U_2^{(n)}}{\partial y_1}\bigg|_{y_1=-\gamma-0},$$

$$\lambda_1 \frac{\partial U_2^{(n)}}{\partial y_1}\bigg|_{y_1=\gamma-0} = \lambda_2 \frac{\partial U_2^{(n)}}{\partial y_1}\bigg|_{y_1=\gamma+0}. \tag{6.30}$$

Since the function $U_2^{(n)}$ is even with respect to y_1, it suffices to consider the second equations in (6.29) and (6.30). Substituting (6.27) into the boundary condition (6.25), the latter becomes

$$\frac{\partial U_2^{(n)}}{\partial y_2}\bigg|_{y_1=0} = \begin{cases} -(1-2\gamma)\left(1-\dfrac{\lambda_2}{\lambda_1}\right), & |y_1|<\gamma, \\ 2\gamma\dfrac{\lambda_1}{\lambda_2}\left(1-\dfrac{\lambda_2}{\lambda_1}\right), & \gamma<|y_1|<\tfrac{1}{2}. \end{cases} \tag{6.31}$$

In Appendix A of this book a special system of functions is developed in terms of which the solution to (6.28) can be conveniently expanded. The functions are continuous and mutually orthogonal on the segment $[-\tfrac{1}{2}, \tfrac{1}{2}]$ and undergo a jump of prescribed magnitude in their first derivatives at $y_1 = \pm\gamma$. Remembering (6.26), the solution, even with respect to y_1, may thus be represented as

$$U_2^{(n)}(y) = \tfrac{1}{2}A_0 + \sum_{n=1}^{\infty} A_n e^{-p_n y_2} \begin{cases} \cos(p_n y_1), & 0 \leqslant y_1 < \gamma, \\ \dfrac{\cos(p_n\gamma)}{\cos(p_n(\tfrac{1}{2}-\gamma))}\cos(p_n(\tfrac{1}{2}-y_1)), & \gamma<y_1<\tfrac{1}{2}, \end{cases} \tag{6.32}$$

where A_n are constants (of which A_0 may be set equal to zero), and p_n are the roots of the equation

$$\sin(p_n(\tfrac{1}{2}-\gamma))\cos(p_n\gamma) + \frac{\lambda_1}{\lambda_2}\cos(p_n(\tfrac{1}{2}-\gamma))\sin(p_n\gamma) = 0. \tag{6.33}$$

It is left as an exercise for the reader to show by substitution that solution (6.32) satisfies the conditions

$$U_2^{(n)}(\gamma-0, y_2) = U_2^{(n)}(\gamma+0, y_2),$$

$$\lambda_1 \frac{\partial U_2^{(n)}}{\partial y_1}\bigg|_{y_1=\gamma-0} = \lambda_2 \frac{\partial U_2^{(n)}}{\partial y_1}\bigg|_{y_1=\gamma+0}.$$

To determine the constants A_n, substitute (6.32) into (6.31), which gives

$$\sum_{n=1}^{\infty} p_n A_n \left\{ \begin{matrix} \cos(p_n y_1), & 0 \leqslant y_1 < \gamma \\ \dfrac{\cos(p_n\gamma)}{\cos(p_n(\tfrac{1}{2}-\gamma))}\cos(p_n(\tfrac{1}{2}-y_1)), & \gamma<y_1<\tfrac{1}{2} \end{matrix} \right\}$$

$$= \begin{cases} (1-2\gamma)\left(1-\dfrac{\lambda_2}{\lambda_1}\right), & 0 \leqslant y_1 < \gamma, \\ -2\gamma\dfrac{\lambda_1}{\lambda_2}\left(1-\dfrac{\lambda_2}{\lambda_1}\right), & \gamma < y_1 < \frac{1}{2} \end{cases}. \tag{6.34}$$

Using formula (A.20) of Appendix A now yields, for $\delta = \lambda_1/\lambda_2$ and $\mu = 1$,

$$A_n = -\frac{2}{p_n^2 C_n}\left(1-\frac{\lambda_2}{\lambda_1}\right)\left[(1-2\gamma)\,C_n^{(1)}\sin(p_n\gamma) + 2\gamma\frac{\lambda_1}{\lambda_2}\sin(p_n(p_n(\tfrac{1}{2}-\gamma))\right], \tag{6.35}$$

where

$$C_n^{(1)} = \sin(p_n(\tfrac{1}{2}-\gamma))\sin(p_n\gamma) - \frac{\lambda_1}{\lambda_2}\cos(p_n(\tfrac{1}{2}-\gamma))\cos(p_n\gamma), \tag{6.36}$$

$$C_n = \left(\tfrac{1}{2}-\gamma+\frac{\lambda_1}{\lambda_2}\gamma\right)\cos(p_n(\tfrac{1}{2}-\gamma))\cos(p_n\gamma)$$

$$-\left(\gamma+\frac{\lambda_1}{2\lambda_2}-\frac{\lambda_1}{\lambda_2}\gamma\right)\sin(p_n(\tfrac{1}{2}-\gamma))\sin(p_n\gamma). \tag{6.37}$$

We now proceed to apply the two-dimensional local analysis to the problem (4.32)–(4.33) for a unit cell taken in the form of a square with a circular hole of radius $r_0 < \frac{1}{2}$, as in Fig. 6.3. The task at hand is to solve the equation

$$\frac{\partial^2 U_\gamma}{\partial y_1^2} + \frac{\partial^2 U_\gamma}{\partial y_2^2} = 0 \quad \text{in } \mathscr{Y}, \tag{6.38}$$

subject to the condition

$$\left.\frac{\partial U_\gamma}{\partial r}\right|_{r=r_0} = \begin{cases} \cos\theta, & \gamma = 1, \\ \sin\theta, & \gamma = 2, \end{cases} \tag{6.39}$$

where r and θ are polar coordinates with origin at $y_1 = 0$, $y_2 = 0$. Clearly the functions $U_\gamma(y_1, y_2)$ must be 1-periodic with respect to both its arguments.

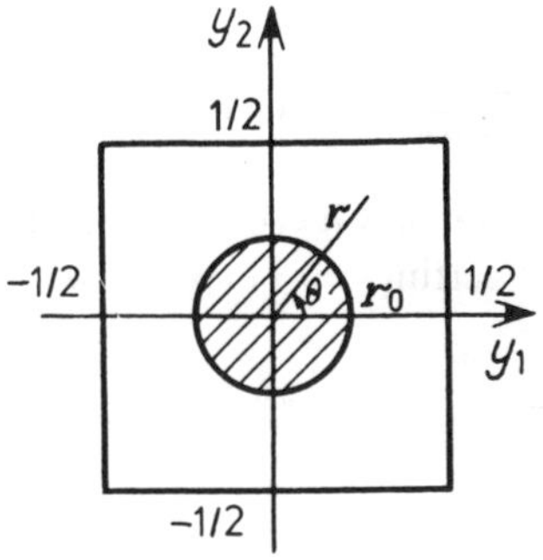

Fig. 6.3 *Unit cell containing a circular hole.*

Introducing a complex variable $z = y_1 + iy_2$, the (doubly periodic) harmonic function $U_1(y_1, y_2)$ can be expanded in a series of the type

$$U_1 = \mathrm{Re}\left\{A_0\left[z - \frac{1}{\pi}\zeta(z)\right] + \sum_{k=0}^{\infty} A_{2k+4} r_0^{2k+4} \frac{\wp^{(2k+1)}(z)}{(2k+3)!}\right\}, \tag{6.40}$$

where $\wp(z)$ is Weierstrass's elliptic function; $\zeta(z)$ is Weierstrass's zeta function, satisfying the quasi-periodicity conditions

$$\zeta(z+1) - \zeta(z) = \delta_1, \qquad \zeta(z+i) - \zeta(z) = \delta_2;$$

the constants A_0 and A_{2k+4} are real; and the constants δ_1 and δ_2 are related by

$$i\delta_1 - \delta_2 = 2\pi i$$

(Legendre's relation) and

$$i\delta_1 + \delta_2 = 0,$$

so that $\delta_1 = +\pi$ and $\delta_2 = -\pi i$. The above-mentioned periodicity property of $U_1(y_1, y_2)$ is readily deduced from the quasi-periodicity of $\zeta(z)$ and the double periodicity of Weierstrass's elliptic function and its derivatives.

From (6.40), using $\zeta'(z) = -\wp(z)$,

$$\frac{\partial U_1}{\partial r} = \mathrm{Re}\left\{A_0\left[1 - \frac{1}{\pi}\wp(z)\right]e^{i\theta} + e^{i\theta}\sum_{k=0}^{\infty} A_{2k+4} r_0^{2k+4} \frac{\wp^{(2k+2)}(z)}{(2k+3)!}\right\}. \tag{6.41}$$

Now use will be made of the expansions (Grigolyuk and Fil'shtinskii 1970)

$$\wp(z) = \frac{1}{z^2} + \sum_{j=0}^{\infty} r_{j,0} z^{2j}, \tag{6.42}$$

$$\frac{\wp^{(2k+2)}(z)}{(2k+3)!} = \frac{1}{z^{2k+4}} + \sum_{j=0}^{\infty} z_{j,k+1} z^{2j}, \tag{6.43}$$

where

$$r_{j,k} = \frac{(2j+2k+1)!\,g_{j+k+1}}{(2j)!(2k+1)!\,2^{2j+2k+2}},$$

$$g_{j+k+1} = \sum_{m,n}{}' \frac{1}{T_{m,n}^{2j+2k+2}}, \qquad T_{m,n} = \tfrac{1}{2}(m + in),$$

where a prime on $\sum_{m,n}$ indicates that $m = n = 0$ is excluded in the summation. Substituting (6.42) and (6.43) into (6.41) and setting $z = r\exp(i\theta)$, we find

$$\left.\frac{\partial U_1}{\partial r}\right|_{r=r_0} = \left[A_0\left(1 + \frac{1}{\pi r_0^2}\right) + \sum_{k=0}^{\infty} A_{2k+4} r_0^{2k+4} r_{0,k+1}\right]\cos\theta$$
$$+ \sum_{j=1}^{\infty}\left[A_0 \frac{1}{\pi} r_0^{2j} r_{j,0} + A_{2j+2} + \sum_{k=0}^{\infty} A_{2k+4} r_0^{2k+2j+4} r_{j,k+1}\right]\cos(2j+1)\theta, \tag{6.44}$$

which when combined with the boundary condition (6.39), gives an infinite system of algebraic equations for the constants A_0 and A_{2k}:

$$A_0\left(1+\frac{1}{\pi r_0^2}\right)+\sum_{k=1}^{\infty} A_{2k+2}r_0^{2k+2}r_{0,k}=1, \tag{6.45}$$

$$A_0\frac{1}{\pi r_0^2}r_0^{2j+2}r_{j,0}+A_{2j+2}+\sum_{k=1}^{\infty} A_{2k+2}r_0^{2k+2+2j}r_{j,k}=0. \tag{6.46}$$

Eliminating A_0 from (6.45) and (6.46) results in the following system of equations for $A_{2k+2}(k=1,2,3,\ldots)$:

$$A_{2j+2}+\sum_{k=1}^{\infty} A_{2k+2}r_0^{2k+2j+2}\left(r_{j,k}-\frac{r_0^2}{\pi r_0^2+1}r_{j,0}r_{0,k}\right)=-\frac{r_0^{2j+2}r_{j,0}}{(\pi r_0^2+1)} \quad (j=1,2,\ldots). \tag{6.47}$$

Using (6.43) and the well-known result (Grigolyuk and Fil'shtinskii 1970)

$$\zeta(z)=\frac{1}{z}-\sum_{j=1}^{\infty}\frac{r_{j,0}}{(2j+1)}z^{2j+1}, \tag{6.48}$$

(6.40) becomes

$$\begin{aligned}U_1=\mathrm{Re}\Bigg\{&\left[A_0+\sum_{k=1}^{\infty}A_{2k+2}r_0^{2k+2}\frac{r_{\mathrm{i},k-1}}{k(2k+1)}\right]z\\&+\sum_{j=1}^{\infty}\left[\frac{A_0}{\pi}\frac{r_{j,0}}{(2j+1)}+(2j+2)\sum_{k=1}^{\infty}A_{2k+2}\frac{r_0^{2k+2}r_{j+1,k-1}}{2k(2k+1)}\right]z^{2j+1}\\&-A_0\frac{1}{\pi z}-\sum_{k=1}^{\infty}A_{2k+2}\frac{r_0^{2k+2}}{(2k+1)}\frac{1}{z^{2k+1}}\Bigg\}.\end{aligned} \tag{6.49}$$

The function $U_2(y_1, y_2)$, doubly periodic in its arguments, now is defined by analogy with (6.40) as

$$U_2=-\mathrm{Re}\left\{A_0\left[\mathrm{i}z-\frac{1}{\pi}\zeta(\mathrm{i}z)\right]+\sum_{k=0}^{\infty}A_{2k+4}\frac{r_0^{2k+4}\wp^{(2k+1)}(\mathrm{i}z)}{(2k+3)!}\right\}, \tag{6.50}$$

where again A_0 and A_{2k+4} are real constants.

Taking the derivative $\partial U_2/\partial r$ and making use of the expansions (6.42) and (6.43), we obtain, for $z=r_0\exp(\mathrm{i}\theta)$,

$$\begin{aligned}\left.\frac{\partial U_2}{\partial r}\right|_{r=r_0}=&\left[A_0\left(1+\frac{1}{\pi r_0^2}\right)+\sum_{k=1}^{\infty}A_{2k+2}r_0^{2k+2}r_{0,k}\right]\sin\theta\\&+\sum_{j=1}^{\infty}(-1)^j\left[A_0\frac{1}{\pi}r_0^{2j}r_{j,0}+A_{2j+2}+\sum_{k=1}^{\infty}A_{2k+2}r_0^{2k+2}r_{j,k}\right]\sin(2j+1)\theta.\end{aligned} \tag{6.51}$$

This is now substituted into the boundary condition (6.39) to give the system

(6.45)–(6.46) for determining the constants A_0 and A_{2k+2}; thus

$$U_2 = -\operatorname{Re}\left\{ i\left[A_0 + \sum_{k=1}^{\infty} A_{2k+2} r_0^{2k+2} \frac{r_{1,k-1}}{k(2k+1)} \right] z + i \sum_{k=1}^{\infty} (-1)^j \left[\frac{A_0 r_{j,0}}{\pi(2j+1)} + (2j+2) \sum_{k=1}^{\infty} A_{2k+2} \frac{r_0^{2k+2} r_{j+1,k-1}}{2k(2k+1)} \right] z^{2j+1} + A_0 \frac{i}{\pi z} + i \sum_{k=1}^{\infty} A_{2k+2} \frac{r_0^{2k+2}(-1)^k}{(2k+1)} \frac{1}{z^{2k+1}} \right\}. \tag{6.52}$$

Using the expansions (6.49) and (6.52) along with (4.38), the effective coefficients involved in the averaged equation (4.37) are

$$q_{11} = \frac{1}{(1-\pi r_0^2)} \int_{\mathscr{Y}} \left(1 - \frac{\partial U_1}{\partial y_1}\right) dy_1 dy_2 = 1 + \frac{r_0}{(1-\pi r_0^2)} \int_0^{2\pi} U_1(r_0, \theta) \cos\theta \, d\theta,$$

$$q_{22} = \frac{1}{(1-\pi r_0^2)} \int_{\mathscr{Y}} \left(1 - \frac{\partial U_2}{\partial y_2}\right) dy_1 dy_2 = 1 + \frac{r_0}{(1-\pi r_0^2)} \int_0^{2\pi} U_2(r_0, \theta) \sin\theta \, d\theta, \tag{6.53}$$

$$q_{12} = q_{21} = 0,$$

or, substituting (6.49) and (6.52) into (6.53),

$$q_{11} = q_{22} = 1 - A_0 + \frac{\pi r_0^2}{(1-\pi r_0^2)} \sum_{k=1}^{\infty} A_{2k+2} r_0^{2k+2} \frac{r_{1,k-1}}{k(2k+1)}, \quad q_{12} = q_{21} = 0. \tag{6.54}$$

The equality $q_{11} = q_{22}$ is a clear consequence of the symmetry of the unit cell with respect to the y_1 and y_2 axes.

3 Elasticity of Regular Composite Structures

The asymptotic methods developed for boundary value problems described by partial differential equations with rapidly oscillating coefficients are relevant to the study of the elastic behaviour of composite materials with a regular structure.

The method of homogenization we have outlined in the preceding section was introduced into classical linear elasticity theory in the 1970s by the pioneering work of Duvaut and co-workers (Duvaut 1976, 1977, Duvaut and Metellus 1976, Artola and Duvaut 1977) and Pobedrya and Gorbachev (1977). It has been shown (Duvaut 1976, Sanchez-Palencia 1980) that for a highly heterogeneous (composite) medium with a periodic structure, the solution of the original linear elasticity problem converges to that of the homogenized problem in the limit $\varepsilon \rightarrow 0$, ε being the (relative) dimension of the periodicity cell of the system. Useful estimates for the remainder of the asymptotic ε-expansions of the theory of elasticity have been given by Bakhvalov and Panasenko (1984); further work in this direction has been reviewed by Kalamkarov *et al.* (1987a).

7. HOMOGENIZATION OF THE LINEAR ELASTICITY PROBLEM

Consider an anisotropic elastic body of regular structure occupying a bounded region Ω in R^3 space with a smooth boundary $\partial\Omega = \partial_1\Omega \cup \partial_2\Omega (\partial_1\Omega \cap \partial_2\Omega = 0)$. We assume that the region Ω is made up by the periodic repetition of the unit cell Y in the form of a parallelepiped with dimensions $\varepsilon Y_i (i = 1, 2, 3)$. The equations of motion and the boundary conditions of the linear elasticity problem of interest may be written in the form

$$\frac{\partial \sigma_{ij}^{(\varepsilon)}}{\partial x_j} + P_i = \rho^{(\varepsilon)} \frac{\partial^2 u_i}{\partial t^2} \quad \text{in } \Omega, \tag{7.1}$$

$$u_i^{(\varepsilon)} = 0 \quad \text{on } \partial_1\Omega, \qquad \sigma_{ij}^{(\varepsilon)} n_j = 0 \quad \text{on } \partial_2\Omega, \tag{7.2}$$

$$u_i^{(\varepsilon)} = \frac{\partial u_i^{(\varepsilon)}}{\partial t} = 0 \quad \text{when } t = 0. \tag{7.3}$$

In the above,

$$\sigma_{ij}^{(\varepsilon)} = c_{ijkl}\left(\frac{\boldsymbol{x}}{\varepsilon}\right) e_{kl}^{(\varepsilon)}, \qquad e_{kl}^{(\varepsilon)} = \frac{1}{2}\left(\frac{\partial u_k^{(\varepsilon)}}{\partial x_l} + \frac{\partial u_l^{(\varepsilon)}}{\partial x_k}\right),$$

the quantities P_i are independent of ε and represent the imposed body forces, and

$\rho^{(\varepsilon)} = \rho(\boldsymbol{y})$ is the mass density function of the material. The coefficients $c_{ijkl}(\boldsymbol{y})$ satisfying the symmetry and ellipticity conditions

$$\begin{gathered} c_{ijkl}(\boldsymbol{y}) = c_{jikl}(\boldsymbol{y}) = c_{ijlk}(\boldsymbol{y}) = c_{klij}(\boldsymbol{y}), \\ \begin{cases} c_{ijkl}(\boldsymbol{y})\eta_{ij}\eta_{kl} \geqslant \kappa\eta_{mn}\eta_{mn} \\ \text{for any symmetric matrix } \|\eta_{mn}\|, \kappa = \text{const.}, \kappa > 0, \end{cases} \end{gathered} \tag{7.4}$$

are the definite Y-periodic functions of $\boldsymbol{y} = \boldsymbol{x}/\varepsilon$ which are defined on the unit cell and may be piecewise-constant in practical applications. This is illustrated in Fig. 7.1, where the unit cell consists of two dissimilar materials, one for the matrix and the other for fibres, and the functions $c_{ijkl}(\boldsymbol{y})$ are defined by

$$c_{ijkl}(\boldsymbol{y}) = \begin{cases} c_{ijkl}^{(\mathrm{I})} & \text{in } Y_{\mathrm{I}}, \\ c_{ijkl}^{(\mathrm{M})} & \text{in } Y_{\mathrm{M}}, \end{cases} \tag{7.5}$$

with $c_{ijkl}^{(\mathrm{I})}$ and $c_{ijkl}^{(\mathrm{M})}$ constant. Note also that at the interface between the matrix and fibre materials, the conditions

$$[\![\sigma_{ij}^{(\varepsilon)} n_j]\!] = 0, \qquad [\![u_i^{(\varepsilon)}]\!] = 0 \tag{7.6}$$

must be satisfied, where $[\![A]\!]$ indicates a jump in the value of A.

In the spirit of the two-scale expansion method, and following Sanchez-Palencia (1980) and Bakhvalov and Panasenko (1984), we begin by writing the series representations

$$u_i^{(\varepsilon)} = u_i^{(0)}(\boldsymbol{x}, t) + \varepsilon u_i^{(1)}(\boldsymbol{x}, \boldsymbol{y}, t) + \cdots, \tag{7.7}$$

$$e_{ij}^{(\varepsilon)} = e_{ij}^{(0)}(\boldsymbol{x}, \boldsymbol{y}, t) + \varepsilon e_{ij}^{(1)}(\boldsymbol{x}, \boldsymbol{y}, t) + \cdots, \tag{7.8}$$

$$\sigma_{ij}^{(\varepsilon)} = \sigma_{ij}^{(0)}(\boldsymbol{x}, \boldsymbol{y}, t) + \varepsilon \sigma_{ij}^{(1)}(\boldsymbol{x}, \boldsymbol{y}, t) + \cdots, \tag{7.9}$$

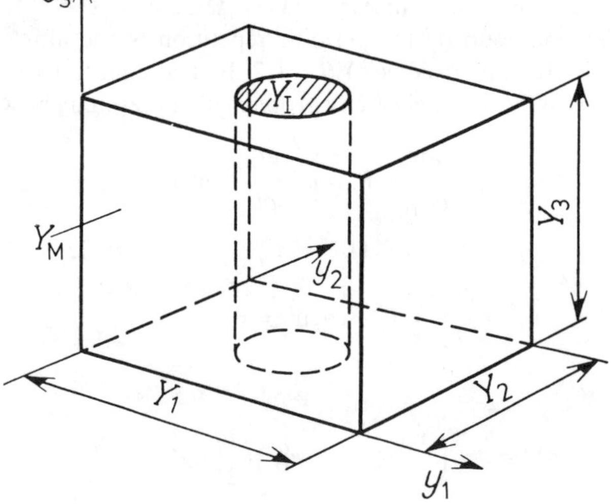

Fig. 7.1 *Unit cell of a fibre composite.*

for the solutions of the problem, where

$$\begin{aligned} e_{ij}^{(0)}(x,y,t) &= \frac{1}{2}\left(\frac{\partial u_i^{(0)}}{\partial x_j}+\frac{\partial u_j^{(0)}}{\partial x_i}\right)+\frac{1}{2}\left(\frac{\partial u_i^{(1)}}{\partial y_j}+\frac{\partial u_j^{(1)}}{\partial y_i}\right), \\ e_{ij}^{(1)}(x,y,t) &= \frac{1}{2}\left(\frac{\partial u_i^{(1)}}{\partial x_j}+\frac{\partial u_j^{(1)}}{\partial x_i}\right)+\frac{1}{2}\left(\frac{\partial u_i^{(2)}}{\partial y_j}+\frac{\partial u_j^{(2)}}{\partial y_i}\right), \end{aligned} \tag{7.10}$$

$$\begin{aligned} \sigma_{ij}^{(0)}(x,y,t) &= c_{ijkh}(y)\frac{\partial u_k^{(0)}}{\partial x_h}+c_{ijkh}(y)\frac{\partial u_k^{(1)}}{\partial y_h}, \\ \sigma_{ij}^{(1)}(x,y,t) &= c_{ijkh}(y)\frac{\partial u_k^{(1)}}{\partial x_h}+c_{ijkh}(y)\frac{\partial u_k^{(2)}}{\partial y_h}. \end{aligned} \tag{7.11}$$

Using (7.7) and (7.9) in (7.1) and retaining the terms of $O(\varepsilon^{-1})$ and $O(\varepsilon^0)$ we find

$$\frac{\partial \sigma_{ij}^{(0)}}{\partial y_j}=0, \tag{7.12}$$

$$\frac{\partial \sigma_{ij}^{(0)}}{\partial x_j}+\frac{\partial \sigma_{ij}^{(1)}}{\partial y_j}+P_i=\rho^{(\varepsilon)}\frac{\partial^2 u_i^{(0)}}{\partial t^2}. \tag{7.13}$$

Equation (7.11) is now substituted into (7.12) to give

$$\frac{\partial}{\partial y_j}\left(c_{ijkh}(y)\frac{\partial u_k^{(1)}(x,y,t)}{\partial y_h}\right)=-\frac{\partial u_k^{(0)}(x,t)}{\partial x_l}\frac{\partial c_{ijkl}(y)}{\partial y_j}. \tag{7.14}$$

Now if we consider x and t as parameters in this last equation, then writing

$$u_n^{(1)}(x,y,t)=N_n^{kl}(y)\frac{\partial u_k^{(0)}(x,t)}{\partial x_l} \tag{7.15}$$

we obtain the following local problems for determining the Y-periodic functions $N_n^{kl}(y)$ $(n,k,l=1,2,3)$:

$$\frac{\partial}{\partial y_j}\left(c_{ijnh}(y)\frac{\partial N_n^{kl}(y)}{\partial y_h}\right)=-\frac{\partial c_{ijkl}(y)}{\partial y_j}. \tag{7.16}$$

Introducing the notation

$$\tau_{ij}^{kl}(y)=c_{ijnh}(y)\frac{\partial N_n^{kl}(y)}{\partial y_h}, \tag{7.17}$$

(7.16) can be written in the form

$$\frac{\partial \tau_{ij}^{kl}(y)}{\partial y_j}=-\frac{\partial c_{ijkl}(y)}{\partial y_j}, \tag{7.18}$$

which is familiar from the ordinary theory of elasticity.

Upon substituting (7.15) we find

$$\sigma_{ij}^{(0)}(\boldsymbol{x},\boldsymbol{y},t)=\left(c_{ijkl}(\boldsymbol{y})+c_{ijnh}\frac{\partial N_n^{kl}(\boldsymbol{y})}{\partial y_h}\right)\frac{\partial u_k^{(0)}(\boldsymbol{x},t)}{\partial x_l}=(c_{ijkl}(\boldsymbol{y})+\tau_{ij}^{kl}(\boldsymbol{y}))\frac{\partial u_k^{(0)}(\boldsymbol{x},t)}{\partial x_l}, \tag{7.19}$$

$$\sigma_{ij}^{(1)}(\boldsymbol{x},\boldsymbol{y},t)=c_{ijnh}(\boldsymbol{y})N_n^{kl}(\boldsymbol{y})\frac{\partial^2 u_k^{(0)}(\boldsymbol{x},t)}{\partial x_l \partial x_h}+c_{ijkh}(\boldsymbol{y})\frac{\partial u_k^{(2)}}{\partial y_h}. \tag{7.20}$$

The equations of motion (equilibrium) of the homogenized body can now be derived from (7.13) which, taken together with (7.20), may be considered as a system of equations to determine the functions $u_k^{(2)}(\boldsymbol{x},\boldsymbol{y},t)$ in Y. Since the functions $\sigma_{ij}^{(1)}$ are Y-periodic in $\boldsymbol{y}$, it follows that the equations

$$\frac{\partial \sigma_{ij}^{(1)}}{\partial y_j}=\rho^{(\varepsilon)}\frac{\partial^2 u_i^{(0)}}{\partial t^2}-P_i-\frac{\partial \sigma_{ij}^{(0)}}{\partial x_j} \tag{7.21}$$

will have a unique solution if

$$\frac{1}{|Y|}\int_Y\left[\rho^{(\varepsilon)}\frac{\partial^2 u_i^{(0)}}{\partial t^2}-P_i-\frac{\partial \sigma_{ij}^{(0)}}{\partial x_j}\right]\mathrm{d}\boldsymbol{y}=0, \tag{7.22}$$

where the integral is performed over the volume $|Y|$ of the unit cell of the problem. Note that (7.22) is a consequence of the relation

$$\int_Y \frac{\partial \sigma_{ij}^{(1)}}{\partial y_j}\mathrm{d}\boldsymbol{y}=\int_{\partial Y}\sigma_{ij}^{(1)}n_j\,\mathrm{d}S_Y=0, \tag{7.23}$$

where n_j is the unit vector in the outward normal direction to the surface ∂Y and the integral has opposite signs on opposite sides of this surface.

The volume averaging procedure applied to the right-hand side of (7.21) now leads to the following macroscopic equations of motion:

$$\frac{\partial\langle\sigma_{ij}^{(0)}\rangle}{\partial x_j}+P_i=\langle\rho\rangle\frac{\partial^2 u_i^{(0)}(\boldsymbol{x},t)}{\partial t^2}, \tag{7.24}$$

where

$$\langle\sigma_{ij}^{(0)}\rangle=\frac{1}{|Y|}\int_Y \sigma_{ij}^{(0)}\,\mathrm{d}\boldsymbol{y}=\langle C_{ijkl}\rangle\frac{\partial u_k^{(0)}(\boldsymbol{x},t)}{\partial x_l} \tag{7.25}$$

and

$$\langle C_{ijkl}\rangle=\frac{1}{|Y|}\int_Y (c_{ijkl}(\boldsymbol{y})+\tau_{ij}^{kl}(\boldsymbol{y}))\,\mathrm{d}\boldsymbol{y} \tag{7.26}$$

are the effective elastic constants of the composite.

The above method for treating elasticity theory problems corresponds to the so-called zeroth-order approximation in which one starts by solving problem (7.16) for the local functions $N_n^{kl}(\boldsymbol{y})$ and then proceeds to calculate the effective moduli (7.26). This is followed by the application of the effective moduli method to the homogenized problem, that is, the solution of (7.24)–(7.26) satisfying the conditions

$$u_i^{(0)} = 0 \quad \text{on } \partial_i\Omega, \qquad \langle \sigma_{ij}^{(0)} \rangle n_j = 0 \quad \text{on } \partial_2\Omega \tag{7.27}$$

is found, and then the stress tensor components are calculated using formula (7.19). We thus see that we have only to solve the homogenized problem (7.24)–(7.27) to approximately predict the microstresses and microdisplacements in the unit cell. If the number of cells is sufficiently large in the system, formula (7.19) estimates quite closely the state of stress in an individual cell of the composite; for laminated elastic composites, this fact has been verified by comparison with exact results due to Gorbachev (1979).

Pobedrya (1985) discusses in considerable detail the application of this theory to the mechanics of a physically non-linear, non-homogeneous deformable solid with a regular structure. The mechanical behaviour of the solid is examined in terms of the displacement and stress variables, and the relationship between the stress and (infinitesimal) strain tensors is given in the very general operator form

$$\sigma_{ij} = F_{ij}(\boldsymbol{x}, \text{grad } \mathbf{u}), \tag{7.28}$$

in which $\mathbf{u}$ is the displacement vector, and the presence of the coordinate $\boldsymbol{x}$ reflects the non-homogeneity of the medium.

We now propose to discuss what modifications must be introduced into the general homogenization scheme when boundary effects are, or may be, important. Consider a two-dimensional elasticity problem for a half-plane with a periodic non-homogeneity and with specified stresses at the boundary (cf. Fig. 7.2). The equations of equilibrium are given in the form

$$\frac{\partial \sigma_{i\alpha}^{(\varepsilon)}}{\partial x_\alpha} = 0, \quad |x_1| < \infty, \; x_2 > 0, \tag{7.29}$$

and the appropriate elasticity relations are written as

$$\sigma_{i\alpha}^{(\varepsilon)}(\boldsymbol{x}) = c_{i\alpha k\beta}(\boldsymbol{y}) \frac{\partial u_k^{(\varepsilon)}(\boldsymbol{x})}{\partial x_\beta}, \tag{7.30}$$

with the functions $c_{i\alpha k\beta}(\boldsymbol{y})$ being 1-periodic in the variables

$$\boldsymbol{y} = \boldsymbol{x}/\varepsilon, \quad \boldsymbol{x} = (x_1, x_2) \quad (\alpha, \beta = 1, 2;\; i, k = 1, 2, 3).$$

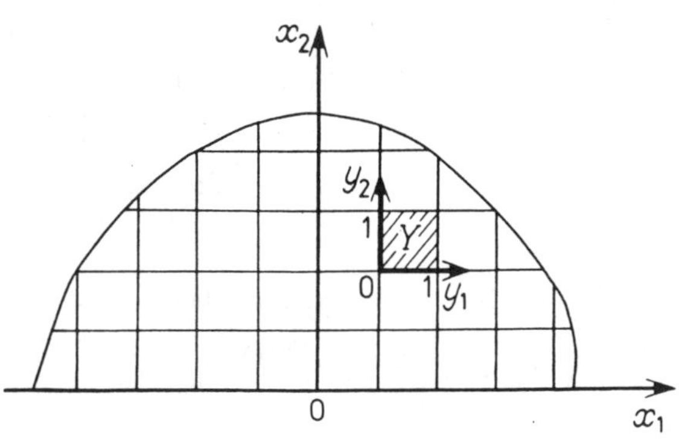

Fig. 7.2 *Half-plane with a periodic non-homogeneity and its unit cell.*

The stress boundary conditions are stated by

$$\sigma_{i2}(x_1, 0) = p_i(x_1). \tag{7.31}$$

To get started we represent the displacements and stress fields by expansions of the form

$$u_k^{(\varepsilon)} = u_k^{(0)}(x) + \varepsilon u_k^{(1)}(x, y) + \cdots, \tag{7.32}$$

$$\sigma_{i\alpha}^{(\varepsilon)} = \sigma_{i\alpha}^{(0)}(x, y) + \varepsilon \sigma_{i\alpha}^{(1)}(x, y) + \cdots, \tag{7.33}$$

where

$$\sigma_{i\alpha}^{(0)}(x, y) = c_{i\alpha k\beta}(y)\left(\frac{\partial u_k^{(0)}(x)}{\partial x_\beta} + \frac{\partial u_k^{(1)}(x, y)}{\partial y_\beta}\right), \tag{7.34}$$

$$\sigma_{i\alpha}^{(1)}(x, y) = c_{i\alpha k\beta}(y)\left(\frac{\partial u_k^{(1)}(x, y)}{\partial x_\beta} + \frac{\partial u_k^{(2)}(x, y)}{\partial y_\beta}\right). \tag{7.35}$$

Using (7.33) in (7.29) and (7.31) yields the following set of problems:

$$\begin{cases} \dfrac{\partial \sigma_{i\alpha}^{(0)}(x, y)}{\partial y_\alpha} = 0, & (7.36) \\ \sigma_{i2}^{(0)}|_{x_2=0} = p_i(x_1), & (7.37) \end{cases}$$

$$\begin{cases} \dfrac{\partial \sigma_{i\alpha}^{(0)}(x, y)}{\partial x_\alpha} + \dfrac{\partial \sigma_{i\alpha}^{(1)}(x, y)}{\partial y_\alpha} = 0, & (7.38) \\ \sigma_{i2}^{(1)}|_{x_2=0} = 0. & (7.39) \end{cases}$$

Combining (7.34) and (7.35) gives

$$\begin{cases} \dfrac{\partial}{\partial y_\alpha}\left[c_{i\alpha k\beta}(y)\dfrac{\partial u_k^{(1)}(x, y)}{\partial y_\beta}\right] = -\dfrac{\partial c_{i\alpha k\beta}(y)}{\partial y_\alpha}\dfrac{\partial u_k^{(0)}(x)}{\partial x_\beta}, & (7.40) \\ c_{i2k\beta}(y)|_{y_2=0}\left.\dfrac{\partial u_k^{(1)}(x, y)}{\partial y_\beta}\right|_{y_2=0} = -c_{i2k\beta}(y)|_{y_2=0}\left.\dfrac{\partial u_k^{(0)}(x)}{\partial x_\beta}\right|_{x_2=0} + p_i(x_1), & (7.41) \end{cases}$$

$$\begin{cases} \dfrac{\partial}{\partial y_\alpha}\left[c_{i\alpha k\beta}(y)\dfrac{\partial u_k^{(2)}(x, y)}{\partial y_\beta}\right] = -\dfrac{\partial c_{i\alpha k\beta}(y)}{\partial y_\alpha}\dfrac{\partial u_k^{(1)}(x, y)}{\partial x_\beta} - c_{i\alpha k\beta}(y)\left[\dfrac{\partial^2 u_k^{(1)}(x, y)}{\partial y_\beta \partial x_\alpha}\right. \\ \qquad\qquad \left. + \dfrac{\partial^2 u_k^{(1)}(x, y)}{\partial y_\alpha \partial x_\beta} + \dfrac{\partial^2 u_k^{(0)}(x)}{\partial x_\alpha \partial x_\beta}\right], & (7.42) \\ c_{i2k\beta}(y)|_{y_2=0}\left.\dfrac{\partial u_k^{(2)}(x, y)}{\partial y_\beta}\right|_{y_2=0} = -c_{i2k\beta}(y)|_{y_2=0}\dfrac{\partial u_k^{(1)}(x, y)}{\partial x_\beta}. & (7.43) \end{cases}$$

A solution to the problem (7.40)–(7.41) is assumed here in the form

$$u_n^{(1)}(x, y) = N_n^{k\beta}(y)\frac{\partial u_k^{(0)}(x)}{\partial x_\beta} + u_n^{(1,2)}(x, y), \tag{7.44}$$

where the function $N_n^{k\beta}(y)$ is 1-periodic in both y_1 and y_2 and is to be determined from the local problem

$$\frac{\partial}{\partial y_\alpha}\left[c_{i\alpha n\gamma}(\boldsymbol{y})\frac{\partial N_n^{k\beta}(\boldsymbol{y})}{\partial y_\gamma}\right] = -\frac{\partial c_{i\alpha k\beta}(\boldsymbol{y})}{\partial y_\alpha} \tag{7.45}$$

on the periodicity cell, whereas $u_n^{(1,2)}(\boldsymbol{x},\boldsymbol{y})$ is 1-periodic only in y_1 and is solved from the problem

$$\frac{\partial}{\partial y_\alpha}\left[c_{i\alpha k\beta}(\boldsymbol{y})\frac{\partial u_k^{(1,2)}(\boldsymbol{x},\boldsymbol{y})}{\partial y_\beta}\right] = 0, \quad 0 < y_1 < 1, y_2 > 0, \tag{7.46}$$

$$c_{i2k\beta}(\boldsymbol{y})\left.\frac{\partial u_k^{(1,2)}(\boldsymbol{x},\boldsymbol{y})}{\partial y_\beta}\right|_{y_2=0} = -\left[c_{i2k\beta}(\boldsymbol{y}) + c_{i2n\gamma}(\boldsymbol{y})\frac{\partial N_n^{k\beta}(\boldsymbol{y})}{\partial y_\gamma}\right]\Bigg|_{y_2=0}\left.\frac{\partial u_k^{(0)}}{\partial x_\beta}\right|_{x_2=0} + p_i(x_1), \tag{7.47}$$

$$u_k^{(1,2)}(\boldsymbol{x},\boldsymbol{y}) \to 0 \quad \text{as } y_2 \to +\infty. \tag{7.48}$$

Integrating (7.46) over the strip $0 < y_1 < 1$, $y_2 > 0$, using (7.47) and (7.48), and recalling the periodicity of $u_k^{(1,2)}(\boldsymbol{x},\boldsymbol{y})$ in y_1, the following solvability condition for the problem (7.46)–(7.48) is obtained (see Kalamkarov *et al.* 1990):

$$C^*_{i2k\beta}\left.\frac{\partial u_k^{(0)}(\boldsymbol{x})}{\partial x_\beta}\right|_{x_2=0} = p_i(x_1), \tag{7.49}$$

$$C^*_{i2k\beta} = \int_0^1 C_{i2k\beta}(y_1, 0)\,\mathrm{d}y_1, \tag{7.50}$$

$$C_{i2k\beta}(y_1, 0) = \left[c_{i2k\beta}(\boldsymbol{y}) + c_{i2n\gamma}(\boldsymbol{y})\frac{\partial N_n^{k\beta}(\boldsymbol{y})}{\partial y_\gamma}\right]\Bigg|_{y_2=0}. \tag{7.51}$$

We next employ (7.49) to eliminate $p_i(x_1)$ from (7.47), and represent $u_k^{(1,2)}(\boldsymbol{x},\boldsymbol{y})$ in the form

$$u_n^{(1,2)}(\boldsymbol{x},\boldsymbol{y}) = N_n^{(1)k\beta}(\boldsymbol{y})\frac{\partial u_k^{(0)}(\boldsymbol{x})}{\partial x_\beta}, \tag{7.52}$$

where the function $N_n^{(1)k\beta}(\boldsymbol{y})$ is 1-periodic in y_1 and satisfies the equation

$$\frac{\partial}{\partial y_\alpha}\left[c_{i\alpha n\gamma}(\boldsymbol{y})\frac{\partial N_n^{(1)k\beta}(\boldsymbol{y})}{\partial y_\gamma}\right] = 0, \quad 0 < y_1 < 1, y_2 > 0, \tag{7.53}$$

and the conditions

$$c_{i2n\gamma}(\boldsymbol{y})|_{y_2=0}\left.\frac{\partial N_n^{(1)k\beta}(\boldsymbol{y})}{\partial y_\gamma}\right|_{y_2=0} = [C^*_{i2k\beta} - C_{i2k\beta}(y_1, 0)], \tag{7.54}$$

$$N_n^{(1)k\beta}(\boldsymbol{y}) \to 0 \quad \text{as } y_2 \to \infty. \tag{7.55}$$

Substituting (7.44) and (7.52) into (7.38) and (7.43) results in the following problem for determining $u_k^{(2)}(x, y)$:

$$\frac{\partial}{\partial y_\alpha}\left[c_{i\alpha k\beta}(y)\frac{\partial u_k^{(2)}(x,y)}{\partial y_\beta}\right]$$

$$= -\frac{\partial}{\partial y_\alpha}\left[c_{i\alpha n\gamma}(y)N_n^{k\beta}(y)\right]\frac{\partial^2 u_k^{(0)}(x)}{\partial x_\beta \partial x_\gamma} - \left[c_{i\alpha k\beta}(y) + c_{i\alpha n\gamma}(y)\frac{\partial N_n^{k\beta}(y)}{\partial y_\gamma}\right]\frac{\partial^2 u_k^{(0)}(x)}{\partial x_\alpha \partial x_\beta}$$

$$-\frac{\partial}{\partial y_\alpha}[c_{i\alpha n\gamma}(y)N_n^{(1)k\beta}(y)]\frac{\partial^2 u_k^{(0)}(x)}{\partial x_\beta \partial x_\gamma} - c_{i\gamma n\alpha}(y)\frac{\partial N_n^{(1)k\beta}(y)}{\partial y_\gamma}\frac{\partial^2 u_k^{(0)}(x)}{\partial x_\alpha \partial x_\beta}, \tag{7.56}$$

$$c_{i2k\beta}(y)|_{y_2=0}\left.\frac{\partial u_k^{(2)}(x,y)}{\partial y_\beta}\right|_{y_2=0} = -c_{i2n\gamma}(y)|_{y_2=0}[N_n^{k\beta}(y) + N_n^{(1)k\beta}(y)]_{y_2=0}\left.\frac{\partial^2 u_k^{(0)}(x)}{\partial x_\beta \partial x_\gamma}\right|_{x_2=0}. \tag{7.57}$$

By analogy with (7.44), the solution of (7.56) is given in the form

$$u_k^{(2)}(x,y) = u_k^{(2,1)}(x,y) + u_k^{(2,2)}(x,y), \tag{7.58}$$

where the function $u_k^{(2,1)}(x, y)$ is 1-periodic in y_1 and y_2 and solves the local problem:

$$\frac{\partial}{\partial y_\alpha}\left[c_{i\alpha k\beta}(y)\frac{\partial u_k^{(2,1)}(x,y)}{\partial y_\beta}\right] = -\frac{\partial}{\partial y_2}[c_{i\alpha n\gamma}(y)N_n^{k\beta}(y)]\frac{\partial^2 u_k^{(0)}(x)}{\partial x_\beta \partial x_\gamma}$$

$$-\left[c_{i\alpha n\gamma}(y)\frac{\partial N_n^{k\beta}(y)}{\partial y_\gamma} + c_{i\alpha k\beta}(y)\right]\frac{\partial^2 u_k^{(0)}(x)}{\partial x_\alpha \partial x_\beta}. \tag{7.59}$$

whereas the function $u_k^{(2,2)}(x, y)$ is periodic only in y_1 and satisfies (a) equation (7.59) in the region $0 < y_1 < 1$, $y_2 > 0$ (with $N_n^{k\beta}$ replaced by $N_n^{(1)k\beta}$), and (b) the conditions

$$c_{i2k\beta}(y)\left.\frac{\partial u_k^{(2,2)}(x,y)}{\partial y_\beta}\right|_{y_2=0} = -c_{i2k\beta}(y)\left.\frac{\partial u_k^{(2,1)}(x,y)}{\partial y_\beta}\right|_{y_2=0}$$

$$-c_{i2n\gamma}(y)|_{y_2=0}(N_n^{k\beta}(y) + N_n^{(1)k\beta}(y))|_{y_2=0}\left.\frac{\partial^2 u_k^{(0)}}{\partial x_\beta \partial x_\gamma}\right|_{x_2=0}, \tag{7.60}$$

$$u_k^{(2,2)}(x,y) \to 0 \quad \text{as } y_2 \to \infty. \tag{7.61}$$

The solvability condition for the local problem (7.59) now gives the following averaged equation for $u_k^{(0)}(x)$ in the half-plane $x_2 > 0$:

$$\langle C_{i\alpha k\beta}\rangle \frac{\partial^2 u_k^{(0)}(x)}{\partial x_\alpha \partial x_\beta} = 0, \tag{7.62}$$

where the quantities

$$\langle C_{i\alpha k\beta}\rangle = \int_Y \left[c_{i\alpha k\beta}(y) + c_{i\alpha n\gamma}(y)\frac{\partial N_n^{k\beta}(y)}{\partial y_\gamma}\right] dy_1\, dy_2 \tag{7.63}$$

are the effective elastic moduli.

We thus conclude that, in the framework of the effective moduli method, the above half-plane elasticity problem reduces to that of solving equations (7.62) subject to the conditions (7.49). To perform this solution we need only to calculate the local functions $N_n^{k\beta}(y)$ from (7.45), but to determine the zeroth-order displacement and stress fields of the problem, the local functions $N_n^{(1)k\beta}(y)$ must be found from the system (7.53)–(7.55). The displacements and stresses are then calculated from

$$u_n = u_n^{(0)}(x) + \varepsilon(N_n^{k\beta}(y) + N_n^{(1)k\beta}(y))\frac{\partial u_k^{(0)}(x)}{\partial x_\beta} + \cdots, \tag{7.64}$$

$$\sigma_{i\alpha} = \left[c_{i\alpha k\beta}(y) + c_{i\alpha n\gamma}(y)\left(\frac{\partial N_n^{k\beta}(y)}{\partial y_\gamma} + \frac{\partial N_n^{(1)k\beta}(y)}{\partial y_\gamma}\right)\right]\frac{\partial u_k^{(0)}(x)}{\partial x_\beta} + \cdots. \tag{7.65}$$

The above discussion goes through equally well if, instead of stresses, we consider material displacements as being specified at the boundary $x_2 = 0$. So if we write

$$u_i^{(\varepsilon)}|_{x_2=0} = v_i(x_1) \tag{7.66}$$

the homogenized problem reduces to the solution in the plane $x_2 > 0$ of equation (7.62) subject to the condition

$$u_i^{(0)}|_{x_2=0} = v_i(x_1). \tag{7.67}$$

The stress and displacement fields will again be determined by (7.64) and (7.65), but the boundary layer type functions $N_n^{(2)k\beta}(y)$ will have to be found from the problem

$$\frac{\partial}{\partial y_\alpha}\left[c_{i\alpha n\gamma}(y)\frac{\partial N_n^{(2)k\beta}(y)}{\partial y_\gamma}\right] = 0, \quad 0 < y_1 < 1,\ y_2 > 0, \tag{7.68}$$

$$N_n^{(2)k\beta}(y)|_{y_2=0} = -N_n^{k\beta}(y)|_{y_2=0} + h_n^{(2)k\beta}, \tag{7.69}$$

$$N_n^{(2)k\beta}(y) \to 0 \quad \text{as } y_2 \to \infty, \tag{7.70}$$

and must be 1-periodic in y_1.

As shown by Panasenko (1979) (see also Bakhvalov and Panasenko 1984) the constants $h_n^{(2)k\beta}$ in (7.69) can be determined in a unique manner from the solvability condition for the problem (7.68)–(7.70). To this end, we first solve the problem (7.68) under the condition

$$\tilde{N}_n^{(2)k\beta}(y)|_{y_2=0} = -N_n^{k\beta}(y)|_{y_2=0} \tag{7.71}$$

in the class of functions which are bounded and 1-periodic in y_1. This problem is solved in a unique manner and the constants $\tilde{h}_n^{(2)k\beta}$ have the property

$$\lim_{y_2\to\infty} \tilde{N}_n^{(2)k\beta}(y) = \tilde{h}_n^{(2)k\beta} \tag{7.72}$$

so that, setting

$$h_n^{(2)k\beta} = -\tilde{h}_n^{(2)k\beta}, \tag{7.73}$$

it can be seen that

$$N_n^{(2)k\beta} = \tilde{N}_n^{(2)k\beta} - \tilde{h}_n^{(2)k\beta} \tag{7.74}$$

is a unique solution to the problem (7.68)–(7.70). We note also that because of the differentiation involved, the constants $h_n^{(2)k\beta}$ will be absent from the resulting local stress formula (7.65).

If the boundary conditions are of mixed type, for example if

$$\begin{aligned} \sigma_{l2}|_{x_2=0} &= p_l(x_1), \quad l = 1,3, \\ u_2|_{x_2=0} &= v_2(x_1), \end{aligned} \tag{7.75}$$

then the homogenized problem reduces to that of solving (7.62) in the half-plane $x_2 > 0$ under the conditions

$$\begin{aligned} C^*_{l2k\beta} \frac{\partial u_k^{(0)}}{\partial x_\beta}\bigg|_{x_2=0} &= p_l(x_1), \quad l = 1, 3, \\ u_2^{(0)}|_{x_2=0} &= v_2(x_1), \end{aligned} \tag{7.76}$$

and the local functions $N_n^{(3)k\beta}(y)$ entering the displacement and stress expressions, equations (7.64) and (7.65) respectively, are 1-periodic in y_1 and have to be found from the problem (see Kalamkarov *et al.* 1990):

$$\frac{\partial}{\partial y_\alpha}\left[c_{i\alpha n\gamma}(y)\frac{\partial N_n^{(3)k\beta}(y)}{\partial y_\gamma}\right] = 0, \quad 0 < y_1 < 1, \; y_2 = 0, \tag{7.77}$$

$$c_{l2n\gamma}\frac{\partial N_n^{(3)k\beta}(y)}{\partial y_\gamma}\bigg|_{y_2=0} = C^*_{l2k\beta} - C_{l2k\beta}(y_1, 0), \quad l = 1, 3, \tag{7.78}$$

$$N_2^{(3)k\beta}(y)|_{y_2=0} = -N_2^{k\beta}(y)|_{y_2=0} + h_2^{(3)k\beta}, \tag{7.79}$$

$$N_n^{(3)k\beta}(y) \to 0 \quad \text{as } y_2 \to \infty. \tag{7.80}$$

The constants $h_2^{(3)k\beta}$ in (7.79) are determined using the same procedure as in the problem (7.68)–(7.70) and are absent again in the final expressions for the local stress distribution.

In Appendix B of this book (see also Kalamkarov *et al.* 1991b) the special generalized integral transforms and series are derived, by means of which the solutions of the boundary value problems for periodic laminated composites can be represented in explicit analytical form.

8. LAMINATED COMPOSITES: EFFECTIVE PROPERTIES AND FRACTURE CRITERION

We have seen in the preceding section that to calculate the effective moduli $\langle C_{ijkl}\rangle$ within the framework of the zeroth-order approximation requires that the local functions

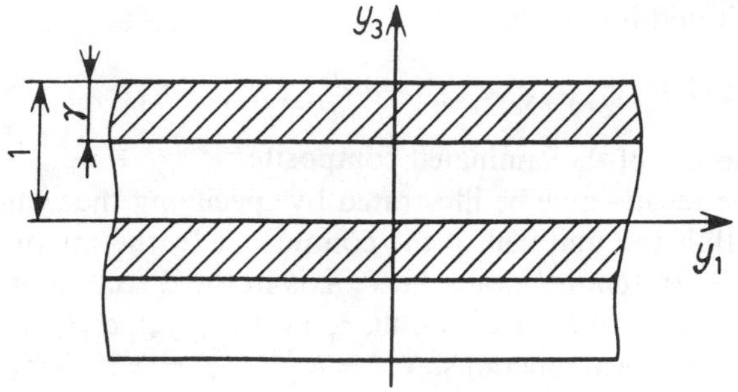

Fig. 8.1 *Layered composite material.*

$N_n^{kl}(y)$ be determined from the local problem (7.16) on the unit cell. This problem is solved without difficulty for the case of a laminated composite made up of non-homogeneous layers (or laminae) periodically repeating themselves along a certain direction. If we take this direction to be along the x_3 axis as in Fig. 8.1, the (true) elastic moduli c_{ijkl} will depend on $y_3 = x_3/\varepsilon$, and equations (7.16) take the form

$$\frac{\mathrm{d}}{\mathrm{d}y_3}\left(c_{i3n3}(y_3)\frac{\mathrm{d}N_n^{kl}}{\mathrm{d}y_3}\right) = -\frac{\mathrm{d}c_{i3kl}(y_3)}{\mathrm{d}y_3}, \quad 0 \leqslant y_3 \leqslant 1, \tag{8.1}$$

which by integration gives

$$c_{i3n3}(y_3)\frac{\mathrm{d}N_n^{kl}}{\mathrm{d}y_3} = -c_{i3kl}(y_3) + A_{ikl}, \tag{8.2}$$

where the A_{ikl} are constants of integration. From (8.2),

$$\frac{\mathrm{d}N_n^{kl}}{\mathrm{d}y_3} = -c_{n3i3}^{-1}(y_3)c_{i3kl}(y_3) + c_{n3i3}^{-1}(y_3)A_{ikl}, \tag{8.3}$$

where the 3×3 matrix c_{n3i3}^{-1} is inverse to the matrix c_{i3n3}, i.e. $c_{k3i3}^{-1}c_{i3n3} = \delta_{kn}$. Equation (8.3) is again integrated to give

$$\langle c_{n3i3}^{-1}\rangle A_{ikl} = \langle c_{n3q3}^{-1}c_{q3kl}\rangle, \tag{8.4}$$

where

$$\langle \cdots \rangle = \int_0^1 (\cdots)\,\mathrm{d}y_3,$$

and where use has been made of the fact that the function $N_n^{kl}(y_3)$ is 1-periodic in y_3, that is, $N_n^{kl}(0) = N_n^{kl}(1)$.

$$A_{ikl} = \langle c_{i3n3}^{-1}\rangle^{-1}\langle c_{n3q3}^{-1}c_{q3kl}\rangle, \tag{8.5}$$

giving

$$\tau_{ij}^{kl}(y_3) = c_{ijn3}(y_3)\frac{\mathrm{d}N_n^{kl}}{\mathrm{d}y_3} = -c_{ijn3}(y_3)c_{n3p3}^{-1}(y_3)c_{p3kl}(y_3)$$
$$+ c_{ijn3}(y_3)c_{n3s3}^{-1}(y_3)\langle c_{s3p3}^{-1}\rangle^{-1}\langle c_{p3q3}^{-1}c_{q3kl}\rangle, \tag{8.6}$$

which when substituted into (7.26) yields

$$\langle C_{ijkl}\rangle = \langle c_{ijkl}\rangle - \langle c_{ijn3}c^{-1}_{n3p3}c_{p3kl}\rangle + \langle c_{ijn3}c^{-1}_{n3s3}\rangle\langle c^{-1}_{s3p3}\rangle^{-1}\langle c^{-1}_{p3q3}c_{q3kl}\rangle \tag{8.7}$$

for the effective moduli of the laminated composite.

Use of the above results may be illustrated by specifying the symmetry of the anisotropic medium within the unit cell of the composite. In the important special case of the hexagonal symmetry 6mm, taking the x_3 axis in the direction of the symmetry axis, the medium is characterized by five moduli, $c_{11}(y_3)$, $c_{12}(y_3)$, $c_{13}(y_3)$, $c_{33}(y_3)$ and $c_{44}(y_3)$, and the set of non-zero local functions, $N_1^{13} = N_2^{23}$, $N_3^{11} = N_3^{22}$, N_3^{33}, are found from the equations

$$\begin{aligned} &\frac{dN_3^{13}}{dy_3} = -1 + \frac{A_{113}}{c_{44}(y_3)}, \qquad \frac{dN_3^{11}}{dy_3} = -\frac{c_{13}(y_3)}{c_{33}(y_3)} + \frac{A_{311}}{c_{33}(y_3)}, \\ &\frac{dN_3^{33}}{dy_3} = -1 + \frac{A_{333}}{c_{33}(y_3)}, \end{aligned} \tag{8.8}$$

which are readily obtained from (8.3). It follows by integrating (8.8) that

$$A_{113} = \frac{1}{\langle c_{44}^{-1}\rangle}, \qquad A_{311} = \frac{\langle c_{13}c_{33}^{-1}\rangle}{\langle c_{33}^{-1}\rangle}, \qquad A_{333} = \frac{1}{\langle c_{33}^{-1}\rangle}, \tag{8.9}$$

and the effective moduli are found to be, in matrix notation,

$$\begin{aligned} &\tilde{c}_{11} = \tilde{c}_{22} = \langle c_{11}\rangle - \langle c_{13}^2 c_{33}^{-1}\rangle + \frac{\langle c_{13}c_{33}^{-1}\rangle^2}{\langle c_{33}^{-1}\rangle}, \\ &\tilde{c}_{33} = \frac{1}{\langle c_{33}^{-1}\rangle}, \qquad \tilde{c}_{23} = \tilde{c}_{13} = \frac{\langle c_{13}c_{33}^{-1}\rangle}{\langle c_{33}^{-1}\rangle}, \\ &\tilde{c}_{12} = \langle c_{12}\rangle - \langle c_{13}^2 c_{33}^{-1}\rangle + \frac{\langle c_{13}c_{33}^{-1}\rangle^2}{\langle c_{33}^{-1}\rangle}, \\ &\tilde{c}_{44} = \frac{1}{\langle c_{44}^{-1}\rangle}, \qquad \tilde{c}_{55} = \tilde{c}_{44}, \\ &\tilde{c}_{66} = \langle c_{66}\rangle = \tfrac{1}{2}[\langle c_{11}\rangle - \langle c_{12}\rangle]. \end{aligned} \tag{8.10}$$

If we set $c_{11} = c_{33}$, $c_{13} = c_{12}$ and $c_{66} = c_{44}$ in the above equations, these yield the effective moduli for the case when the layer material is of cubic symmetry (with three independent elastic moduli). If, furthermore, $c_{11} = c_{33} = \lambda + 2G$, $c_{13} = c_{12} = \lambda$ and $c_{66} = c_{44} = G$, the effective moduli of a composite with isotropic laminae are obtained:

$$\begin{aligned} &\tilde{c}_{11} = \tilde{c}_{22} = \langle \lambda + 2G\rangle - \left\langle \frac{\lambda^2}{\lambda + 2G}\right\rangle + \left\langle \frac{\lambda}{\lambda + 2G}\right\rangle^2 \frac{1}{\langle(\lambda + 2G)^{-1}\rangle}, \\ &\tilde{c}_{33} = \frac{1}{\langle(\lambda + 2G^{-1}\rangle}, \qquad \tilde{c}_{23} = \tilde{c}_{13} = \left\langle \frac{\lambda}{\lambda + 2G}\right\rangle \frac{1}{\langle(\lambda + 2G)^{-1}\rangle}, \end{aligned}$$

$$\tilde{c}_{12} = \langle\lambda\rangle - \left\langle\frac{\lambda^2}{\lambda+2G}\right\rangle + \left\langle\frac{\lambda}{\lambda+2G}\right\rangle^2 \frac{1}{\langle(\lambda+2G)^{-1}\rangle},$$

$$\tilde{c}_{44} = \frac{1}{\langle G^{-1}\rangle}, \qquad \tilde{c}_{55} = \frac{1}{\langle G^{-1}\rangle}, \qquad \tilde{c}_{66} = \langle G\rangle, \tag{8.11}$$

where λ and G are the Lamé contants. It is seen that a homogenized laminated composite composed of isotropic layers acts like a transversely isotropic material possessing five independent elastic moduli $\tilde{c}_{11}$, $\tilde{c}_{33}$, $\tilde{c}_{12}$, $\tilde{c}_{23} = \tilde{c}_{13}$ and $\tilde{c}_{44} = \tilde{c}_{55}$.

In the important practical case of a two-phase composite material (see Fig. 8.2), for homogeneous and isotropic components, the effective moduli are given by

$$\begin{aligned}
\tilde{c}_{11} = \tilde{c}_{22} &= (\lambda_1 + 2G_1)\gamma + (\lambda_2 + 2G_2)(1-\gamma) - \left[\frac{\lambda_1^2}{\lambda_1 + 2G_1}\gamma + \frac{\lambda_2^2}{\lambda_2 + 2G_2}(1-\gamma)\right] \\
&\quad + \left[\frac{\lambda_1\gamma}{\lambda_1 + 2G_1} + \frac{\lambda_2(1-\gamma)}{\lambda_2 + 2G_2}\right]^2 [(\lambda_1 + 2G_1)^{-1}\gamma + (\lambda_2 + 2G_2)^{-1}(1-\gamma)]^{-1}, \\
\tilde{c}_{33} &= [(\lambda_1 + 2G_1)^{-1}\gamma + (\lambda_2 + 2G_2)(1-\gamma)]^{-1}, \\
\tilde{c}_{13} = \tilde{c}_{23} &= \left[\frac{\lambda_1}{\lambda_1 + 2G_1}\gamma + \frac{\lambda_2}{\lambda_2 + 2G_2}(1-\gamma)\right][(\lambda_1 + 2G_1)^{-1}\gamma + (\lambda_2 + 2G_2)^{-1}(1-\gamma)]^{-1}, \\
c_{12} &= [\lambda_1\gamma + \lambda_2(1-\gamma)] - \left[\frac{\lambda_1^2}{\lambda_1 + 2G_1}\gamma + \frac{\lambda_2^2}{\lambda_2 + 2G}(1-\gamma)\right] \\
&\quad + \left[\frac{\lambda_1\gamma}{\lambda_1 + 2G_1} + \frac{\lambda_2(1-\gamma)}{\lambda_2 + 2G_2}\right]^2 [(\lambda_1 + 2G_1)^{-1}\gamma + (\lambda_2 + 2G_2)^{-1}(1-\gamma)], \\
\tilde{c}_{44} &= [G_1^{-1}\gamma + G_2^{-1}(1-\gamma)]^{-1}, \qquad \tilde{c}_{55} = \tilde{c}_{44}, \qquad \tilde{c}_{66} = [(G_1\gamma + G_2(1-\gamma)].
\end{aligned} \tag{8.12}$$

The simple formulae we have derived for the effective moduli of composite materials make it possible to gain a useful insight into the manner in which the geometrical and mechanical properties of the individual components affect the strength properties of the composite as a whole; also, and no less important, the overall strength criterion can be deduced from knowledge of the individual macroscopic strength criteria.

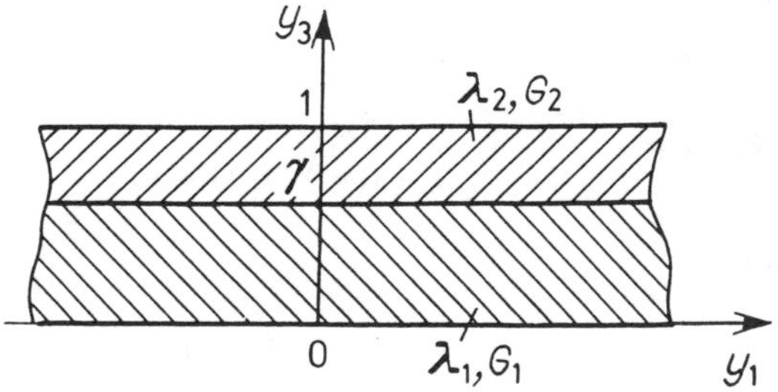

Fig. 8.2 *Unit cell of a two-component laminated composite material with homogeneous isotropic components.*

Following Gorbachev and Pobedrya (1985), we begin by introducing the stress concentration tensor $A_{ijpq}(y)$ which relates $\langle \sigma_{ij}^{(0)} \rangle$ to the layer stresses $\sigma_{ij}^{(0)}$. Making use of (7.25) we determine the averaged strain components,

$$e_{kl}^{(0)} = \tfrac{1}{2}(\partial u_k^{(0)}(\boldsymbol{x})/\partial x_l + \partial u_l^{(0)}(\boldsymbol{x})/\partial x_k),$$

and substitute them into (7.19) to obtain:

$$\sigma_{ij}(\boldsymbol{x}, \boldsymbol{y}) \cong \sigma_{ij}^{(0)}(\boldsymbol{x}, \boldsymbol{y}) = A_{ijpq}\langle \sigma_{pq}^{(0)} \rangle, \tag{8.13}$$

where

$$A_{ijpq}(\boldsymbol{y}) = (c_{ijkl}(\boldsymbol{y}) + \tau_{ij}^{kl}(\boldsymbol{y}))\langle C_{klpq} \rangle^{-1}, \tag{8.14}$$

with tensors $\langle C_{ijkl} \rangle$ and $(\langle C_{klpq} \rangle)^{-1}$ mutually inverse in the sense $\langle C_{ijkl} \rangle \langle C_{klpq} \rangle^{-1} = \tfrac{1}{2}(\delta_{ip}\delta_{jk} + \delta_{iq}\delta_{jp})$.

The stress concentration tensor components can now be rewritten with the use of (8.6) to give

$$\begin{aligned} A_{ijpq}(y_3) = [&c_{ijkl}(y_3) - c_{ijn3}(y_3)c_{n3p3}^{-1}(y_3)c_{p3kl}(y_3) \\ &+ c_{ijn3}(y_3)c_{n3s3}^{-1}(y_3)\langle c_{s3p3}^{-1} \rangle^{-1} \langle c_{p3q3}^{-1} c_{q3kl} \rangle] \langle C_{klpq} \rangle^{-1}. \end{aligned} \tag{8.15}$$

If the material of layers is isotropic (cf. (1.86)),

$$c_{ijkl}(y_3) = \frac{E(y_3)}{1 + \nu(y_3)} \left[\tfrac{1}{2}(\delta_{ik}\delta_{jl} + \delta_{il}\delta_{jk}) + \frac{\nu(y_3)}{1 - 2\nu(y_3)} \delta_{ij}\delta_{kl} \right], \tag{8.16}$$

where $E(y_3)$ denotes Young's modulus and $\nu(y_3)$ Poisson's ratio, it is easily shown that

$$A_{i3pq} = A_{3ipq} = \tfrac{1}{2}(\delta_{ip}\delta_{3q} + \delta_{iq}\delta_{3p}) \tag{8.17}$$

and that the remaining components are given by

$$\begin{aligned} A_{1111} = A_{2222} &= \frac{E(y_3)}{1 - \nu^2(y_3)} \frac{\langle E/(1 - \nu^2) \rangle - \nu(y_3) \langle E\nu/(1 - \nu^2) \rangle}{\langle E/(1 - \nu) \rangle \langle E/(1 + \nu) \rangle}, \\ A_{1122} = A_{2211} &= \frac{E(y_3)}{1 - \nu^2(y_3)} \frac{\nu(y_3) \langle E/(1 - \nu^2) \rangle - \langle E\nu/(1 - \nu^2) \rangle}{\langle E/(1 - \nu) \rangle \langle E/(1 + \nu) \rangle}, \\ A_{1212} &= \tfrac{1}{2} \frac{E(y_3)/(1 + \nu(y_3))}{\langle E/(1 - \nu) \rangle}, \\ A_{1133} = A_{2233} &= \frac{\nu(y_3)}{1 - \nu(y_3)} - \frac{E(y_3)}{1 - \nu(y_3)} \frac{\langle \nu/(1 - \nu) \rangle}{\langle E/(1 - \nu) \rangle}. \end{aligned} \tag{8.18}$$

It will be noted that

$$A_{1212} = \tfrac{1}{2}(A_{1111} - A_{1122}).$$

As a fracture criterion for an individual composite component, a very general form due to Goldenblat and Kopnov (1968) (see also Malmeister *et al.* 1980) will be used, namely

$$P_{ij}(y)\sigma_{ij} + P_{ijkl}(y)\sigma_{ij}\sigma_{kl} + \cdots = 1, \tag{8.19}$$

where P_{ij}, $P_{ijkl}\dots$ are the stress strength tensors of the material. Substituting (8.13) and applying the mean operator, this becomes

$$P^*_{ij}\langle\sigma^{(0)}_{ij}\rangle + P^*_{ijkl}\langle\sigma^{(0)}_{ij}\rangle\langle\sigma^{(0)}_{kl}\rangle + \ldots = 1, \tag{8.20}$$

where P^*_{ij}, $P^*_{ijkl}\dots$ are the effective stress strength tensors defined by

$$\begin{aligned} P^*_{ij} &= \langle P_{mn} A_{mnij}\rangle, \\ P^*_{ijkl} &= \langle P_{mnpq} A_{mnij} A_{pqkl}\rangle, \ldots . \end{aligned} \tag{8.21}$$

As a special case of the individual fracture criterion we take the form

$$P_{ijkl}(y)\sigma_{ij}\sigma_{kl} = 1, \tag{8.22}$$

which expresses the well-known shape variation energy criterion and in which

$$P_{ijkl}(y) = \frac{1}{2\tau_b^2(y_3)}\left[\tfrac{1}{2}(\delta_{ik}\delta_{jl} + \delta_{il}\delta_{jk}) - \tfrac{1}{3}\delta_{ij}\delta_{kl}\right], \tag{8.23}$$

with τ_b denoting the shear strength of the material. The non-zero strength tensor components P^*_{ijkl} are

$$\begin{aligned}
P^*_{1111} &= P^*_{2222} = \frac{1}{6}\left\langle \frac{1}{\tau_b^2}(A_{1111} - A_{1122})^2 + \frac{1}{\tau_b^2}A^2_{1111} + \frac{1}{\tau_b^2}A^2_{1122}\right\rangle, \\
P^*_{1133} &= P^*_{2233} = -\frac{1}{6}\left\langle \frac{1}{\tau_b^2}(1 - A_{1133})(A_{1111} + A_{1122})\right\rangle, \\
P^*_{1122} &= -\frac{1}{6}\left\langle \frac{1}{\tau_b^2}(A_{1111} - A_{1122})^2 - \frac{2}{\tau_b^2}A_{1111}A_{1122}\right\rangle, \\
P^*_{3333} &= \frac{1}{3}\left\langle \frac{1}{\tau_b^2}(1 - A_{1133})^2\right\rangle, \qquad P^*_{1313} = P^*_{2323} = \frac{1}{4}\left\langle \frac{1}{\tau_b^2}\right\rangle, \\
P^*_{1212} &= \left\langle \frac{1}{\tau_b^2}A^2_{1212}\right\rangle.
\end{aligned} \tag{8.24}$$

Following Gorbachev and Pobedrya (1985) we introduce at this point the so-called engineering strength properties $\sigma_L, \sigma_F, \tau_L, \tau_F$ and σ_p, related to the tensor components P^*_{ijkl} by the equations

$$\begin{aligned}
&\sigma_F = (P^*_{3333})^{-1/2}, \qquad \sigma_L = (P^*_{1111})^{-1/2}, \qquad \tau_F = (1/2)(P^*_{1313})^{-1/2}, \\
&\tau_L = (1/2)(P^*_{1212})^{-1/2}, \qquad \sigma_p = (P^*_{3333} + 4P^*_{1111} + 4P^*_{1133} - 4P^*_{1212})^{-1/2},
\end{aligned} \tag{8.25}$$

where σ_L (resp. σ_F) is the strength for stretching or compressing along (resp. perpendicular to) the layer; τ_L (resp. τ_F) is the strength for shear along (resp. perpendicular to) the layer; σ_p is the uniform compression strength. With the quantities so defined, the fracture criterion

$$P^*_{ijkl}\langle\sigma^{(0)}_{ij}\rangle\langle\sigma^{(0)}_{kl}\rangle = 1 \tag{8.26}$$

for a laminated composite can be rewritten as

$$\frac{1}{\sigma_L^2}(\langle\sigma^{(0)}_{11}\rangle^2 + \langle\sigma^{(0)}_{22}\rangle^2) + \frac{1}{\sigma_F^2}\langle\sigma^{(0)}_{33}\rangle^2 + \frac{1}{\tau_L^2}\langle\sigma^{(0)}_{12}\rangle^2 + \frac{1}{\tau_F^2}(\langle\sigma^{(0)}_{13}\rangle^2 + \langle\sigma^{(0)}_{23}\rangle^2)$$
$$+\left(\frac{2}{\sigma_L^2} - \frac{1}{\tau_L^2}\right)\langle\sigma^{(0)}_{11}\rangle\langle\sigma^{(0)}_{22}\rangle + \frac{1}{2}\left(\frac{1}{\sigma_p^2} + \frac{1}{\tau_L^2} - \frac{1}{\sigma_F^2} - \frac{4}{\sigma_L^2}\right)(\langle\sigma^{(0)}_{11}\rangle + \langle\sigma^{(0)}_{22}\rangle)\langle\sigma^{(0)}_{33}\rangle = 1. \tag{8.27}$$

It should be understood that the average stress components $\langle\sigma^{(0)}_{ij}\rangle$ in this last equation are determined by the effective modulus method, that is by solving the equations (7.24)–(7.26) under appropriate boundary conditions. We note also that (8.27) may as well be used as a composite material plasticity criterion if we replace τ_b by the simple shear yield strength and consider σ_L, σ_F, τ_L, τ_F and σ_p as the engineering yield strength properties.

The effective elastic moduli of a laminated composite have also been derived by Duvaut (1976) and Pobedrya (1984). Pobedrya (1977) applies the homogenization method to the static elasticity problem (formulated in terms of stresses) and calculates the components of the effective compliance tensor for a composite system; in a later paper (Pobedrya 1984) expressions for the effective thermal conductivity tensor may be found. With regard to the dynamic elasticity problem for composite materials, an important point to be made is that in the asymptotic expansions involved, higher order terms in ε must be retained. As discussed by Karimov (1986) in connection with the free vibrations of a finite-thickness composite layer, the fourth-order terms retained in the asymptotic expansions (7.7)–(7.9) reveal the so-called wave filter effect, in which free vibrations prove to be non-existent in some frequency range(s).

9. EFFECTIVE CHARACTERISTICS OF UNIDIRECTIONAL FIBRE COMPOSITES

The two-dimensional local problems arising on the unit cell of such a composite are readily amenable to treatment by the analytical methods of the theory of (double periodic) functions of a complex variable. The unit cell in this case is obtained by cutting the material in a plane normal to the fibre direction and may take, for example, the form of a square with a circular inclusion corresponding to the circular cross-section of the fibre (see Fig. 7.1). If both the matrix and fibre materials are homogeneous and the fibres are aligned along the x_3 axis, the elastic tensor of the composite will be a piecewise-constant, doubly periodic function of y_1 and y_2, and the local problem at hand will be stated as

$$\frac{\partial \tau_{\alpha j}^{kl}(\mathrm{f})}{\partial y_\alpha}=0 \quad \text{in } Y_\mathrm{f},$$

$$\frac{\partial \tau_{\alpha j}^{kl}(\mathrm{M})}{\partial y_\alpha}=0 \quad \text{in } Y_\mathrm{M}, \tag{9.1}$$

$$\tau_{\alpha j}^{kl}(\mathrm{f})=c_{\alpha j n\beta}^{(\mathrm{f})}\frac{\partial N_n^{kl}(\mathrm{f})}{\partial y_\beta}, \qquad \tau_{\alpha j}^{kl}(\mathrm{M})=c_{\alpha j n\beta}^{\mathrm{M}}\frac{\partial N_n^{kl}(\mathrm{M})}{\partial y_\beta}$$

$$(\alpha,\beta=1,2;\ j,\ k,\ l,\ n=1,2,3). \tag{9.2}$$

We therefore specify that the solution be doubly periodic in y_1 and y_2 and we also require that the following perfect contact conditions be satisfied on the boundary Γ:

$$N_n^{kl}(\mathrm{f})|_\Gamma = N_n^{kl}(\mathrm{M})|_\Gamma, \tag{9.3}$$

$$(\tau_{\alpha j}^{kl}(\mathrm{f})+c_{\alpha jkl}^{(\mathrm{f})})n_\alpha|_\Gamma=(\tau_{\alpha j}^{kl}(\mathrm{M})+c_{\alpha jkl}^{(\mathrm{M})})n_\alpha|_\Gamma, \tag{9.4}$$

where $c_{\alpha jkl}^{(\mathrm{f})}$ and $c_{\alpha jkl}^{(\mathrm{M})}$ are the elastic moduli tensors of the fibre and matrix materials, respectively, and n_α are the components of the unit vector normal to $\Gamma (\alpha=1,2)$.

Exactly how many problems of the type (9.1)–(9.2) must be solved in each particular case depends, of course, on the symmetry properties of the matrix and fibre materials (note that the double-index superscript kl takes the values 11, 22, 33, 23, 31, 12). If both are isotropic, the problem decomposes into four similar plane strain problems for $\tau_{\alpha\beta}^{11}$, $\tau_{\alpha\beta}^{22}$, $\tau_{\alpha\beta}^{33}$ and $\tau_{\alpha\beta}^{12}$, and two anti-plane problems for $\tau_{\alpha 3}^{23}$ and $\tau_{\alpha 3}^{13}$. In the case of homogeneous isotropic materials, the plane strain problems are of the form

$$\frac{\partial \tau_{11}^{kl}}{\partial y_1}+\frac{\partial \tau_{12}^{kl}}{\partial y_2}=0, \qquad \frac{\partial \tau_{12}^{kl}}{\partial y_1}+\frac{\partial \tau_{22}^{kl}}{\partial y_2}=0, \tag{9.5}$$

$$\tau_{11}^{kl}=c_{11}(y)\frac{\partial N_1^{kl}}{\partial y_1}+c_{12}(y)\frac{\partial N_2^{kl}}{\partial y_2},$$

$$\tau_{22}^{kl}=c_{12}(y)\frac{\partial N_1^{kl}}{\partial y_1}+c_{11}(y)\frac{\partial N_2^{kl}}{\partial y_2}, \tag{9.6}$$

$$\tau_{12}^{kl}=\tfrac{1}{2}(c_{11}(y)-c_{12}(y))\left(\frac{\partial N_2^{kl}}{\partial y_1}+\frac{\partial N_1^{kl}}{\partial y_2}\right)$$

where

$$c_{11}(y)=\begin{cases} c_{11}^{(\mathrm{f})}=\lambda_\mathrm{f}+2G_\mathrm{f}, & y\in Y_\mathrm{f},\\ c_{11}^{(\mathrm{M})}=\lambda_\mathrm{M}+2G_\mathrm{M}, & y\in Y_\mathrm{M},\end{cases}$$

$$c_{12}(y)=\begin{cases} c_{12}^{(\mathrm{f})}=\lambda_\mathrm{f}, & y\in Y_\mathrm{f},\\ c_{12}^{(\mathrm{M})}=\lambda_\mathrm{M}, & y\in Y_\mathrm{M},\end{cases}$$

where λ_M, G_M, λ_f and G_f are the Lamé constants of the matrix and fibre materials, and $kl=$

11, 22, 33, 12. The problems for determining the local functions $\tau_{\alpha 3}^{23}$ and $\tau_{\alpha 3}^{13}$ are, in this case,

$$\frac{\partial \tau_{13}^{kl}}{\partial y_1} + \frac{\partial \tau_{23}^{kl}}{\partial y_2} = 0, \tag{9.7}$$

$$\tau_{13}^{kl} = c_{44}(y)\frac{\partial N_3^{kl}}{\partial y_1}, \qquad \tau_{23}^{kl} = c_{44}(y)\frac{\partial N_3^{kl}}{\partial y_2}, \tag{9.8}$$

where $kl = 23$ and 13 and

$$c_{44}(y) = \begin{cases} c_{44}^{(\mathrm{f})} = G_{\mathrm{f}}, & y \in Y_{\mathrm{f}}, \\ c_{44}^{(\mathrm{M})} = G_{\mathrm{M}}, & y \in Y_{\mathrm{M}}. \end{cases}$$

The solution of the systems (9.5)–(9.6) and (9.7)–(9.8) must be doubly periodic in y_1 and y_2 and must satisfy conditions (9.3) and (9.4) on the fibre–matrix interface. If both the matrix and fibre materials are homogeneous and isotropic, the boundary conditions take the form

$$[\![N_n^{kl}]\!]_\Gamma = 0, \tag{9.9}$$

$$[\![(\tau_{11}^{11} + c_{11})n_1 + \tau_{12}^{11} n_2]\!]_\Gamma = 0, \qquad [\![\tau_{12}^{11} n_1 + (\tau_{22}^{11} + c_{12})n_2]\!]_\Gamma = 0, \tag{9.10}$$

$$[\![(\tau_{11}^{22} + c_{12})n_1 + \tau_{12}^{22} n_2]\!]_\Gamma = 0, \qquad [\![\tau_{12}^{11} n_1 + (\tau_{22}^{22} + c_{11})n_2]\!]_\Gamma = 0, \tag{9.11}$$

$$[\![(\tau_{11}^{33} + c_{12})n_1 + \tau_{12}^{33} n_2]\!]_\Gamma = 0, \qquad [\![\tau_{12}^{33} n_1 + (\tau_{22}^{33} + c_{12})n_2]\!]_\Gamma = 0, \tag{9.12}$$

$$[\![(\tau_{11}^{12} n_1 + (\tau_{12}^{12} + c_{66})n_2]\!]_\Gamma = 0, \qquad [\![(\tau_{12}^{12} + c_{66})n_1 + \tau_{22}^{12} n_2]\!]_\Gamma = 0, \tag{9.13}$$

where $c_{66} = (1/2)(c_{11} - c_{12})$ and $[\![A]\!]$ indicates a jump in the value of the quantity A at the boundary Γ.

It should be noted here that, with one possible exception of circular fibre cross-section, no analytical methods may be expected to be applicable to the local problems (9.5)–(9.6) or (9.7)–(9.8), and in most cases more or less elaborate numerical techniques, such as the Finite Element Method, must be employed.

This numerical work is, of course, simplified if some symmetry elements occur either in the geometry of the unit cell, or in the constituent properties, or both, because a local problem for the whole unit cell may in this case be reduced to a boundary value problem for only part of the cell. As an example of such a reduction, consider the system (9.5)–(9.6) for $kl = 11$ for a composite with isotropic components and with the unit cell shown in Fig. 7.1 ($Y_1 = Y_2 = 1$). We assume $c_{11}(y)$ to be even functions of y_1 and y_2 and consider the system of equations

$$\frac{\partial \tau_{11}^{11}}{\partial y_1} + \frac{\partial \tau_{12}^{11}}{\partial y_2} = -\frac{\partial c_{11}}{\partial y_1}, \qquad \frac{\partial \tau_{12}^{11}}{\partial y_1} + \frac{\partial \tau_{22}^{11}}{\partial y_2} = -\frac{\partial c_{12}}{\partial y_2}, \tag{9.14}$$

$$\tau_{11}^{11} = c_{11}(y)\frac{\partial N_1^{11}}{\partial y_1} + c_{12}(y)\frac{\partial N_2^{11}}{\partial y_2}, \qquad \tau_{12}^{11} = c_{12}(y)\frac{\partial N_1^{11}}{\partial y_1} + c_{11}(y)\frac{\partial N_2^{11}}{\partial y_2},$$

$$\tau_{12}^{11} = \tfrac{1}{2}(c_{11}(y) - c_{12}(y))\left(\frac{\partial N_2^{11}}{\partial y_1} + \frac{\partial N_1^{11}}{\partial y_2}\right). \tag{9.15}$$

From (9.14) we observe that the local functions τ_{11}^{11} and τ_{22}^{11} are even, and τ_{12}^{11} is odd in y_1 and y_2, and from (9.15) it is seen that N_1^{11} (N_2^{11}) is even in y_2 (y_1) and odd in y_1 (y_2). Since the functions N_1^{11}, N_2^{11} and τ_{12}^{11} must be continuous at the boundary of the unit cell and across the lines $y_1 = 0$ and $y_2 = 0$, it then follows that the problem of interest in this case is the plane boundary value problem (9.14)–(9.15) in the regions $0 < y_1 < \frac{1}{2}$, $0 < y_2 < \frac{1}{2}$, and the boundary conditions to be satisfied are

$$\begin{aligned} N_1^{11} &= 0, \tau_{12}^{11} = 0 \quad \text{if } y_1 = 0, y_1 = \tfrac{1}{2}, 0 < y_2 < \tfrac{1}{2}, \\ N_2^{11} &= 0, \tau_{12}^{11} = 0 \quad \text{if } y_2 = 0, y_2 = \tfrac{1}{2}, 0 < y_1 < \tfrac{1}{2}. \end{aligned} \tag{9.16}$$

Note that the derivatives $\partial c_{11}/\partial y_1$ and $\partial c_{12}/\partial y_2$ in the right-hand side of equations (9.14) exhibit a delta function behaviour as a result of the discontinuities in the coefficients c_{11} and c_{12}.

The solutions of the local problems (9.5)–(9.8) for a composite with homogeneous isotropic components have been obtained by Mol'kov and Pobedrya (1985, 1988), who employ the Muskhelishvili complex potentials to estimate the components of the effective elastic moduli tensor. Numerical methods for treating unit cell local problems are discussed by Pobedrya (1979), Sheshenin (1980) and Leont'ev (1984), among others.

We now proceed to the analytical treatment of the two-dimensional local problems set on the unit cell shown in Fig. 7.1 ($Y_1 = Y_2 = 1$). Two of these problems, (9.7) and (9.8), are relatively simple and reduce to the determination of doubly periodic functions $N_3^{kl}(y)$ ($kl = 23, 13$) satisfying Laplace's equation in the regions $Y_{\rm f}$ and $Y_{\rm M}$ and conditions (9.13) at the boundary Γ. This is easily achieved by representing the harmonic functions $N_3^{kl}(y)$ by the series expansion (6.40) in the region $Y_{\rm M}$ and by a Taylor series expansion in powers of $z = y_1 + {\rm i}y_2$ in the region $Y_{\rm f}$; the coefficients in these expansions will be found from the infinite systems of simultaneous algebraic equations resulting from conditions (9.13).

With regard to the local problems (9.5) and (9.6), to which we turn next, it has been found that the theory of doubly periodic functions of a complex variable provides effective means for their treatment. The case when $kl = 11$ will be used to illustrate this approach. We begin by representing the biharmonic stress functions in the usual manner in terms of the analytical functions $\varphi(z)$ and $\psi(z)$ for $z = y_1 + {\rm i}y_2$ and express the local functions τ_{11}^{11}, τ_{22}^{11}, τ_{12}^{11}, N_1^{11} and N_2^{11} in the form

$$\begin{aligned} \tau_{11}^{11} + \tau_{22}^{11} &= 2[\varphi'(z) + \overline{\varphi'(z)}], \\ \tau_{22}^{11} - \tau_{11}^{11} + 2{\rm i}\tau_{12}^{11} &= 2[\bar{z}\varphi''(z) + \psi'(z)], \end{aligned} \tag{9.17}$$

$$2G(N_1^{11} + {\rm i}N_2^{11}) = \kappa\varphi(z) - z\overline{\varphi'(z)} - \overline{\psi(z)}, \tag{9.18}$$

where we have written

$$G = \begin{cases} G_{\rm f}, & z \in Y_{\rm f}, \\ G_{\rm M}, & z \in Y_{\rm M}, \end{cases}$$

$$\kappa = \frac{\lambda + 3G}{\lambda + G} = \begin{cases} \kappa_{\rm f}, & z \in Y_{\rm f}, \\ \kappa_{\rm M}, & z \in Y_{\rm M}. \end{cases}$$

If we introduce polar coordinates $(r = |z|, \theta)$ as shown in Fig. 6.3, the boundary conditions (9.10) at $\Gamma(r=r_0)$ take the form

$$[\tfrac{1}{2}(\tau_{11}^{11}+\tau_{22}^{11})-\tfrac{1}{2}e^{2i\theta}(\tau_{22}^{11}-\tau_{11}^{11}+2i\tau_{12}^{11})]_{r=r_0-0}$$
$$=[\tfrac{1}{2}(\tau_{11}^{11}+\tau_{22}^{11})-\tfrac{1}{2}e^{2i\theta}(\tau_{22}^{11}-\tau_{11}^{11}+2i\tau_{12}^{11})]_{r=r_0+0}$$
$$+\tfrac{1}{2}(c_{11}^{(M)}+c_{12}^{(M)}-c_{11}^{(f)}-c_{12}^{(f)})+\tfrac{1}{2}(c_{11}^{(M)}-c_{12}^{(M)}-c_{11}^{(f)}+c_{12}^{(f)})e^{2i\theta}. \tag{9.19}$$

Making use of (9.17) and introducing the notation

$$\Phi_f(z)=\varphi_f'(z), \qquad \Psi_f(z)=\psi_f'(z),$$
$$\Phi_M(z)=\varphi_M'(z), \qquad \Psi_M(z)=\psi_M'(z),$$

for the analytical functions in the fibre and matrix regions (Y_f and Y_M, respectively), equation (9.19) becomes

$$\Phi_f(t)+\overline{\Phi_f(t)}-e^{2i\theta}(\bar{t}\Phi_f'(t)+\Psi_f(t))$$
$$=\Phi_M(t)+\overline{\Phi_M(t)}-e^{2i\theta}(\bar{t}\Phi_M'(t)+\Psi_M(t))+\tfrac{1}{2}(c_{11}^{(M)}+c_{12}^{(M)}-c_{11}^{(f)}-c_{12}^{(f)})$$
$$+\tfrac{1}{2}(c_{11}^{(M)}-c_{12}^{(M)}-c_{11}^{(f)}+c_{12}^{(f)})e^{2i\theta}, \tag{9.20}$$

where $t=r_0\exp(i\theta)$ is an arbitrary point on Γ.

To proceed further it is necessary to specialize condition (9.9) by expressing the continuity of the functions N_1^{11} and N_2^{11} across the contour $\Gamma(r=r_0)$. Using (9.18) we obtain:

$$[\kappa_f\varphi_f(t)-t\overline{\varphi_f'(t)}-\overline{\psi_f(t)}]=\frac{G_M}{G_f}[\kappa_M\varphi_M(t)-t\overline{\varphi_M'(t)}-\overline{\psi_M(t)}]. \tag{9.21}$$

This is now differentiated along the direction s tangent to Γ giving

$$[-\kappa_f\overline{\Phi_f(t)}+\Phi_f(t)-e^{2i\theta}(\bar{t}\Phi_f'(t)+\Psi_f(t))]$$
$$=\frac{G_M}{G_f}[-\kappa_M\overline{\Phi_M(t)}+\Phi_M(t)-e^{2i\theta}(\bar{t}\Phi_M'(t)+\Psi_M(t))], \tag{9.22}$$

where it is recalled that

$$\frac{d}{ds}=-\frac{\partial}{\partial y_1}\sin\theta+\frac{\partial}{\partial y_2}\cos\theta,$$
$$\frac{\partial}{\partial y_1}=\frac{\partial}{\partial z}+\frac{\partial}{\partial\bar{z}}, \qquad \frac{\partial}{\partial y_2}=i\left(\frac{\partial}{\partial z}-\frac{\partial}{\partial\bar{z}}\right).$$

In view of the symmetry of the unit cell with respect to the y_1 and y_2 axes, the functions $\Phi_M(z)$ and $\Psi_M(z)$ can be represented in the respective forms:

$$\Phi_M(z)=\alpha_0+\sum_{k=0}^{\infty}\alpha_{2k+2}r_0^{2k+2}\frac{\wp^{(2k)}(z)}{(2k+1)!}, \tag{9.23}$$

$$\Psi_M(z)=\beta_0+\sum_{k=0}^{\infty}\beta_{2k+2}\frac{r_0^{2k+2}\wp^{(2k)}(z)}{(2k+1)!}-\sum_{k=0}^{\infty}\alpha_{2k+2}\frac{r_0^{2k+2}Q^{(2k+1)}(z)}{(2k+1)!}. \tag{9.24}$$

where α_{2k} and β_{2k} $(k=0,1,2,\ldots)$ are real constants; $\wp(z)$ is the doubly periodic Weierstrass elliptic function with periods $\omega_1=1$ and $\omega_2=\mathrm{i}$; the special meromorphic function $Q(z)$ is defined by

$$Q(z)=\sum_{m,n}{}'\left\{\frac{\bar{P}}{(z-P)}-2z\frac{\bar{P}}{P^3}-\frac{\bar{P}}{P^2}\right\} \tag{9.25}$$

for $P=m+\mathrm{i}n$ and $\bar{P}=m-\mathrm{i}n$ and satisfies the relations

$$Q^{(k)}(z+1)-Q^{(k)}(z)=\wp^{(k)}(z),$$
$$Q^{(k)}(z+\mathrm{i})-Q^{(k)}(z)=-\mathrm{i}\wp^{(k)}(z)\quad(k=1,2,\ldots).$$

The functions $\Phi_{\mathrm{M}}(z)$ and $\Psi_{\mathrm{M}}(z)$, as given by (9.23) and (9.24), ensure a symmetrical doubly periodic distribution of the local stresses τ_{11}^{11}, τ_{22}^{11} and τ_{12}^{11} and it can be shown that the constants α_0 and β_0 are related to α_2 and β_2 by

$$\alpha_0=\frac{\pi}{2}\beta_2 r_0^2,\qquad \beta_0=(\gamma_1+\pi)\alpha_2 r_0^2,\qquad \gamma_1=2Q(\tfrac{1}{2})-\wp(\tfrac{1}{2}). \tag{9.26}$$

To obtain the derivatives $\wp^{(2k)}(z)$ and $Q^{(2k+1)}(z)$ involved in (9.23) and (9.24) we use the well-known Laurent expansions of these functions about the point $z=0$:

$$\frac{\wp^{(2k)}(z)}{(2k+1)!}=\frac{1}{z^{2k+2}}+\sum_{j=0}^{\infty}r_{j,k}z^{2j},\quad k=0,1,2,\ldots, \tag{9.27}$$

$$\frac{Q^{(2k+1)}(z)}{(2k+1)!}=\sum_{j=0}^{\infty}s_{j,k}z^{2j},\quad k=0,1,2,\ldots, \tag{9.28}$$

where

$$r_{j,k}=\frac{(2j+2k+1)!}{(2j)!(2k+1)!}\frac{g_{j+k+1}}{2^{2j+2k+2}},\quad r_{0,0}=0,$$

$$g_{j+k+1}=\sum_{m,n}{}'(P/2)^{-2j-2k-2},$$

$$s_{j,k}=\frac{(2j+2k+2)!}{(2j)!(2k+2)!}\frac{\rho_{j+k+1}}{2^{2j+2k+2}},\quad s_{0,0}=0,$$

$$\rho_{j+k+1}=\sum_{m,n}{}'(\bar{P}/2)(P/2)^{-2j-2k-3}.$$

Returning to (9.23) and (9.24), the Laurent expansions for Φ_{M} and Ψ_{M} are then given by

$$\Phi_{\mathrm{M}}(z)=\alpha_0+\sum_{k=0}^{\infty}\alpha_{2k+2}\frac{r_0^{2k+2}}{z^{2k+2}}+\sum_{k=0}^{\infty}\sum_{j=0}^{\infty}\alpha_{2k+2}r_0^{2k+2}r_{j,k}z^{2j}, \tag{9.29}$$

$$\Psi_{\mathrm{M}}(z)=\beta_0+\sum_{k=0}^{\infty}\beta_{2k+2}\frac{r_0^{2k+2}}{z^{2k+2}}+\sum_{k=0}^{\infty}\sum_{j=0}^{\infty}\beta_{2k+2}r_0^{2k+2}r_{j,k}z^{2j}$$
$$-\sum_{k=0}^{\infty}\sum_{j=0}^{\infty}(2k+2)\alpha_{2k+2}r_0^{2k+2}s_{j,k}z^{2j}. \tag{9.30}$$

Since the functions $\Phi_f(z)$ and $\Psi_f(z)$ are both regular in Y_f we are justified in representing them by the Taylor series

$$\Phi_f(z) = \sum_{k=0}^{\infty} a_{2k} z^{2k}, \qquad \Psi_f(z) = \sum_{k=0}^{\infty} b_{2k} z^{2k}, \tag{9.31}$$

with a_{2k} and b_{2k} real constants.

Substituting the expansions (9.29)–(9.31) into conditions (9.20) and (9.22) on Γ and equating coefficients of equal powers of $\exp(i\theta)$ yields the following infinite system of algebraic equations for the constants α_{2k} and β_{2k+2} $(k = 2, 3, \ldots)$:

$$(2k+1)\alpha_{2k} - \beta_{2k+2} + \sum_{j=1}^{\infty} \alpha_{2j+2} A_{k,j}^{(1)} + \sum_{j=2}^{\infty} \beta_{2j+2} B_{k,j}^{(1)}$$
$$= \frac{(\kappa_f - \kappa_M G_M/G_f)}{2\Delta_\alpha \Delta_\alpha^*} (c_{11}^{(M)} - c_{12}^{(M)} - c_{11}^{(f)} + c_{12}^{(f)}), \tag{9.32}$$

$$-\frac{(\kappa_M G_M/G_f + 1)}{(G_M/G_f - 1)} \alpha_{2k} + \sum_{j=1}^{\infty} \alpha_{2j+2} A_{k,j}^{(2)} + \sum_{j=2}^{\infty} \beta_{2j+2} B_{kj}^{(2)}$$
$$= \frac{\gamma_{k,0} r_0^{2k}}{2\Delta_\alpha \Delta_\alpha^*} (c_{11}^{(M)} - c_{12}^{(M)} - c_{11}^{(f)} + c_{12}^{(f)})$$
$$+ \frac{r_{k-1,0} r_0^{2k} (1 - \kappa_f)}{4\Delta_\beta} (c_{11}^{(M)} + c_{12}^{(M)} - c_{11}^{(f)} - c_{12}^{(f)}) \quad (k = 2, 3, \ldots), \tag{9.33}$$

where

$$A_{k,j}^{(1)} = \frac{(\kappa_f - \kappa_M G_M/G_f)}{(\kappa_f + G_M/G_f)} \left[r_{k,j} + r_0^2 r_{k,0} \frac{(G_M/G_f - 1)}{\Delta_\alpha \Delta_\alpha^*} \gamma_{1,j} \right] r_0^{2j+2k+2},$$

$$B_{k,j}^{(1)} = (\kappa_f - \kappa_M G_M/G_f) \frac{(G_M/G_f - 1)}{\Delta_\alpha \Delta_\alpha^*} r_{k,0} r_{0,j} r_0^{2k+2j+4},$$

$$A_{k,j}^{(2)} = \left[\gamma_{k,j} - \frac{(G_M/G_f - 1)}{\Delta_\alpha \Delta_\alpha^*} r_0^2 \gamma_{k,0} \gamma_{1,j} \right.$$
$$\left. + \frac{1}{\Delta_\beta} \left((1 - \kappa_M) \frac{G_M}{G_f} - 1 + \kappa_f \right) r_0^2 r_{k-1,0} r_{0,j} \right] r_0^{2j+2k},$$

$$B_{k,j}^{(2)} = \left[-r_{k-1,j} + \frac{(G_M/G_f - 1)}{\Delta_\alpha \Delta_\alpha^*} r_0^2 \gamma_{k,0} \gamma_{0,j} \right] r_0^{2k+2j},$$

$$\gamma_{k,0} = r_0^2 r_{k,0} - (2k-2) r_{k-1,0} + 2 s_{k-1,0}$$
$$- \frac{(\kappa_f - \kappa_M G_M/G_f)}{(\kappa_f + G_M/G_f)} r_0^6 r_{1,0} r_{k-1,1} - 3 r_0^2 r_{k-1,1},$$

$$\gamma_{k,j} = r_0^2 r_{k,j} - (2k-2) r_{k-1,j} + (2j+2) s_{k-1,j} - \frac{(\kappa_f - \kappa_M G_M/G_f)}{(\kappa_f + G_M/G_f)} r_0^6 r_{1,j} r_{k-1,1},$$

$$\Delta_\alpha = -1 - \kappa_M \frac{G_M}{G_f} - \left(\frac{G_M}{G_f} - 1\right)(r_0^4 r_{1,0} - (\gamma_1 + \pi) r_0^2),$$

$$\Delta_\beta = \frac{G_M}{G_f}\left((1-\kappa_M)\frac{\pi r_0^2}{2} - 1\right) - \tfrac{1}{2}(1-\kappa_f)(\pi r_0^2 - 1),$$

$$\Delta_\alpha^* = 1 - \frac{3}{\Delta_\alpha}(G_M/G_f - 1) r_0^4 r_{0,1} - \frac{1}{\Delta_\alpha}(G_M/G_f - 1)\frac{(\kappa_f - \kappa_M G_M/G_f)}{(\kappa_f + G_M/G_f)} r_0^8 r_{1,0} r_{0,1}.$$

The remaining coefficients in the expansions (9.29)–(9.31) can be calculated from

$$\alpha_2 = -\frac{(G_M/G_f - 1)}{\Delta_\alpha \Delta_\alpha^*}\sum_{j=1}^{\infty} \gamma_{1,j}\alpha_{2j+2} z_0^{2j+2} + \frac{(G_M/G_f - 1)}{\Delta_\alpha \Delta_\alpha^*}\sum_{j=2}^{\infty} \beta_{2j+2} r_0^{2j+2} r_{0,j}$$
$$- (2\Delta_\alpha \Delta_\alpha^*)^{-1}(c_{11}^{(M)} - c_{12}^{(M)} - c_{11}^{(f)} + c_{12}^{(f)}), \tag{9.34}$$

$$\beta_2 = \frac{(1-\kappa_f)}{4\Delta_\beta}(c_{11}^{(M)} + c_{12}^{(F)} - c_{11}^{(f)} - c_{12}^{(f)}) - \frac{1}{\Delta_\beta}[(1-\kappa_M) G_M/G_f$$
$$- (1-\kappa_f)] \sum_{j=1}^{\infty} \alpha_{2j+2} r_0^{2j+2} r_{0,j}, \tag{9.35}$$

$$\beta_4 = \left[3 + \frac{(\kappa_f - \kappa_M G_M/G_f)}{(\kappa_f + G_M/G_f)} r_0^4 r_{1,0}\right]\alpha_2 + \frac{(\kappa_f - \kappa_M G_M/G_f)}{(\kappa_f + G_M/G_f)}\sum_{j=1}^{\infty} \alpha_{2j+2} r_0^{2j+4} r_{1,j}, \tag{9.36}$$

$$a_0 = \tfrac{1}{2}(\pi r_0^2 - 1) + \sum_{j=1}^{\infty} \alpha_{2j+2} r_0^{2j+2} r_{0,j} + \tfrac{1}{4}(c_{11}^{(M)} + c_{12}^{(M)} - c_{11}^{(f)} - c_{12}^{(f)}), \tag{9.37}$$

$$a_2 r_0^2 = \frac{(1+\kappa_M) G_M/G_f}{(\kappa_f + G_M/G_M)}\left[r_0^4 r_{1,0}\alpha_2 + \sum_{j=1}^{\infty} \alpha_{2j+2} r_0^{2j+4} r_{1,j}\right], \tag{9.38}$$

$$a_{2k} r_0^{2k} = r_0^{2k+2} r_{k,0}\alpha_2 + (2k+1)\alpha_{2k} - \beta_{2k+2} + \sum_{j=1}^{\infty} \alpha_{2j+2} r_0^{2j+2k+2} r_{k,j} \quad (k = 2, 3, \ldots), \tag{9.39}$$

$$b_0 = -a_2 r_0^2 - (1 - (\gamma_1 + \pi) r_0^2 + r_0^4 r_{1,0})\alpha_2 - \sum_{j=1}^{\infty} (r_0^2 r_{1,j}$$
$$+ (2j+2) s_{0,j})\alpha_{2j+2} r_0^{2j+2} + r_0^4 r_{0,1}\beta_4 + \sum_{j=2}^{\infty} \beta_{2j+2} r_0^{2j+2} r_{0,j}$$
$$- \tfrac{1}{2}(c_{11}^{(M)} - c_{12}^{(M)} - c_{11}^{(f)} + c_{12}^{(f)}), \tag{9.40}$$

$$b_{2k-2} r_0^{2k-2} = -(2k-1) a_{2k} r_0^{2k} - (r_0^{2k+2} r_{k,0} - (2k-2) r_0^{2k} r_{k-1,0}$$
$$+ 2 s_{k-1,0} r_0^{2k})\alpha_2 + r_0^{2k} r_{k-1,0}\beta_2 - \alpha_{2k} - \sum_{j=1}^{\infty} (r_0^2 r_{k,j} - (2k-2) r_{k-1,j}$$
$$+ (2j+2) s_{k-1,j})\alpha_{2j+2} r_0^{2j+2k} + \sum_{j=1}^{\infty} \beta_{2j+2} r_0^{2j+2k} r_{k-1,j} \quad (k = 2, 3, \ldots). \tag{9.41}$$

All the coefficients in (9.29)–(9.31) having been determined, formula (9.17) can now be employed to calculate the local stresses τ_{11}^{11}, τ_{22}^{11} and τ_{12}^{11}, and the solution of the local problem (9.5)–(9.6) for $kl = 11$ is thus completed. The cases $kl = 22, 33, 12$, which we have to consider next, differ little if at all from the case $kl = 11$ and are treated in very much the same way. Thus we have provided all the information necessary to determine the effective elastic moduli

$$\tilde{c}_{ijkl} = \int_Y (c_{ijkl}(y) + \tau_{ij}^{kl}(y))\,\mathrm{d}y_1\,\mathrm{d}y_2, \tag{9.42}$$

where $Y = Y_f \bigcup Y_M$ denotes the region occupied by the unit cell of the (unidirectional fibre) composite. It is useful to point out that, in view of the symmetry of the above local problems,

$$\int_Y \tau_{12}^{kl}\,\mathrm{d}y_1\,\mathrm{d}y_2 = 0 \quad (kl = 11, 22, 33),$$

$$\int_Y \tau_{11}^{12}\,\mathrm{d}y_1\,\mathrm{d}y_2 = \int_Y \tau_{22}^{12}\,\mathrm{d}y_1\,\mathrm{d}y_2 = 0, \qquad \int_Y \tau_{13}^{23}\,\mathrm{d}y_1\,\mathrm{d}y_2 = \int_Y \tau_{23}^{13}\,\mathrm{d}y_1\,\mathrm{d}y_2 = 0,$$

and it is therefore readily shown that, changing to the two-index notation, the effective elastic matrix of such a composite is of the form

$$\tilde{c}_{\alpha\beta} = \begin{pmatrix} \tilde{c}_{11} & \tilde{c}_{12} & \tilde{c}_{13} & 0 & 0 & 0 \\ \tilde{c}_{12} & \tilde{c}_{11} & \tilde{c}_{13} & 0 & 0 & 0 \\ \tilde{c}_{13} & \tilde{c}_{13} & \tilde{c}_{33} & 0 & 0 & 0 \\ 0 & 0 & 0 & \tilde{c}_{44} & 0 & 0 \\ 0 & 0 & 0 & 0 & \tilde{c}_{44} & 0 \\ 0 & 0 & 0 & 0 & 0 & \tilde{c}_{66} \end{pmatrix}, \tag{9.43}$$

where

$$\tilde{c}_{11} = \int_Y (c_{11}(y) + \tau_{11}^{11}(y))\mathrm{d}y_1\,\mathrm{d}y_2,$$

$$\tilde{c}_{12} = \int_Y (c_{12}(y) + \tau_{11}^{22}(y))\mathrm{d}y_1\,\mathrm{d}y_2,$$

$$\tilde{c}_{13} = \int_Y (c_{12}(y) + \tau_{11}^{33}(y))\mathrm{d}y_1\,\mathrm{d}y_2,$$

$$\tilde{c}_{33} = \int_Y (c_{11}(y) + \tau_{33}^{33}(y))\mathrm{d}y_1\,\mathrm{d}y_2, \tag{9.44}$$

$$= \int_Y \left[c_{11}(y) + \frac{c_{12}(y)}{c_{11}(y) + c_{12}(y)}(\tau_{11}^{33} + \tau_{22}^{33}) \right] \mathrm{d}y_1\,\mathrm{d}y_2$$

$$\tilde{c}_{44} = \int_Y (c_{44}(y) + \tau_{23}^{23}(y))\mathrm{d}y_1\,\mathrm{d}y_2,$$

$$\tilde{c}_{66} = \int_Y (c_{66}(y) + \tau_{12}^{12}(y))\mathrm{d}y_1\,\mathrm{d}y_2.$$

Note that the local function $\tau_{33}^{33}(y)$ entering the $\tilde{c}_{33}$ expression is defined by

$$\tau_{33}^{33} = c_{12}\left(\frac{\partial N_1^{33}}{\partial y_1} + \frac{\partial N_2^{33}}{\partial y_2}\right) \tag{9.45}$$

or, using (9.6),

$$\tau_{33}^{33} = \frac{c_{12}}{c_{12}+c_{11}}(\tau_{11}^{33} + \tau_{22}^{33}). \tag{9.46}$$

10. PLANE ELASTICITY PROBLEM FOR A PERIODIC COMPOSITE WITH A CRACK

It is common practice when performing the stress analysis of a composite with an ideally smooth macrocrack that the non-homogeneous composite medium being studied be replaced with a homogeneous anisotropic medium whose response is assumed to be equivalent to that of the actual composite in a certain averaged sense (see, for example, Cherepanov 1983). The main attraction of the method is that the calculation of the averaged stress field in such a composite reduces to an elasticity theory problem for a homogeneous anisotropic material with a mathematical cut within it.

If the material has a periodic structure, the averaged (or effective) properties of the equivalent medium can be estimated by means of the homogenization method, which also gives asymptotically correct results for the local structure of the stress field in the composite. Kalamkarov *et al.* (1990) adopted this approach in their analysis of the stress field arising in the neighbourhood of a macrocrack in a laminated composite with a periodic structure; the stress intensity factors of the composite are expressed by the authors in terms of the constituent properties and the parameters determining the position of the crack in the material. In this section, the application of this approach will be discussed in which the boundary effect occurring in the vicinity of the crack is taken into account by introducing additional solutions of the boundary layer type discussed in Section 7. For the sake of simplicity, only the plane formulation will be considered.

We are thus interested in the state of stress of a periodically non-homogeneous (composite) medium containing a rectilinear macrocrack the length of which is much greater than the unit cell dimensions. It is assumed that the elastic medium possesses a doubly periodic inhomogeneity in the (x_1, x_2) plane and that the rims of the (tunnel) crack are parallel to the boundary of the unit cell as illustrated in Fig. 10.1. The problem so stated is relevant, for example, to the study of a fibre composite with a tunnel crack lying in the plane normal to the fibre direction; or of a laminated composite with the plane crack normal or parallel to the laminae.

Let the rectilinear crack lie in the interface between two unit cells of the (unbounded) composite, and let its rims be subjected to a prescribed set of self-balanced normal and tangential loads. To determine the state of stress and strain in the neighbourhood of such a crack, an asymptotic method will be applied to the equations of elasticity in the periodically non-homogeneous half-plane $x_2 > 0$ under mixed boundary conditions on $x_2 = 0$. These latter specify the stresses σ_{i2} $(i = 1, 2, 3)$ on the segment $|x_1| < a$, and the stresses σ_{12}, σ_{32} and displacement u_2 at $|x_1| > a$ (see Fig. 10.2). To satisfy these types of

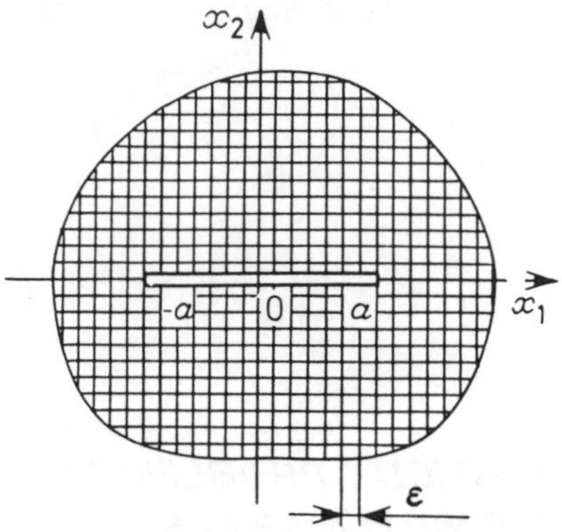

Fig. 10.1 *Macroscopic crack in a composite material.*

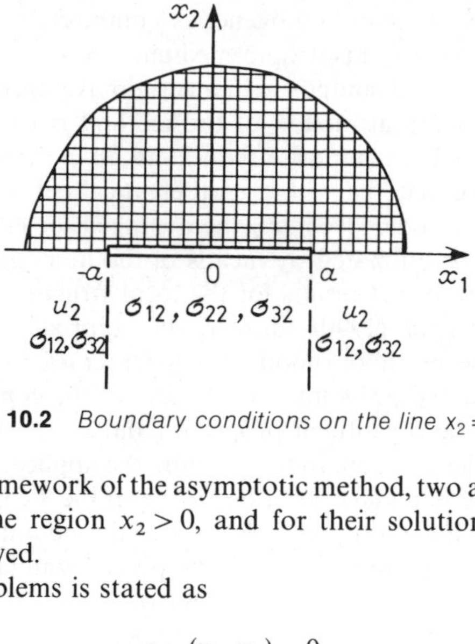

Fig. 10.2 *Boundary conditions on the line* $x_2 = 0$.

conditions within the framework of the asymptotic method, two auxiliary plane problems will be considered in the region $x_2 > 0$, and for their solution the results derived in Section 7 will be employed.

The first of these problems is stated as

$$\sigma_{i\alpha,\alpha}(x_1, x_2) = 0, \tag{10.1}$$

$$\sigma_{i\alpha} = c_{i\alpha k\beta}(y_1, y_2)u_{k,\beta}, \tag{10.2}$$

where the elastic coefficients $c_{i\alpha k\beta}\,(y_1, y_2)$ are 1-periodic functions of the variables $y_\alpha = x_\alpha/\varepsilon$; a comma (vertical line) denotes partial differentiation with respect to x_1, x_2 (y_1, y_2); Latin indices range from 1 to 3 and Greek (indices) from 1 to 2.

The conditions to be satisfied at the boundary of the half-plane are

$$\sigma_{i2}(x_1, 0) = p_i(x_1), \quad i = 1, 2, 3. \tag{10.3}$$

In Section 7 the asymptotic analysis of the problem (10.1)–(10.3) yielded formulas (7.64) and (7.65) from which the displacement and stress fields can be estimated to leading terms in ε. The local functions $N_n^{k\beta}(\boldsymbol{y})$ and boundary layer functions $N_n^{(1)k\beta}(\boldsymbol{y})$ involved in these formulas are determined from the problems (7.45) and (7.53)–(7.55), respectively.

In the second auxiliary problem for the half-plane $x_2 > 0$, the (mixed) boundary conditions at $x_2 = 0$ are of the form

$$\begin{aligned} \sigma_{l2}(x_1, 0) &= p_l(x_1), \quad l = 1, 3, \\ u_2(x_1, 0) &= v_2(x_1). \end{aligned} \tag{10.4}$$

In this case it has been shown by Kalamkarov *et al.* (1990) (the same procedure used in Section 7) that the displacements and stresses are given (to the same accuracy) by formulas very similar to (7.64) and (7.65) in which the functions $N_n^{(1)k\beta}(y)$ are replaced by boundary layer type functions $N_n^{(3)k\beta}(y)$ obtained from the problem (7.77)–(7.80).

With the above results at our disposal we can now proceed to the stress analysis of a rectilinear crack of normal rupture passing in the interface between two rectangular unit cells in a composite material. If we assume that there are no tangential stresses on the rims of the crack ($\sigma_{12} = \sigma_{23} = 0$) and $\sigma_{22} = p_2(x_1)$ at $x_2 = \pm 0$, $|x_1| < a$, then, because of the symmetry of the stress state with respect to the x_1 axis, we may reduce the problem to one concerning the upper half-plane $x_2 > 0$ with boundary conditions of the form (cf. Fig. 10.2)

$$\begin{aligned} \sigma_{l2} &= 0 && \text{when } x_2 = 0, \ |x_1| < \infty, \ l = 1, 3, \\ \sigma_{22} &= p_2(x_1) && \text{when } x_2 = 0, \ |x_1| < a, \\ u_2 &= 0 && \text{when } x_2 = 0, \ |x_1| > a. \end{aligned} \tag{10.5}$$

Referring to Fig. 10.3, several characteristic regions should be distinguished in the non-homogeneous half-plane subjected to these conditions. To begin with, there is region IV in the vicinity of the points $x_1 = \pm a$, $x_2 = 0$, which has the form of a rectangle with dimensions $\tilde{\varepsilon}_1 + \tilde{\varepsilon}_2$ and $\tilde{\varepsilon}$ and contains a finite (and sufficiently small) number of unit cells within itself. The state of stress (strain) in this region must be determined directly from the solution of the elasticity problem rather that by applying the asymptotic process of the homogenization method. In the unbounded strip $0 < x_2 < \tilde{\varepsilon}$, $|x_1| < \infty$, of which region IV is a part, region I ($0 < x_2 < \tilde{\varepsilon}$, $x_1 < a - \tilde{\varepsilon}_1$) and region II ($0 < x_2 < \tilde{\varepsilon}$, $x_1 > a + \tilde{\varepsilon}_2$) may be recognized, in which the asymptotic solutions of, respectively, the first and the second of the auxiliary problems just discussed may be utilized. We should remark here that the functions $N_n^{(1)k\beta}(y)$ and $N_n^{(3)k\beta}(y)$ determined from problems (7.53)–(7.55) and

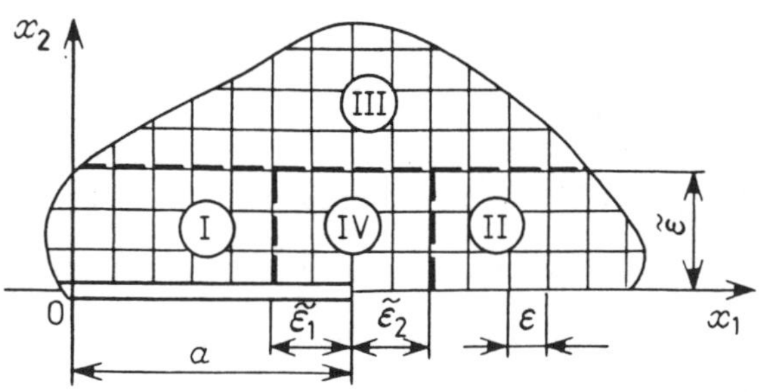

Fig. 10.3 *Auxiliary regions I–IV in the vicinity of a macrocrack in a composite material of periodic structure.*

(7.77)–(7.80) are boundary layer type solutions and it is therefore possible to set

$$N_n^{(1)k\beta}(y)=0, \qquad N_n^{(3)k\beta}(y)=0$$

in region III ($x_2>\tilde{\varepsilon}$). The magnitude of $\tilde{\varepsilon}$ is thus found from the conditions $N_n^{(1)k\beta}(y_1,\tilde{\varepsilon})\approx 0$ and $N_n^{(3)k\beta}(y_1,\tilde{\varepsilon})\approx 0$.

The magnitudes of $\tilde{\varepsilon}_1$ and $\tilde{\varepsilon}_2$ may be obtained by noting that the stress perturbation resulting from the change in boundary conditions in the vicinity of the tip of the crack is localized within region IV and does not spread. The characteristic dimension of the perturbed area can therefore be estimated from knowledge of the asymptotic expressions for the stress field in the vicinity of the crack. Note also that, for the sake of convenience, the values of $\tilde{\varepsilon}, \tilde{\varepsilon}_1$ and $\tilde{\varepsilon}_2$ should be selected so as to make the boundaries of region IV coincide with those of the unit cells of the composite (see Fig. 10.3).

With the boundary layer solutions available, stresses on the two vertical sides ($x_1=a-\tilde{\varepsilon}_1,\ 0<x_2<\tilde{\varepsilon};\ x_1=a+\tilde{\varepsilon}_2,\ 0<x_2<\tilde{\varepsilon}$) and the upper horizontal side ($a-\tilde{\varepsilon}_1<x_1<a+\tilde{\varepsilon}_2, x_2=\tilde{\varepsilon}$) of region IV can be specified. Since the boundary conditions of the segment $a-\tilde{\varepsilon}_1<x_1<a+\tilde{\varepsilon}_2$, $x_2=0$ are known (see equations (10.5)), in principle it is possible to solve the plane elasticity problem in region IV—for example by numerical methods. The problem is

$$\begin{aligned}
&\sigma_{\alpha\beta,\beta}=0, && \sigma_{\alpha\beta}=c_{\alpha\beta\lambda\mu}(y)u_{\lambda,\mu},\\
&\sigma_{12}=0 && \text{when } x_2=0,\ a-\tilde{\varepsilon}_1<x_1<a+\tilde{\varepsilon}_2,\\
&\sigma_{22}=p_2(x_1) && \text{when } x_2=0,\ a-\tilde{\varepsilon}_1<x_1<a,\\
&u_2=0 && \text{when } x_2=0,\ a<x_1<a+\tilde{\varepsilon}_2,
\end{aligned} \tag{10.6}$$

$$\begin{aligned}
&\sigma_{\mu 1}=[c_{\mu 1\lambda\beta}(y)+c_{\mu 1n\gamma}(y)(N_{n|\gamma}^{\lambda\beta}+N_{n|\gamma}^{(1)\lambda\beta})]u_{\lambda,\beta}^{(0)}\\
&\qquad \text{when } x_1=a-\tilde{\varepsilon}_1,\ 0<x_2<\tilde{\varepsilon},\\
&\sigma_{\mu 1}=[c_{\mu 1\lambda\beta}(y)+c_{\mu 1n\gamma}(y)(N_{n|\gamma}^{\lambda\beta}+N_{n|\gamma}^{(3)\lambda\beta})]u_{\lambda,\beta}^{(0)}\\
&\qquad \text{when } x_1=a+\tilde{\varepsilon}_2,\ 0<x_2<\tilde{\varepsilon},\\
&\sigma_{\mu 2}=[c_{\mu 2\lambda\beta}(y)+c_{\mu 2n\gamma}(y)N_{n|\gamma}^{\lambda\beta}]u_{\lambda,\beta}^{(0)}\\
&\qquad \text{when } x_2=\tilde{\varepsilon},\ a-\tilde{\varepsilon}_1<x_1<a+\tilde{\varepsilon}_2.
\end{aligned} \tag{10.7}$$

It should be noted in connection with equations (10.7) that they do not contain the constants involved in conditions (7.79) and that the functions $u_\alpha^{(0)}(\boldsymbol{x})$ are solutions of the following homogenized problem:

$$\begin{aligned}
\langle C^*_{\lambda\alpha\mu\beta}\rangle u_{\mu,\alpha\beta}^{(0)}&=0 && \text{for } x_2>0,\\
C^*_{12\alpha\beta}u_{\alpha,\beta}^{(0)}&=0 && \text{when } x_2=0,\ |x_1|<\infty,\\
C^*_{22\alpha\beta}u_{\alpha,\beta}^{(0)}&=p_2(x_1) && \text{when } x_2=0,\ |x_1|<a,\\
u_2^{(0)}&=0 && \text{when } x_2=0,\ |x_1|>a,
\end{aligned} \tag{10.8}$$

with coefficients defined by (7.50) and (7.63).

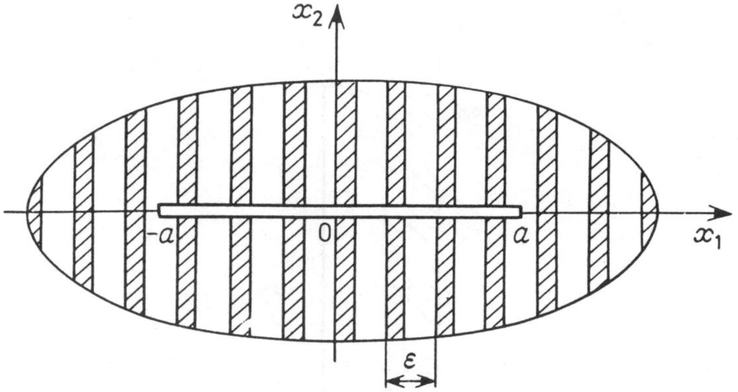

Fig. 10.4 *Macro crack in a laminated two-component composite.*

What makes special the significance of the separation in the composite of a local region IV with known boundary conditions on its perimeter is the amenability of this region to a reasonably rigorous stress analysis for various positions of the crack tip in the unit cell of the composite.

The above approach will now be applied to the problem of a macrocrack in a laminated periodic composite shown in Fig. 10.4, the unit cell of which consists of two (isotropic) layers (see Fig. 10.5) characterized by the parameters E_1, ν_1 and E_2, ν_2.

The local problems (7.45) are solved in an elementary fashion in this case and the non-zero 1-periodic functions are given by

$$N_1^{11} = \begin{cases} \dfrac{(c_{11}^{(2)} - c_{11}^{(1)})(1-\gamma)}{(1-\gamma)c_{11}^{(1)} + \gamma c_{11}^{(2)}} \left(y_1 - \dfrac{\gamma}{2} \right) & \text{when } 0 < y_1 < \gamma, \\ \dfrac{(c_{11}^{(2)} - c_{11}^{(1)})\gamma}{(1-\gamma)c_{11}^{(1)} + \gamma c_{11}^{(2)}} \left(\dfrac{1+\gamma}{2} - y_1 \right) & \text{when } \gamma < y_1 < 1, \end{cases}$$

$$N_1^{22} = \begin{cases} \dfrac{(c_{12}^{(2)} - c_{12}^{(1)})(1-\gamma)}{(1-\gamma)c_{11}^{(1)} + \gamma c_{11}^{(2)}} \left(y_1 - \dfrac{\gamma}{2} \right) & \text{when } 0 < y_1 < \gamma, \\ \dfrac{(c_{12}^{(2)} - c_{12}^{(1)})\gamma}{(1-\gamma)c_{11}^{(1)} + \gamma c_{11}^{(2)}} \left(\dfrac{1+\gamma}{2} - y_1 \right) & \text{when } \gamma < y_1 < 1, \end{cases}$$

$$N_2^{12} = N_2^{21} = N_3^{31} \tag{10.9}$$

$$= \begin{cases} \dfrac{(c_{44}^{(2)} - c_{44}^{(1)})(1-\gamma)}{(1-\gamma)c_{44}^{(1)} + \gamma c_{44}^{(2)}} \left(y_1 - \dfrac{\gamma}{2} \right) & \text{when } 0 < y_1 < \gamma, \\ \dfrac{(c_{44}^{(2)} - c_{44}^{(1)})\gamma}{(1-\gamma)c_{44}^{(1)} + \gamma c_{44}^{(2)}} \left(\dfrac{1+\gamma}{2} - y_1 \right) & \text{when } \gamma < y_1 < 1, \end{cases}$$

where the superscripts (1) and (2) on the elastic properties c_{ik} refer to the layers of the composite. Note that

$$C_{i2k\beta}^{*} = \langle C_{i2k\beta} \rangle \tag{10.10}$$

because the relevant functions $C_{i2k\beta}$ are independent of y_2.

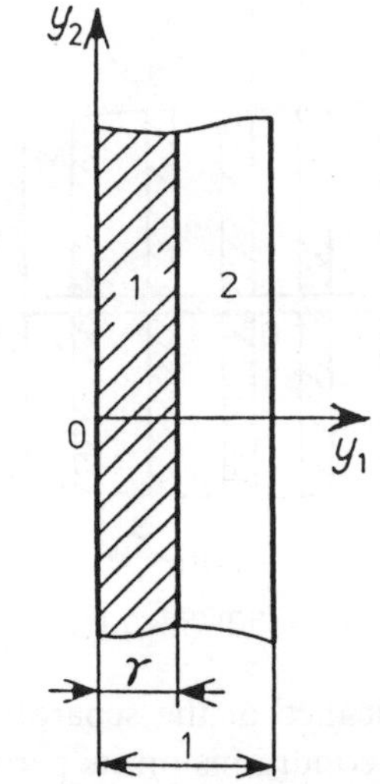

Fig. 10.5 *Unit cell of a laminated two-component composite.*

The solution of the relevant boundary layer problem (7.53)–(7.55) is non-trivial only for the functions

$$N_1^{(1)11}, N_2^{(1)11} \quad \text{and} \quad N_1^{(1)22}, N_2^{(1)22}, \tag{10.11}$$

whereas non-zero solutions of the problem (7.77)–(7.80) only exist for the boundary layer functions

$$N_1^{(3)21} = N_1^{(3)12}(y_1, y_2), \qquad N_2^{(3)21} = N_2^{(3)12}(y_1, y_2), \tag{10.12}$$

a result which is easily deduced by the use of expressions (10.9) and (10.10) for the local functions $N_n^{k\beta}$. The functions (10.11) and (10.12) can be obtained numerically from the appropriate boundary layer problems.

In the special case we consider, the homogenized medium is transversely isotropic, with x_1 as the axis of symmetry, and its effective characteristics are as follows:

$$\begin{aligned}
\tilde{c}_{11} &= \left[\gamma \frac{(1+\nu_1)(1-2\nu_1)}{E_1(1-\nu_1)} + (1-\gamma)\frac{(1+\nu_2)(1-2\nu_2)}{E_2(1-\nu_2)}\right]^{-1}, \\
\tilde{c}_{12} &= \tilde{c}_{11}\left[\gamma \frac{\nu_1}{1-\nu_1} + (1-\gamma)\frac{\nu_2}{1-\nu_2}\right], \\
\tilde{c}_{22} &= \gamma \frac{E_1}{1-\nu_1^2} + (1-\gamma)\frac{E_2}{1-\nu_2^2} + \frac{(\tilde{c}_{12})^2}{\tilde{c}_{11}}, \\
\tilde{c}_{55} &= \left[\gamma \frac{2(1+\nu_1)}{E_1} + (1-\gamma)\frac{2(1+\nu_2)}{E_2}\right]^{-1}.
\end{aligned} \tag{10.13}$$

Using (10.10), it turns out (Liebowitz 1968) that the homogenized problem (10.8) is solved analytically to give

$$
\begin{aligned}
&u_1^{(0)} = 2\,\mathrm{Re}[r_1\varphi(z_1) + r_2\psi(z_2)], \qquad u_2^{(0)} = 2\,\mathrm{Re}[q_1\varphi(z_1) + q_2\psi(z_2)] \\
&\sigma_{11} = 2\,\mathrm{Re}[\mu_1^2\varphi'(z_1) + \mu_2^2\psi'(z_2)], \\
&\sigma_{22} = 2\,\mathrm{Re}[\varphi'(z_1) + \psi'(z_2)], \\
&\sigma_{12} = -2\,\mathrm{Re}[\mu_1\varphi'(z_1) + \mu_2\psi'(z_2)],
\end{aligned}
\tag{10.14}
$$

where

$$
\begin{aligned}
&z_1 = x_1 + \mu_1 x_2, \qquad z_2 = x_1 + \mu_2 x_2, \\
&r_1 = \tilde{a}_{11}\mu_1^2 + \tilde{a}_{12}, \qquad r_2 = \tilde{a}_{11}\mu_2^2 + \tilde{a}_{12}, \\
&q_1 = (\tilde{a}_{12}\mu_1^2 + \tilde{a}_{22})/\mu_1, \qquad q_2 = (\tilde{a}_{12}\mu_2^2 + \tilde{a}_{22})/\mu_2, \\
&\tilde{a}_{11} = \tilde{c}_{22}(\tilde{c}_{11}\tilde{c}_{22} - \tilde{c}_{12}^2)^{-1}, \qquad \tilde{a}_{12} = -\tilde{c}_{12}(\tilde{c}_{11}\tilde{c}_{22} - \tilde{c}_{12}^2)^{-1}, \\
&\tilde{a}_{22} = \tilde{c}_{11}(\tilde{c}_{11}\tilde{c}_{22} - \tilde{c}_{12}^2)^{-1}, \qquad \tilde{a}_{55} = (2\tilde{c}_{55})^{-1}, \\
&\varphi(z_1) = -\frac{\mu_2}{\mu_2 - \mu_1}\frac{p}{2}\left[\sqrt{z_1^2 - a^2} - z_1\right], \\
&\psi(z_2) = \frac{\mu_1}{\mu_2 - \mu_1}\frac{p}{2}\left[\sqrt{z_2^2 - a^2} - z_2\right],
\end{aligned}
$$

and where $p_2(x_1) = p$, and

$$
\mu_1 = \alpha_1 + \mathrm{i}\beta_1, \qquad \mu_2 = \alpha_2 + \mathrm{i}\beta_2 \quad (\beta_1 > 0, \beta_2 > 0, \beta_1 \neq \beta_2)
$$

are the roots of the characteristic equation

$$
\tilde{a}_{11}\mu^4 + 2(\tilde{a}_{12} + \tilde{a}_{55})\mu^2 + \tilde{a}_{22} = 0.
$$

The final formulation of the elastic problem for region IV of Fig. 10.3 will be obtained from (10.7) by substituting the local functions $N_n^{k\beta}$ given by (10.9) and the numerically calculated functions (10.11) and (10.12).

Although in principle straightforward, the finite elements solution of the boundary layer problem (7.53)–(7.55) has been given close examination by the author of this book in view of the importance of the final result for the construction of the solution for region IV. Numerical computations performed on a 494-element (540-site) grid with an aspect ratio of 1:10 showed that already at a distance as short as only four length units from the lower edge, the functions sought for differed very little from the zero ones, a fact which enables us to reduce the aspect ratio to 1:4 (i.e. to set $\tilde{\varepsilon} = 4\varepsilon$) in subsequent work. A great degree of confidence in the numerical results obtained was provided by the observation that they changed only slightly when a similar calculation using square eight-site elements (that is, with nearly twice as many sites) was carried out. The results were thus substituted into (10.7), and the computation procedure was limited to region IV, this being represented by 376 elements and 419 sites. The tip of the crack was enclosed by a singular finite element, and the displacement field was approximated with the use of the known solution for the crack-containing region. It turned out the even without reducing the grid spacing, the singular finite element scheme provided high accuracy in estimating the stress intensity factors and in modelling the singular stress field in the vicinity of the crack tip.

Numerical results were obtained for $\gamma = 0.667$, equivalent to the 2:1 ratio of the layer thicknesses in the unit cell. The length of the crack was taken to be $2a = 20.333\,\varepsilon$, the tip was assumed to lie in the interior of the wider layer and the values of the parameters $\tilde{\varepsilon}_1$ and $\tilde{\varepsilon}_2$ (see Fig. 10.3) were chosen to be $2.333\,\varepsilon$ and $2.667\,\varepsilon$, respectively. Of the two cases considered in the study, one was obtained from the other by interchanging the layer materials, so that in one case the tip of the crack lay in the stiffer material (Case 1), and in other in the softer material (Case 2). The elastic parameters of the layers were, respectively,

$$E_1 = 1.5\,\text{GPa}, \qquad E_2 = 20\,\text{GPa}$$

and

$$\nu_1 = 0.446, \qquad \nu_2 = 0.3.$$

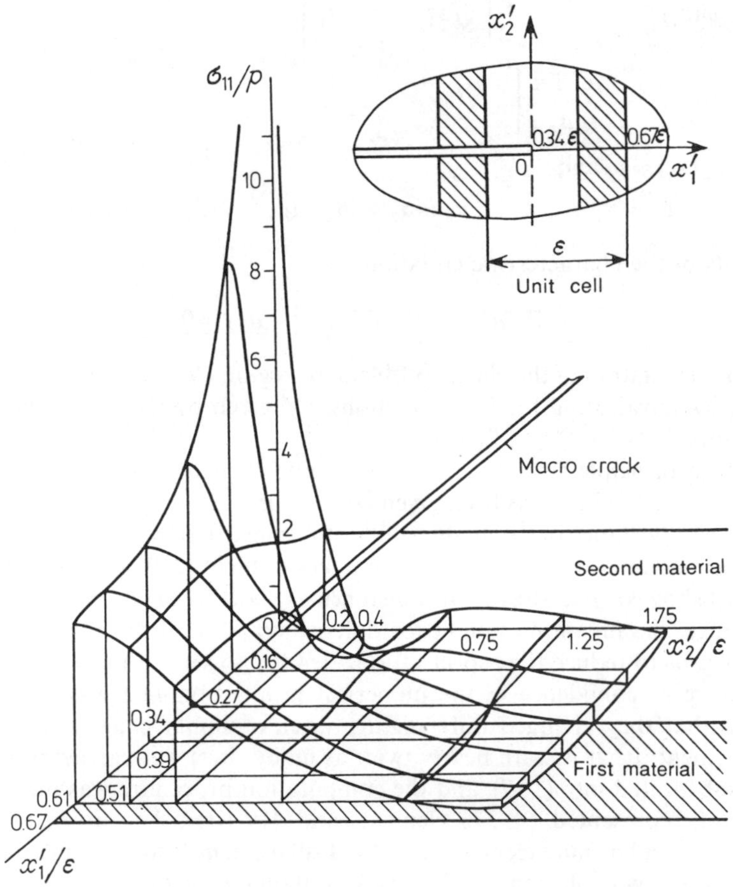

Fig. 10.6 *Local distribution of σ_{11}/p in Case 1.*

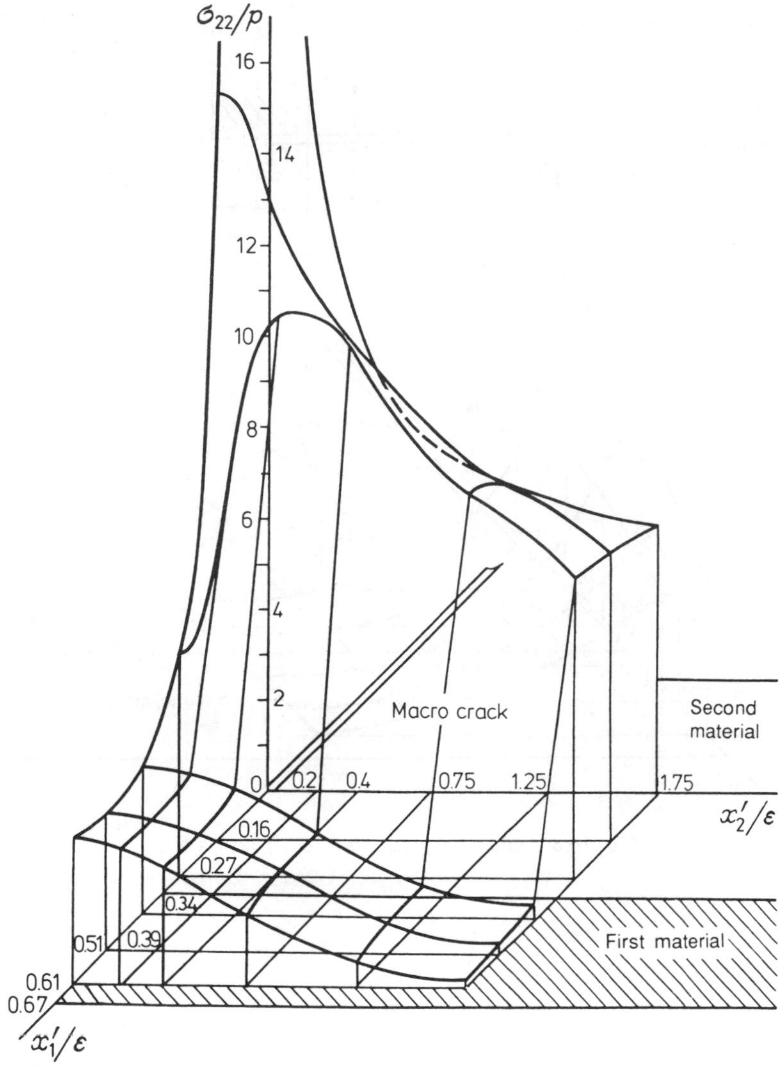

Fig. 10.7 *Local distribution of σ_{22}/p in Case* 1.

In either case, the values of the stress intensity factors, and the local stress fields in the vicinity of the crack tip were calculated.

The stress intensity factor

$$\frac{K_I}{P\sqrt{\pi a}} = \lim_{x_1 \to a+0} [\sqrt{2\pi(x_1 - a)}\sigma_{22}(x_1, 0)]$$

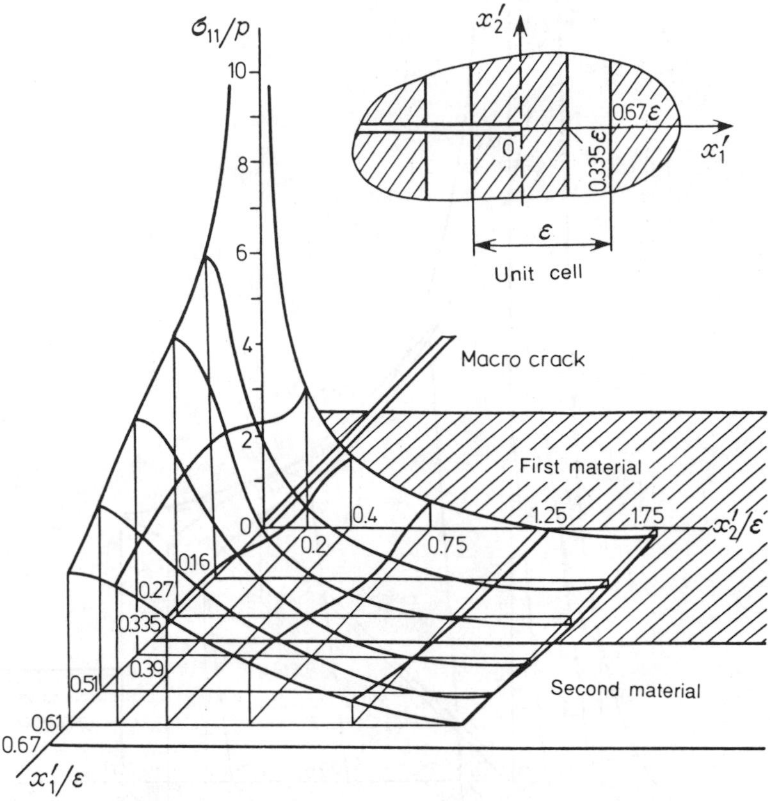

Fig. 10.8 *Local distribution of σ_{11}/p in Case 2*

was found to be about 3.21 in Case 1 and about 1.18 in Case 2. The local stress distributions σ_{11}/p and σ_{22}/p shown in Figs. 10.6 and 10.7 (Case 1) and in Figs. 10.8 and 10.9 (Case 2) are seen to exhibit strong oscillations in the vicinity of the crack tip. The stresses σ_{22}, for example, undergo a marked jump at the interface between the layers, the values in the stiffer material exceeding by a factor of 10 or more those in the softer one (see Figs. 10.7 and 10.9).

The reliability of the results obtained was estimated by fine grid test calculations of the stress intensity factors for the entire plane composite with a central crack and the same geometric and material parameters. The accuracy, in this sense, was found to be within 12%, or probably even better, if the error of the test calculations themselves (unavoidable for highly non-homogeneous media) were taken into account. It was also of interest to compare the above approach with the approximating scheme developed by Parton *et al.* (1988). This latter was found to give strongly underestimated values for the stress intensity factor (1.15 and 0.73 in, respectively, the first and second cases), indicative of the considerable role of the near-crack local boundary effects.

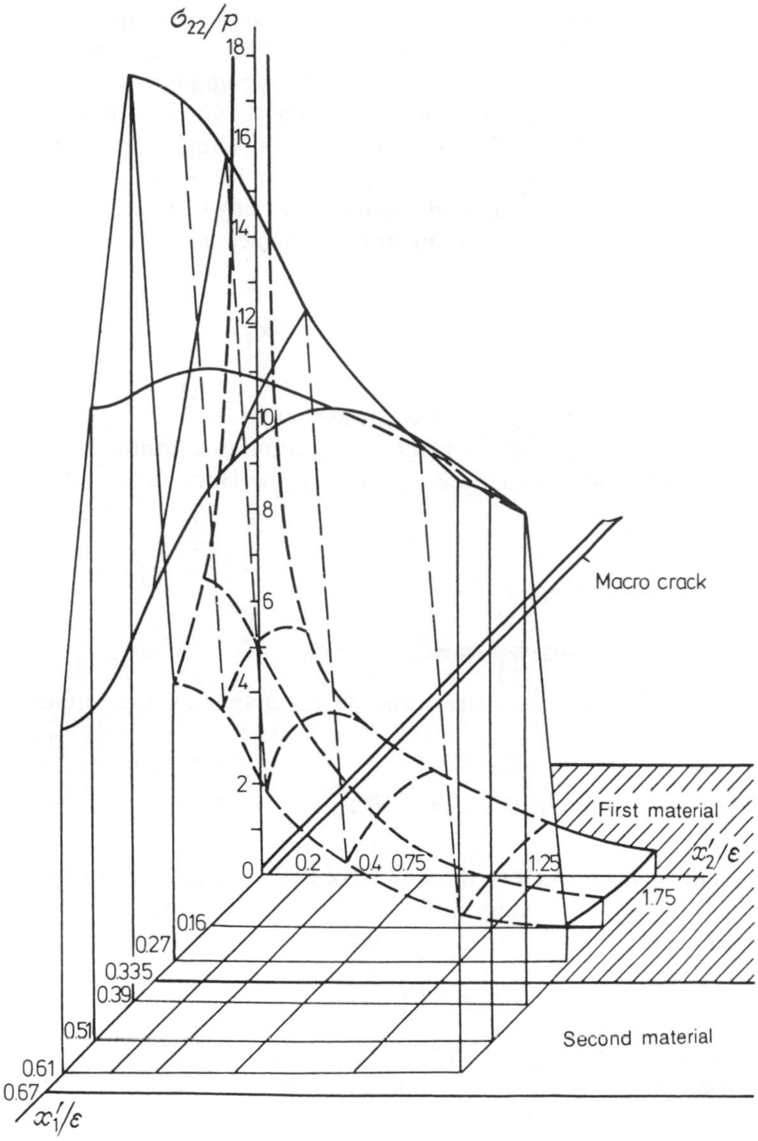

Fig. 10.9 *Local distribution of σ_{22}/p in Case 2.*

11. HOMOGENIZATION OF THE GEOMETRICALLY NON-LINEAR ELASTICITY PROBLEM FOR A PERIODIC COMPOSITE. ELASTIC STABILITY EQUATIONS

The problem of predicting the stability of composite material structures is as important as it is complex. Because of the inherent high non-homogeneity of such materials, the equations of non-linear elasticiy (of geometrically non-linear elasticity in the case of small undercritical deformations) are virtually intractable unless simplifying assumptions are made or the homogenization principles adopted.

In this section we employ the asymptotic method of homogenization in constructing the equations of the geometrically non-linear theory of elasticity for composites with a regular structure. On the basis of these equations, the elastic stability of three-dimensional bodies made of such materials is examined, under the assumption that the functions determining the stress conditions at the moment of buckling vary on a scale much larger than the unit cell dimensions.

For small tensile and shear deformations, relating the body to a curvilinear coordinate system $(\theta_1, \theta_2, \theta_3)$, the nonlinear equations of motion may be represented in the form (cf. Novozhilov 1948, Guz' 1986)

$$\nabla_j(\sigma^{ij} + \sigma^{lj}\nabla_l u^i) + P^{*i} = \rho \ddot{u}^i, \tag{11.1}$$

where ∇_j denotes a covariant derivative; P^{*i} is the body force after the deformation; indices range from 1 to 3; and the generalized stresses are considered to be indistinguishable form the true ones in view of the small deformation assumption.

The mechanical constitutive equations are taken to be Hooke's law (cf. (1.77))

$$\sigma^{ij} = c^{ijkl} e_{kl}, \tag{11.2}$$

in which the strain tensor is defined by

$$2e_{kl} = \nabla_l u_k + \nabla_k u_l + \nabla_k u_m \nabla_l u^m. \tag{11.3}$$

The boundary conditions on the stress and displacements on the surface $s = s_1 \cup s_2$ of the body are

$$(\sigma^{ij} + \sigma^{lj}\nabla_l u^i)n_j \overset{s_1}{=} p^{*i}, \qquad u_i \overset{s_2}{=} u_i^*, \tag{11.4}$$

where $\boldsymbol{n}$ is the unit vector normal to the surface of the non-deformed body; and p^{*i} are the external surface tractions acting in the deformed body. Note that both (11.1) and (11.4) take a more compact form if we introduce

$$t^{ij} = \sigma^{ij} + \sigma^{li}\nabla_l u^j, \tag{11.5}$$

the non-sysmmetric Kirchhoff stress tensor.

Consider the problem (11.1)–(11.4) for a body made from a periodic composite with unit cell Y: $\{-h_i/2 < y_i < h_i/2\}$, $y_i = \theta_i/\varepsilon$, $\varepsilon \ll 1$, $h_1 \sim 1$ $(i = 1, 2, 3)$. The mass density and elastic coefficients are represented in this case by piecewise-smooth periodic functions of the 'rapid' variables $\boldsymbol{y} = (y_1, y_2, y_3)$, with Y as the unit cell.

We solve the problem by asymptotically expanding the solution in powers of the small parameter ε in the following manner:

$$u_i = u_i^{(0)}(\boldsymbol{\theta}, t) + \varepsilon u_i^{(1)}(\boldsymbol{\theta}, t, \boldsymbol{y}) + \varepsilon^2 u_i^{(2)}(\boldsymbol{\theta}, t, \boldsymbol{y}) + \cdots, \tag{11.6}$$

where $u_i^{(m)}(\theta, t, y)$, $m = 1, 2\ldots$, are periodic functions of y with the unit cell Y; $\boldsymbol{\theta} = (\theta_1, \theta_2, \theta_3)$. In accordance with the two-scale expansion method, and remembering that the metric tensor g_{ij} is independent of the 'rapid' coordinates $\boldsymbol{y}$, the operator ∇_i will everywhere be replaced by $\nabla_i + (1/\varepsilon)\delta/\delta y_i$.

From equations (11.2), (11.3), (11.5) and (11.6), and introducing the notation

$\delta u_j/\delta y_i = u_{j|i}$, the leading terms in the stress and strain expressions are found to be given by

$$\begin{aligned} 2e_{kl}^{(0)} &= \nabla_l u_k^{(0)} + \nabla_k u_l^{(0)} + \nabla_k u_m^{(0)} \nabla_l u^{(0)m} + u_{l|k}^{(1)} + u_{k|l}^{(1)} \\ &\quad + u_{m|k}^{(1)} \nabla_l u^{(0)m} + u_{|l}^{(1)m} \nabla_k u_m^{(0)} + u_{m|k}^{(1)} u_{|l}^{(1)m}, \\ \sigma^{(0)ij} &= c^{ijkl} e_{kl}^{(0)}, \\ t^{(0)ij} &= \sigma^{(0)ij} + \sigma^{(0)li}(\nabla_l u^{(0)j} + u_{|l}^{(1)j}). \end{aligned} \tag{11.7}$$

Setting to zero the terms $O(\varepsilon^{-1})$ and $O(\varepsilon^0)$ in (11.1) and (11.4), we obtain:

$$t_{|j}^{(0)ji} = 0, \quad \nabla_j t^{(0)ji} + t_{|j}^{(1)ji} + P^{*i} = \rho \ddot{u}^{(0)i}, \tag{11.8}$$

$$t^{(0)ji} n_j \overset{S_1}{=} p^{*i}, \qquad u_i^{(0)} \overset{S_2}{=} u_i^*. \tag{11.9}$$

Expressions (11.7) for $t^{(0)ij}$ are now substituted into the first of equations (11.8), and the products of three and more rapid coordinate derivatives $\nabla_i u_j^{(0)}$ are discarded because of their smallness. The solution of the resulting expression for $u_k^{(1)}$ may be given in the form

$$u_k^{(1)} = N_k^{mn} \nabla_n u_m^{(0)} + W_k^{mnpq} \nabla_n u_m^{(0)} \nabla_q u_p^{(0)} \tag{11.10}$$

if the $N_k^{mn}(y)$ and $W_k^{mnpq}(y)$ are periodic functions of period Y solving the local problems

$$C_{|i}^{ijmn} = 0, \qquad C^{ijmn} = c^{ijmn} + c^{ijkl} N_{k|l}^{mn}, \tag{11.11}$$

$$\begin{aligned} B_{|i}^{ijmnpq} + C^{ilmn} g^{kj} N_{k|li}^{pq} &= 0, \\ B^{ijmnpq} &= c^{ijkl} W_{k|l}^{mnpq} + \tfrac{1}{2} c^{ijnq} g^{mp} \\ &\quad + c^{ijkn} g^{ms} N_{s|k}^{pq} + \tfrac{1}{2} c^{ijkl} g^{sr} N_{s|k}^{mn} N_{r|l}^{pq}, \end{aligned} \tag{11.12}$$

with the perfect contact continuity conditions

$$[N_k^{mn}] = 0, \qquad [C^{ijmn} n_i^{(k)}] = 0, \tag{11.13}$$

$$\begin{aligned} &[W_k^{mnpq}] = 0, \\ &[(B^{ijmnpq} + C^{ilmn} g^{kj} N_{k|l}^{pq}) n_i^{(k)}] = 0 \end{aligned} \tag{11.14}$$

on surfaces where discontinuities in the material characteristics occur; $n_i^{(k)}$ representing the normal to the discontinuity surface.

The local problem (11.11) and (11.13) coincides with the corresponding local problem obtained in Section 7 in the process of homogenization of linear elasticity equations. The second local problem, equations (11.12) and (11.14), is identical in structure, linear in the unknown functions and incorporates the solution of the first problem as its input data. Either problem has a unique solution, to within a constant term.

Combining (11.7) and (11.10) results in the forms

$$\begin{aligned} \sigma^{(0)ij} &= C^{ijmn} \nabla_n u_m^{(0)} + B^{ijmnpq} \nabla_n u_m^{(0)} \nabla_q u_p^{(0)}, \\ t^{(0)ij} &= \sigma^{(0)ij} + C^{ilmn} \nabla_n u_m^{(0)} \nabla_l u^{(0)j} \\ &\quad + g^{kj} C^{ilmn} N_{k|l}^{pq} \nabla_n u_m^{(0)} \nabla_q u_p^{(0)}, \end{aligned} \tag{11.15}$$

which, when subjected to the unit cell average operation

$$\langle \varphi \rangle = \frac{1}{|Y|} \int_Y \varphi \, \mathrm{d}\boldsymbol{y},$$

yield the pair of equations

$$\langle \sigma^{(0)ij} \rangle = \langle C^{ijmn} \rangle \nabla_n u_m^{(0)} + \langle B^{ijmnpq} \rangle \nabla_n u_m^{(0)} \nabla_q u, \tag{11.16}$$

$$\langle t^{(0)ij} \rangle = \langle \sigma^{(0)ij} \rangle + \langle \sigma^{(0)li} \rangle \nabla_l u^{(0)j}, \tag{11.17}$$

where use has been made of the fact that $\langle C^{ilmn} N_{k|l}^{pq} \rangle = 0$ because of (11.11) and the periodicity in $\boldsymbol{y}$.

Application of the same average operator to the second of equations (11.8) gives

$$\begin{gathered} \nabla_j [\langle \sigma^{(0)lj} \rangle (g_l^i + \nabla_l u^{(0)i})] + P^{*i} = \langle \rho \rangle \ddot{u}^{(0)i}, \\ \langle \sigma^{(0)lj} \rangle (g_l^i + \nabla_l u^{(0)i}) n_j \overset{S_1}{=} p^{*i}, \qquad u_i^{(0)} \overset{S_2}{=} u_i^* \end{gathered} \tag{11.18}$$

after the use of (11.17) and the periodicity in $\boldsymbol{y}$.

Equations (11.16) and (11.18) comprise the homogenized problem of small strain non-linear elasticity and, together with relations (11.15), provide the zeroth approximation to the local structure of the stress field in the material under study. In the limiting case $c^{ijkl} = \text{const.}$ (homogeneous material), we find that

$$C^{ijmn} = c^{ijmn}, \qquad B^{ijmnpq} = \tfrac{1}{2} g^{mp} c^{ijnq}$$

from (11.11) and (11.12), and the problem (11.16), (11.18) reduces to a known one.

We can now proceed to the problem of the elastic stability of three-dimensional bodies made from composite materials with a regular structure. Because our discussion will be limited to the leading term in the expansion (11.6), the superscript (0) will be omitted in the homogenized problem (11.16), (11.18).

Instead, we append it, without parentheses, on functions related to the (unperturbed) state of equilibrium and, as is customary, we describe the perturbed state by the functions $u_i^0 + v_i$, $\sigma^{0ij} + \sigma^{ij}$ and $e_{ij}^0 + e_{ij}$, in which the second terms denote perturbations from the corresponding equilibrium values. We substitute these expressions into (11.16) and (11.18) and make use of the fact that the unperturbed quantities satisfy the equations of equilibrium.

Linearizing with respect to the displacement perturbations v_i then yields

$$\begin{gathered} \nabla_j [\langle \sigma^{lj} \rangle (g_l^i + \nabla_l u^{0i}) + \langle \sigma^{0lj} \rangle \nabla_l v^i] = \langle \rho^0 \rangle \ddot{v}^i, \\ [\langle \sigma^{lj} \rangle (g_l^i + \nabla_l u^{0i}) + \langle \sigma^{0lj} \rangle \nabla_l v^i] n_j \overset{S_1}{=} f^i, \\ v_i \overset{S_2}{=} 0, \end{gathered} \tag{11.19}$$

and from (11.16) it follows that

$$\langle \sigma^{ij} \rangle = \langle C^{ijmn} \rangle \nabla_n v_m + 2 \langle B^{ijmnpq} \rangle \nabla_n u_m^0 \nabla_q v_p. \tag{11.20}$$

Equations (11.19) have been derived under the assumption that the external body forces are independent of displacements. For surface tractions, we have

$$f^i = p^{*i}(u_k^0 + v_k) - p^{*i}(u_k^0). \tag{11.21}$$

For the special case of the follow-up loading, a formula for f^i may be found in the book by Novozhilov (1948).

The equilibrium displacements u_i^0 are determined from the problem (11.16), (11.18) in which all non-linear terms may be omitted because the turning angles are negligibly small in the state of equilibrium. This results in the following linear problem:

$$\begin{gathered} \nabla_j\langle \sigma^{0ij}\rangle + P^i = 0, \qquad \langle \sigma^{0ij}\rangle = C^{ijmn}\rangle \nabla_n u_m^0, \\ \langle \sigma^{0ij}\rangle n_j \overset{S_1}{=} p^i, u_i^0 \overset{S_2}{=} u_i^*. \end{gathered} \tag{11.22}$$

In view of the smallness of $\nabla_l u^{0i}$, relations (11.19) reduce to the problem

$$\begin{gathered} \nabla_j[\langle \sigma^{ij}\rangle + \langle \sigma^{0lj}\rangle \nabla_l v^i] = \langle \rho^o\rangle \ddot{v}^i, \\ (\langle \sigma^{ij}\rangle + \langle \sigma^{0lj}\rangle \nabla_l v^i) n_j \overset{S_1}{=} f^i, \qquad v_i \overset{S_2}{=} 0, \end{gathered} \tag{11.23}$$

which admits of still further simplification by assuming that the turning angles in the perturbed state are much greater than the strain components; see Parton and Kalamkarov (1988b).

It will be understood that, strictly speaking, problems (11.19)–(11.23) do not describe the stability of the original composite medium, but rather relate to the homogenized medium whose effective properties are determined from the local problems (11.11)–(11.14). It is therefore a necessary condition, for the critical loads so obtained to be realistic, that the unit cell be small on the length scale for the variation of the functions determining the stress (strain) state of the composite at the moment of loss of stability.

4 Thermoelasticity of Regular Composite Structures

One of the most important applications of homogenization ideas is to analyse coupled fields in deformable composite media with a regular structure. In this chapter, problems concerned with the coupling of elastic variables to thermal variables are treated.

In particular, the effective properties of thermoelastic laminar and unidirectional fibre composites are derived. The use of the effective property results is illustrated by considering the problem of designing a laminar composite with prescribed stiffness and thermal properties.

12. HOMOGENIZATION OF THE THERMOELASTICITY PROBLEM

Thermoelasticity may perhaps be considered as a classical example of a theory involving the interaction of physical fields each with a different nature. Ene (1983) and Kalamkarov *et al.* (1987a) have applied the method of homogenization to the coupled thermoelasticity analysis of a regularly non-homogeneous composite. Paşa (1983) has proved that the solution of a non-homogeneous problem converges to that of the homogenized coupled thermoelasticity problem in the limit as the period of the structure tends to zero.

Following Kalamkarov *et al.* (1987a), we consider a non-steady-state problem of coupled thermoelasticity for a body made of an anisotropic non-homogeneous material (a composite). Let the position of a typical point of the body be denoted by three coordinates η_1, η_2 and η_3 of a curvilinear system of axes, and let the unit cell of the structure, Y, be defined by the inequalities

$$-h_i/2 < y_i < h_i/2,$$

where $y_i = \eta_i/\varepsilon$, $\varepsilon \ll 1$ and $h_i \sim 1$ $(i = 1,2,3)$.

The linear version of the equation of motion (11.1) is

$$\nabla_j \sigma^{ij} + P^i = \rho \ddot{u}^i, \tag{12.1}$$

where, once again, ∇_j denotes a covariant derivative.

The stress and strain tensors are related to the temperature change by the equation

$$\sigma^{ij} = c^{ijkl} e_{kl} - \beta^{ij}\theta, \qquad \beta^{ij} = c^{ijkl}\alpha^{\mathrm{T}}_{kl}, \tag{12.2}$$

which is known as the Duhamel–Neumann law (cf. (2.14) and (2.16)), the strain tensor being defined by

$$2e_{kl} = \nabla_l u_k + \nabla_k u_l. \tag{12.3}$$

The heat flow vector is expressed in terms of temperature by the Fourier law (cf. (2.16)):

$$q^i = -\lambda^{ij}\nabla_j\theta, \tag{12.4}$$

where λ^{ij} denotes the thermal conductivity components, and the heat balance equation is taken in the form (cf. (2.21))

$$-\nabla_i q^i - T_0\beta^{ij}\dot{e}_{ij} = c_v\dot{\theta} - f, \tag{12.5}$$

where c_v is the volumetric specific heat and f is the density of internal heat sources. As in Section 11 in the preceding chapter, all coefficients in relations (12.1)–(12.5) are considered to be piecewise-smooth periodic functions of the coordinates y_1, y_2 and y_3 with unit cell Y.

Equations (12.1) and (12.5), together with relations (12.2)–(12.4), form a closed system of equations of linear coupled thermoelasticity for an anisotropic non-homogeneous solid. To this system we have to adjoin appropriate boundary and initial conditions. By analogy with (11.4), the placement and traction conditions on the surface $\partial\Omega = \partial_1\Omega \cup \partial_2\Omega$ of the body are taken to be

$$\sigma^{ij}n_j \overset{\partial_1\Omega}{=} P^i, \qquad u_i \overset{\partial_2\Omega}{=} u_i^*, \tag{12.6}$$

and the heat exchange conditions are written as

$$q^j n_j \overset{\partial\Omega}{=} \alpha_S T - q_S^*, \tag{12.7}$$

where α_S represents the heat transfer coefficient and q_S^* is the external heat flow. In the special case of heat exchange of the third kind we have (cf. (2.25))

$$q_S^* = \alpha_S T_S^*, \tag{12.8}$$

where T_S^* is the ambient temperature.

Finally, the initial conditions (at $t = 0$, t being time) are specified in the form

$$u_i = u_i^0, \qquad \dot{u}_i = v_i^0, \qquad T = T^0. \tag{12.9}$$

Since the degree of strain–temperature coupling is usually insignificant in practical applications, it is common practice in problems of the kind we are discussing to neglect the effects of deformation on the temperature distribution in the material. It is known, however, that the inclusion of coupling may even change the qualitative nature of the solution of a dynamic problem, as exemplified by the work of Podstrigach and Shvets (1978) on the thermoelasticity of thin shells. On the other hand, there is evidence that in some polymeric materials thermoelastic coupling may be strong enough to be considered, especially when impact-type loads are applied (Kovalenko 1970). As shown by Lukovkin *et al.* (1983), a jump in temperature due to an impact load may play a crucial role in the fracture of glassy polymers, the reason lying in the specifics of deformation and the low thermal conductivity of such materials rather than in a large thermoelastic coupling. Since composite materials often have polymers as their matrices, clearly the above facts are of special significance in the present context.

The solution of the problem will be sought by assuming asymptotic small-parameter expansions of the form

$$\begin{aligned} u_i &= u_i^{(0)}(\boldsymbol{\eta}, t) + \varepsilon u_i^{(1)}(\boldsymbol{\eta}, t, \boldsymbol{y}) + \varepsilon^2 u_i^{(2)}(\boldsymbol{\eta}, t, \boldsymbol{y}) + \cdots, \\ \theta &= \theta^{(0)}(\boldsymbol{\eta}, t) + \varepsilon \theta^{(1)}(\boldsymbol{\eta}, t, \boldsymbol{y}) + \varepsilon^2 \theta^{(2)}(\boldsymbol{\eta}, t, \boldsymbol{y}) + \cdots, \end{aligned} \tag{12.10}$$

where $u_i^{(m)}(\boldsymbol{\eta}, t, \boldsymbol{y})$ and $\theta^{(m)}(\boldsymbol{\eta}, t, \boldsymbol{y}), m = 1, 2, \ldots,$ are periodic functions of $\boldsymbol{y}$ with unit cell Y.

Without going into the details of the homogenization procedure, the leading terms in the expansions (12.10) are found to be given by

$$\begin{aligned} u_k &= u_k^{(0)}(\boldsymbol{\eta}, t) + \varepsilon[N_k^{mn} \nabla_n u_m^{(0)} + M_k \theta^{(0)}] + \cdots, \\ \theta &= \theta^{(0)}(\boldsymbol{\eta}, t) + \varepsilon W^n \nabla_n \theta^{(0)} + \cdots, \end{aligned} \tag{12.11}$$

where $N_k^{mn}(\boldsymbol{y})$, $M_k(\boldsymbol{y})$ and $W^n(\boldsymbol{y})$ are periodic functions with unit cell Y, of which N_k^{mn} solves the local problem (11.11), (11.13) resulting from the homogenization of the elasticity theory problem, and M_k and W^n are solutions of the following local problems:

$$S^{ij}_{|i} = 0, \qquad S^{ij} = \beta^{ij} - c^{ijkl} M_{k|l}, \tag{12.12}$$

$$\Lambda^{ij}_{|i} = 0, \qquad \Lambda^{ij} = \lambda_{ij} + \lambda^{ik} W^j_{|k}. \tag{12.13}$$

The continuity conditions to be satisfied across the surfaces of discontinuity in material properties are similar to those given by (11.13), namely

$$[\![M_k]\!] = 0, \qquad [\![S^{ij} n_i^{(k)}]\!] = 0, \tag{12.14}$$

$$[\![W^k]\!] = 0, \qquad [\![\Lambda^{ij} n_i^{(k)}]\!] = 0. \tag{12.15}$$

In accordance with (12.11), the leading order terms in the stress tensor and heat flow vector expansions are given by

$$\begin{aligned} \sigma^{(0)ij} &= C^{ijmn} \nabla_n u_m^{(0)} - S^{ij} \theta^{(0)}, \\ q^{(0)i} &= -\Lambda^{ij} \nabla_j \theta^{(0)}. \end{aligned} \tag{12.16}$$

In a manner parallel to that discussed in Section 11, we set to zero the terms of $O(1)$ in equations (12.1) and (12.5) and take a volume average over the unit cell Y, to obtain:

$$\begin{aligned} \nabla_j \langle \sigma^{(0)ij} \rangle + P^i &= \langle \rho \rangle \ddot{u}^{(0)i}, \\ -\nabla_i \langle q^{(0)i} \rangle - T_0 \langle B^{ij} \rangle \nabla_j \dot{u}_i^{(0)} &= \langle C_v \rangle \dot{\theta}^{(0)} - f, \end{aligned} \tag{12.17}$$

where we have introduced

$$B^{ij} = C^{klij}\alpha^{T}_{kl}, \qquad C_v = c_v + T_0\alpha^{T}_{kl}(\beta^{kl} - S^{kl}). \tag{12.18}$$

Averaging (12.16) results in the equations

$$\langle \sigma^{(0)ij} \rangle = \langle C^{ijmn} \rangle \nabla_n u^{(0)}_m - \langle S^{ij} \rangle \theta^{(0)}, \qquad \langle q^{(0)i} \rangle = -\langle \Lambda^{ij} \rangle \nabla_j \theta^{(0)}, \tag{12.19}$$

which are the constitutive relations of the homogenized medium, and the coefficients of which represent the (effective) elastic and thermal properties of the medium. It is important to note that

$$\langle B^{ij} \rangle = \langle S^{ij} \rangle. \tag{12.20}$$

To prove this, combine (11.11), (12.2) and (12.18) to give

$$B^{ij} = \beta^{ij} + \beta^{mn} N^{ij}_{m|n}. \tag{12.21}$$

Using the periodicity in $\boldsymbol{y}$ and the local problem (12.12), observe that

$$\langle \beta^{mn} N^{ij}_{m|n} \rangle = -\langle N^{ij}_m \beta^{mn}_{|n} \rangle = -\langle N^{ij}_m (c^{mnkl} M_{k|l})_{|n} \rangle = \langle c^{mnkl} M_{k|l} N^{ij}_{m|n} \rangle. \tag{12.22}$$

Similarly, from the local problem (11.11), we obtain:

$$-\langle c^{ijkl} M_{k|l} \rangle = \langle M_k c^{ijkl}_{|l} \rangle = -\langle M_k (c^{klmn} N^{ij}_{m|n})_{|l} \rangle = \langle c^{klmn} N^{ij}_{m|n} M_{k|l} \rangle. \tag{12.23}$$

Comparing (12.22) and (12.23) and using the symmetry properties of the elastic modulus tensor, we now find that

$$\langle \beta^{mn} N^{ij}_{m|n} \rangle = -\langle c^{ijkl} M_{k|l} \rangle, \tag{12.24}$$

which proves (12.20) when combined with (12.12) and (12.21).

To obtain the boundary and initial conditions of the homogenized problem, take a volume average of (12.6) and (12.7) and retain only the leading order terms in expansions (12.11) to arrive at

$$\begin{aligned} &\langle \sigma^{(0)ij} \rangle n_j \overset{\partial_1\Omega}{=} p^i, \qquad u_i \overset{\partial_2\Omega}{=} u^*_i, \\ &\langle q^{(0)j} \rangle n_j \overset{\partial\Omega}{=} \langle \alpha_S \rangle T^{(0)} - q^*_S, \\ &u^{(0)}_i = u^0_i, \qquad \dot{u}^{(0)}_i = v^0_i, \qquad T^{(0)} = T^0 \quad \text{when } t = 0. \end{aligned} \tag{12.25}$$

Relations (12.17)–(12.20) and (12.25) describe the homogenized problem of thermoelasticity for the medium being examined. The effective properties of the medium are calculated from the soutions of the local problems (11.11), (11.13), (12.12), (12.14) and (12.13), (12.15), and relations (12.16) provide the zeroth-order approximation to the local structure of the stress and heat flow fields. Note that the solutions of the local problems are only unique to within constant terms, an ambiguity which is easily removed by introducing the conditions

$$\langle N_k^{mn} \rangle = 0 \quad (N_k^{mn} \leftrightarrow M_k \leftrightarrow W^n). \tag{12.26}$$

Averaging (12.11) then yields the formulas

$$\langle u_k \rangle = u_k^{(0)}(\boldsymbol{\eta}, t), \qquad \langle \theta \rangle = \theta^{(0)}(\boldsymbol{\eta}, t), \tag{12.27}$$

clarifying the meaning of the leading order in the expansions (12.11).

If coupling effects may be neglected, the term $\nabla_j \dot{u}_i^{(0)}$ in the heat balance equation (12.17) should be omitted, and the effective specific heat of the system should be written as $\langle C_v \rangle = \langle c_v \rangle$, equivalent to the use of the rule of mixtures.

13. FIBRE COMPOSITES: LOCAL STRESSES AND EFFECTIVE PROPERTIES

We shall be concerned in this section with predicting the mechanical behaviour of unidirectional fibre reinforced composites with a regular structure. The unit cell in this case is obtained by cutting the material along the plane orthogonal to the direction of the fibres and may be represented, in particular, by a square containing an inclusion in the form of the fibre cross-section. If the fibres are aligned along the y_3-axis, then for a homogeneous matrix and fibre materials all the characteristics of the composite will be described by piecewise-smooth, doubly periodic functions of y_1 and y_2. The local problem (11.11) reduces in this case to the system

$$\begin{aligned} \tau_{|\alpha}^{\alpha jmn}(\mathrm{f}) &= 0 \quad \text{in } Y_{\mathrm{f}}, \\ \tau_{|\alpha}^{\alpha jmn}(\mathrm{M}) &= 0 \quad \text{in } Y_{\mathrm{M}}, \\ \tau^{\alpha jmn}(\mathrm{f}) &= c^{\alpha jk\lambda}(\mathrm{f}) N_{k|\lambda}^{mn}(\mathrm{f}), \\ \tau^{\alpha jmn}(\mathrm{M}) &= c^{\alpha jk\lambda}(\mathrm{M}) N_{k|\lambda}^{mn}(\mathrm{M}), \\ &(\alpha, \lambda = 1,2;\ j,k,m,n = 1,2,3). \end{aligned} \tag{13.1}$$

the solutions of which, apart from the double periodicity property, must also satisfy conditions (11.13) of the perfect bond on the interface Γ, that is,

$$\begin{aligned} N_k^{mn}(\mathrm{f})|_\Gamma &= N_k^{mn}(\mathrm{M})|_\Gamma, \\ (\tau^{\alpha jmn}(\mathrm{f}) + c^{\alpha jmn}(\mathrm{f})) n_\alpha^{(k)}|_\Gamma &= (\tau^{\alpha jmn}(\mathrm{M}) + c^{\alpha jmn}(\mathrm{M})) n_\alpha^{(k)}|_\Gamma. \end{aligned} \tag{13.2}$$

where $c^{\alpha jmn}(\mathrm{f})$ and $c^{\alpha jmn}(\mathrm{M})$ are the elasticity tensors of, respectively, the fibre and matrix materials, and $n_\alpha^{(k)}$ is the unit vector in the outward normal direction to the surface Γ; $\alpha = 1,2$.

As already discussed in Section 9, the specific structure of the local problems (13.1)–(13.2) depends on the symmetry properties of the matrix and fibre materials. If both materials are isotropic, for example, we have to solve four similar plane strain problems for $\tau^{\alpha\beta 11}$, $\tau^{\alpha\beta 22}$, $\tau^{\alpha\beta 33}$ and $\tau^{\alpha\beta 12}$ and two antiplane problems for $\tau^{\alpha 323}$ and $\tau^{\alpha 313}$.

It was remarked earlier that, if there are some elements of symmetry in the geometry of the unit cell or in the properties of the constituents of the composite, a unit cell local problem reduces to a boundary value problem for only a part of the cell, with a resultant reduction in the amount of numerical work.

In particular, for a square unit cell $|y_1| < \frac{1}{2}$, $|y_2| < \frac{1}{2}$ with circular or square fibres, the reader is referred to Fig. 13.1 for the boundary conditions to be satisfied by the functions N_i^{kl} $(i, k, l = 1,2,3)$.

A similar analysis can be carried out in the case of unidirectional fibre composites for the local problems (12.12), (12.14) and (12.13), (12.15). Since

$$\langle S^{ij} \rangle = \langle B^{ij} \rangle = \langle \beta^{ij} + \beta^{mn} N^{ij}_{m|n} \rangle \tag{13.3}$$

from (12.20) and (12.21), it follows that the effective thermoelastic coefficients $\langle S^{ij} \rangle$ can be calculated using the solution of the local elasticity problem (13.1)–(13.2).

The method used by Kalamkarov *et al.* (1987a) in their finite element analysis of the local problem (13.1)–(13.2) consists of dividing a quarter of the unit cell of the composite into a set of triangular elements numbering from 350 to 400 for different composite combinations. The good agreement between the effective moduli predicted and the relevant results known from earlier work (Pobedrya 1984, Mol'kov and Pobedrya 1985) encouraged the application of the solutions so obtained to the calculation of the effective properties of various types of fibre composites.

The authors considered, in particular, a system of transversally isotropic carbon and organic fibres of circular cross-section embedded in an isotropic plastic (EDT-10) matrix.

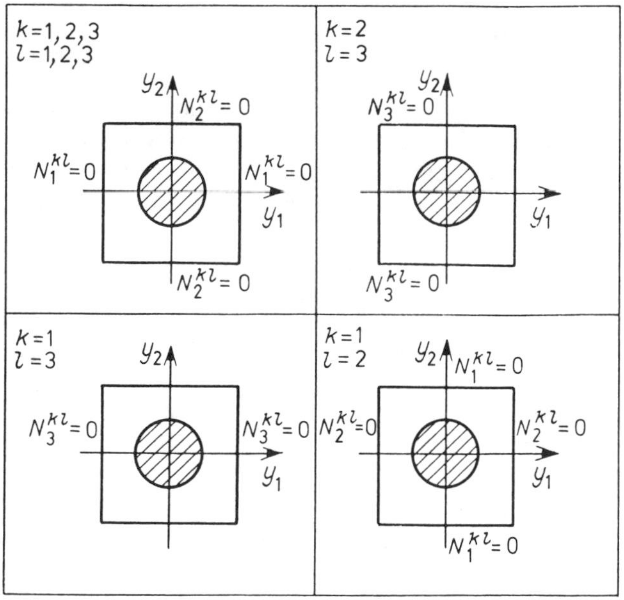

Fig. 13.1 *Boundary conditions on the contour of the unit cell for the functions* N_i^{kl} $(i, k, l = 1, 2, 3)$.

Table 13.1 *Elastic moduli of fibre materials.*

	Organic fibres	Carbon fibres
E_3 (GPa)	127.500	225.600
E_1 (GPa)	3.236	7.751
G_{13} (GPa)	2.354	59.720
ν_{13}	0.29	0.30
ν_{12}	0.17	0.20

The properties of the fibre materials are listed in Table 13.1 reproduced from Maksimov *et al.* (1983), and the characteristics of the matrix material are, from the same source, $E_M = 3.236$ GPa and $\nu_M = 0.42$. The engineering properties of these systems were computed with the use of the local problem solutions and are represented in Table 13.2 together with the corresponding experimental results taken, again, from Maksimov *et al.* (1983). The agreement between the two sets of values may be considered to be quite satisfactory, especially if one notes that, of the experimental values shown, only E_3 had been measured directly. The other four had been computed from approximate formulas the inaccuracy of which accounts at least in part for the discrepancies in the values of E_1 and G_{13}.

The effects of unit cell geometry and fibre volume fraction on the effective properties of a composite were examined by Kalamkarov *et al.* (1987a) by applying their method to an Al–B–Si fibre, epoxy matrix composite, the epoxy being taken either pure or reinforced with a fine elastomer dispersion (which is an effective means for enhancing the impact strength of a material; see, for example, Lukovkin *et al.* 1983). The elastic and thermal properties of the fibre and matrix materials are shown in Table 13.3.

Of the four major geometries considered in the study (see Fig. 13.2), the first three differ in their matrix materials as follows:

(I) a pure epoxy;

(II) a macroscopically isotropic polydispersed medium (see Christensen 1979) formed by spherical rubber inclusions of volume fraction 0.2 embedded in the epoxy;

(III) a transversely isotropic material formed by rubber fibres of volume fraction 0.2 embedded in the epoxy (note the elongated shape of the inclusions).

Table 13.2 *Elastic moduli of unidirectional fibre reinforced composites.*

	Organic plastics (fibre volume fraction 0.48)		Carbon plastic (fibre volume fraction 0.40)	
	Numerical calculation	Experimental values	Numerical calculation	Experimental values
E_3 (GPa)	62.610	62.763 ± 4.630	92.150	87.279 ± 6.982
E_1 (GPa)	3.761	3.727 ± 0.333	7.035	5.884 ± 0.628
G_{13} (GPa)	1.814	1.569 ± 0.177	4.341	3.825 ± 0.451
ν_{13}	0.365	0.37 ± 0.06	0.374	0.38 ± 0.05
ν_{12}	0.489	0.45 ± 0.06	0.506	0.46 ± 0.06

Table 13.3 *Elastic and thermal properties of the fibre and matrix materials (room temperature).*

	E(GPa)	ν	G(GPa)	$\alpha^T(K^{-1})$	$\beta(N/m^2K)$	λ(W/mK)	$c_v(J/m^3K)$
Al–B–Si fibre	73.0	0.23	29.68	5×10^{-6}	6.765×10^5	0.9	1.826×10^6
Epoxy resin	3.43	0.35	1.27	6×10^{-5}	6.852×10^5	0.198	1.533×10^6
Rubber	2.94×10^{-3}	0.4998	9.8×10^{-4}	2×10^{-4}	1.47×10^6	0.22	1.623×10^6
Epoxy with embedded rubber inclusions of volume fraction 0.2	2.22	0.393	7.974×10^{-1}	8.632×10^{-5}	8.938×10^5	0.202	1.551×10^6

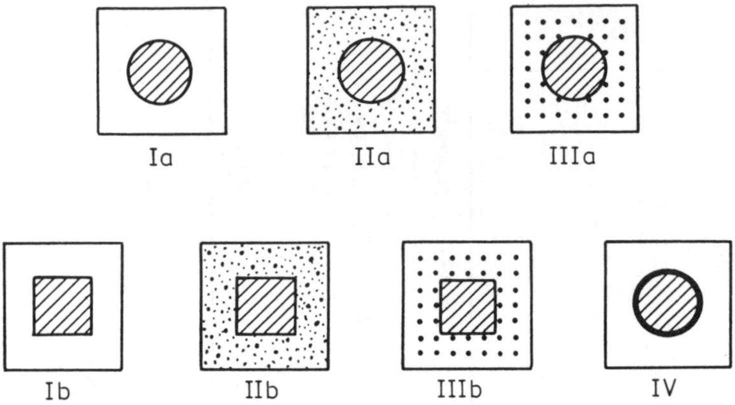

Fig. 13.2 *Unit cells of the unidirectional fibre composites under study.*

In composition IV, the matrix material is again a pure epoxy, but this time there is a thin rubber coating around each fibre; the volume fraction of the rubber in the region outside the fibres is again 0.2.

The results of the numerical calculations are summarized in Table 13.4 in which all the effective properties of the composites are related to the corresponding fibre material properties; see Table 13.3. The notation adopted in the figures is as follows: γ denotes the fibre volume fraction; K_{12} is the bulk elastic modulus for plane deformation normal to the fibre direction,

$$K_{12} = (\tfrac{1}{2})(\langle C^{2222} \rangle + \langle C^{1122} \rangle);$$

and K_f is the bulk modulus for the plane deformation of an isotropic fibre material. On the basis of the data given in Table 13.3, it is found that $K_f = 54.953$ GPa. The compositions are labelled in accordance with Fig. 13.2.

Referring to (12.18), the effective heat capacity is given by the expression

$$\langle C_v \rangle = \langle c_v \rangle + \langle T_0 \alpha^{\mathrm{T}}_{kl}(\beta^{kl} - S^{kl}) \rangle, \tag{13.4}$$

in which the first term corresponds to the rule of mixtures while the second corrects for thermoelastic coupling effects and turns out to be negligibly small. For example, for $\gamma = 0.6$, with constituent properties taken from Table 13.3, the rule of mixtures gives $\langle C_v \rangle = 1.709 \times 10^6$ J/m^{3}K for combinations I(a,b) of Fig. 13.2, whereas the (room temperature) values of the correction term are 0.108 J/m^{3}K and 0.101 J/m^{3}K, that is, seven orders of magnitude less, for, respectively, the circular, composition I(a), and square, composition I(b), cross-sections.

The comparison between the predicted effective properties (shown in Table 13.4) and their Hashin–Shtrikman variational estimates (see (3.63), (3.64)) is illustrated in Figs. 13.3 and 13.4, which show the effective engineering constants of the composition I(a) as functions of the fibre volume fraction γ and in which the curves indicating the Hashin–Shtrikman upper and lower bounds of the corresponding properties are drawn. Also shown are the values computed for composition I(b) (closed squares) and composition IV (closed circles). While for the E_3 property the agreement is all but

Table 13.4 *Calculated values of the elastic and thermal characteristics of unidirectional glass-reinforced plastics (room temperature).*

	I(a) $\gamma = 0.2$	II(a) $\gamma = 0.2$	III(a) $\gamma = 0.2$	I(a) $\gamma = 0.4$	II(a) $\gamma = 0.4$	III(a) $\gamma = 0.4$
E_3/E_f	0.238	0.225	0.230	0.429	0.420	0.424
K_{12}/K_f	0.157	0.146	0.148	0.245	0.232	0.234
ν_{31}/ν_f	1.234	1.400	1.268	1.134	1.170	1.152
G_{32}/G_f	0.123	0.104	0.109	0.212	0.188	0.194
G_{12}/G_f	0.120	0.102	0.102	0.200	0.180	0.180
$\langle \Lambda^{11} \rangle / \lambda_f$	0.316	0.321	0.321	0.427	0.431	0.431
$\langle \Lambda^{33} \rangle / \lambda_f$	0.376	0.380	0.380	0.532	0.535	0.535
$\langle S^{11} \rangle / \beta_f$	1.012	1.294	1.272	1.010	1.265	1.244
$\langle S^{33} \rangle / \beta_f$	1.011	1.274	1.223	1.009	1.224	1.184

	I(a) $\gamma = 0.6$	I(b) $\gamma = 0.6$	II(a) $\gamma = 0.6$	II(b) $\gamma = 0.6$	III(a) $\gamma = 0.6$	III(b) $\gamma = 0.6$	IV $\gamma = 0.6$
E_3/E_f	0.619	0.620	0.613	0.614	0.616	0.617	0.616
K_{12}/K_f	0.347	0.422	0.331	0.405	0.333	0.408	0.333
ν_{31}/ν_f	1.081	1.060	1.104	1.077	1.092	1.068	1.090
G_{32}/G_f	0.316	0.391	0.282	0.361	0.291	0.368	0.223
G_{12}/G_f	0.285	0.366	0.261	0.342	0.261	0.342	0.275
$\langle \Lambda^{11} \rangle / \lambda_f$	0.556	0.603	0.560	0.607	0.560	0.607	0.560
$\langle \Lambda^{33} \rangle / \lambda_f$	0.688	0.688	0.690	0.690	0.690	0.690	0.690
$\langle S^{11} \rangle / \beta_f$	1.009	1.008	1.231	1.205	1.213	1.189	1.119
$\langle S^{33} \rangle / \beta_f$	1.007	1.006	1.173	1.162	1.145	1.134	1.097

	I(a) $\gamma = 0.7$	I(b) $\gamma = 0.7$	II(a) $\gamma = 0.7$	II(b) $\gamma = 0.7$	III(a) $\gamma = 0.7$	III(b) $\gamma = 0.7$
E_3/E_f	0.715	0.715	0.710	0.710	0.712	0.712
K_{12}/K_f	0.415	0.496	0.395	0.476	0.398	0.480
ν_{31}/ν_f	1.061	1.044	1.078	1.057	1.070	1.050
G_{32}/G_f	0.389	0.463	0.347	0.427	0.358	0.436
G_{12}/G_f	0.338	0.428	0.309	0.401	0.309	0.401
$\langle \Lambda^{11} \rangle / \lambda_f$	0.635	0.682	0.639	0.685	0.639	0.685
$\langle \Lambda^{33} \rangle / \lambda_f$	0.766	0.766	0.767	0.767	0.767	0.767
$\langle S_{11} \rangle / \beta_f$	1.008	1.007	1.209	1.181	1.192	1.166
$\langle S_{33} \rangle / \beta_f$	1.006	1.005	1.145	1.133	1.122	1.111

perfect; it is seen from the figures that in other properties discrepancies up to a factor of two occur.

On the basis of the analysis of Table 13.4, the following general conclusions can be made:

(1) With constituent volume fractions equal in any two composites, the geometric structure generally influences the effective properties. For example, the properties E_3,

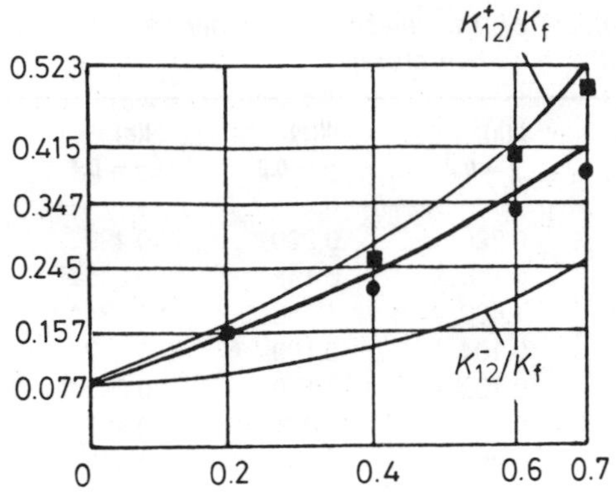

Fig. 13.3 *Effective bulk elastic modulus K_{12} for plane deformation and the Hashin–Shtrikman estimates $K^{\pm}_{12}$.*

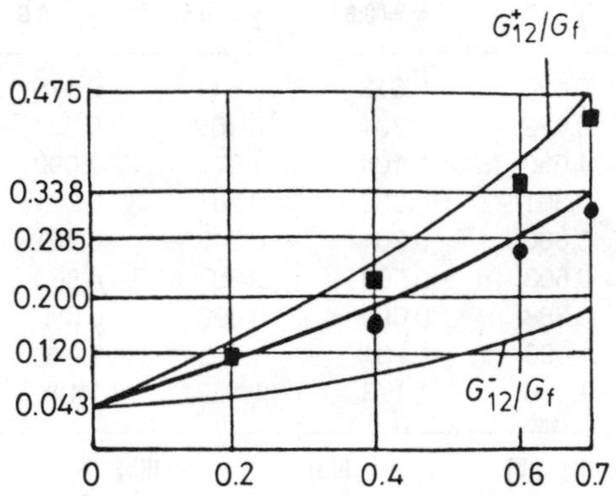

Fig. 13.4 *Effective shear modulus G_{12} and the Hashin–Shtrikman estimates $G^{\pm}_{12}$.*

K_{12} and G_{32} increase, and ν_{31}, $\langle S^{11}\rangle$ and $\langle S^{33}\rangle$ decrease in passing from composition II to composition III in Fig. 13.2. The properties G_{12} and $\langle \Lambda^{11}\rangle$ remain the same, though.

(2) At $\gamma = 0.6$, compositions II, III and IV are identical in terms of the volume fractions of their three components. The properties E_3, K_{12} and $\langle \Lambda^{11}\rangle$ coincide for compositions III(a) and IV, but in the latter case a higher value of G_{12} and lower values of ν_{31}, G_{32}, $\langle S^{11}\rangle$ and $\langle S^{33}\rangle$ are observed. Also, G_{12} in IV is greater than in II(a).

(3) For equal constituent volume contents, the shape of the fibre cross-section is a factor controlling the effective properties of the composite. The values of K_{12}, G_{32}, G_{12} and $\langle \Lambda^{11}\rangle$, for example, increase from circular to square cross-section (that is, from compositions I–III(a) to I–III(b) in Fig. 13.2). For ν_{31}, $\langle S^{11}\rangle$ and $\langle S^{33}\rangle$, the reverse is the case.

To summarize: the results obtained from the solutions of the local problems make it possible to elucidate the effects of both component content and unit cell geometry on the mechanical behaviour of a composite, thereby enabling the prediction and optimization of the properties of the composite.

Note also that, on the basis of the numerical solutions of the local problems, the distributions of stress over the unit cell of the composite can be determined.

With knowledge of the distributions of the functions C^{ijmn} and S^{ij} and of the solution of the homogenized problem, formula (12.16) will give, with a high degree of accuracy, the distribution of local stresses for various loading conditions and given temperature gradients. In particular, if the temperature is constant and macroscopic deformation is uniform ($e_{11}^{(0)} = 1$), the functions C^{1111}, C^{2211}, C^{3311} and C^{1211} coincide with the respective stress components $\sigma_{11}^{(0)}$, $\sigma_{22}^{(0)}$, $\sigma_{33}^{(0)}$ and $\sigma_{12}^{(0)}$. This means that, apart from the first three components, this type of deformation also gives rise to the fourth, $\sigma_{12}^{(0)}$, which, as seen from Fig. 13.5, is less than, but of the same order of magnitude as, the others. We note also, again with reference to Fig. 13.5, that maximum stresses occur in the vicinity of the fibre–matrix interface.

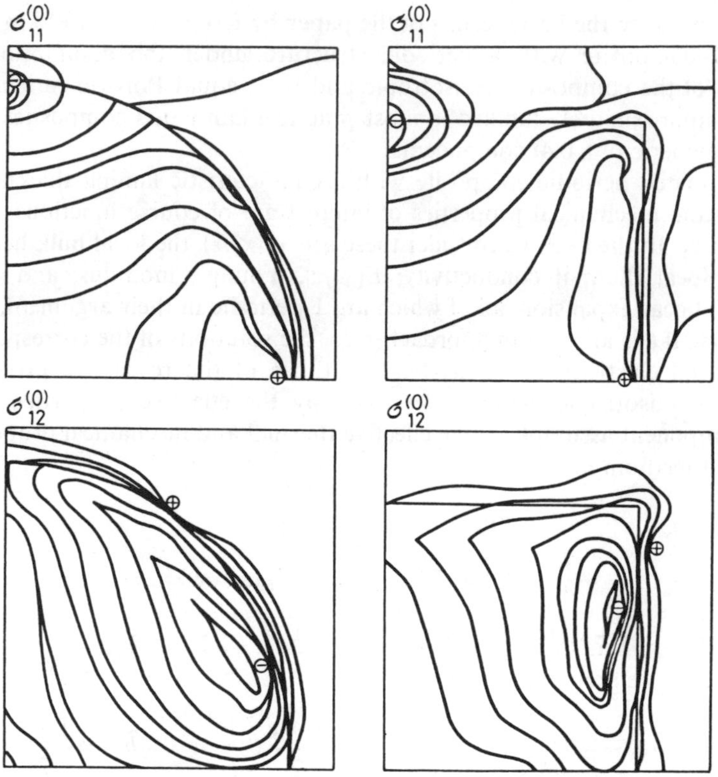

Fig. 13.5 *Local stress distributions for the case of macro deformation* $e_{11}^{(0)} = 1$ *(compositions Ia, Ib,* $\gamma = 0.6$*). Extremum stress values (in GPa):*

$\sigma_{11}^{+} = 78.675$, $\sigma_{11}^{+} = 90.501$;

$\sigma_{11}^{-} = -93.078$, $\sigma_{11}^{-} = -116.177$;

$\sigma_{12}^{+} = 0.865$, $\sigma_{12}^{+} = 5.144$;

$\sigma_{12}^{-} = -25.436$, $\sigma_{12}^{-} = -61.232$.

Once the maximum stress regions having been located, and the values of the maxima calculated, the strength criteria known for the homogeneous component materials can be employed to determine the failure conditions for the entire composite system.

14. LAMINATED COMPOSITE WITH PRESCRIBED THERMOELASTIC PROPERTIES. COMPOSITE MATERIAL DESIGN

The problem we consider here is concerned with the design of laminated composites and may be formulated as follows: given the prescribed overall properties and a set of local component properties of the composite, secure the former by varying the latter. The solution of the problem has been given by Kolpakov and Rakin (1986) on the basis of the earlier work of Pontryagin *et al.* (1976) and Alekseev *et al.* (1979). In the present section we reproduce the basic results of the paper by Kolpakov and Rakin derived for a laminated composite with a periodic structure under the assumption that the components of the composite are isotropic and have equal Poisson ratios. The latter condition is approximately fulfilled in most practical laminated composites with metal ($\nu \cong \frac{1}{3}$) or polymeric ($\nu \cong 0.4$) components.

For a laminated periodic composite with a characteristic lamina thickness ε, $\varepsilon \ll 1$, the thermal and mechanical properties of interest are of course functions of only one variable, say x_1. In the case we consider these are: $c_v(x_1/\varepsilon)$, the local bulk heat capacity; $\lambda(x_1/\varepsilon)$, the local thermal conductivity; $E(x_1/\varepsilon)$, Young's modulus; and $\alpha(x_1/\varepsilon)$, the coefficient of linear expansion, all of which are 1-periodic in their argument, x_1 varying from zero to ε. If ε is allowed to approach zero, the solutions of the corresponding heat conduction and elasticity problems tend to those of the respective problems for a homogeneous anisotropic material described by the effective properties. Under the isotropic component assumption, the effective thermal and mechanical properties of the homogenized medium are

$$\tilde{c}_v = \langle c_v \rangle, \tag{14.1}$$

$$\tilde{\lambda}_{11} = \langle 1/\lambda \rangle^{-1}, \qquad \tilde{\lambda}_{22} = \tilde{\lambda}_{33} = \langle \lambda \rangle, \tag{14.2}$$

$$\tilde{h}_{11} = \frac{(1+\nu)(1-2\nu)}{1-\nu}\left\langle \frac{1}{E} \right\rangle + \frac{2\nu^2}{1-\nu}\frac{1}{\langle E \rangle}, \qquad \tilde{h}_{22} = \tilde{h}_{33} = \frac{1}{\langle E \rangle}, \tag{14.3}$$

$$\tilde{h}_{12} = \tilde{h}_{13} = \tilde{h}_{23} = -\frac{\nu}{\langle E \rangle}, \qquad \tilde{h}_{55} = \tilde{h}_{66} = \frac{1+\nu}{2}\left\langle \frac{1}{E} \right\rangle, \qquad \tilde{h}_{44} = \frac{1+\nu}{2}\frac{1}{\langle E \rangle},$$

$$\tilde{\alpha}_{11} = \frac{1+\nu}{1-\nu}\langle \alpha \rangle - \frac{2\nu}{1-\nu}\frac{\langle E\alpha \rangle}{\langle E \rangle}, \qquad \tilde{\alpha}_{22} = \tilde{\alpha}_{33} = \frac{\langle E\alpha \rangle}{\langle E \rangle}. \tag{14.4}$$

In the above,

$$\langle f \rangle = \frac{1}{\varepsilon}\int_0^{\varepsilon} f(x_1/\varepsilon)\mathrm{d}x_1 = \int_0^1 f(t)\mathrm{d}t,$$

$\tilde{\lambda}_{11}$, $\tilde{\lambda}_{22}$ and $\tilde{\lambda}_{33}$ are the (only) non-zero elements of the heat conductivity tensor; $\tilde{h}_{\alpha\beta}$ is the elastic compliances tensor (in matrix notation, $\alpha, \beta = 1,2,\ldots,6$); and $\tilde{\alpha}_{11}$, $\tilde{\alpha}_{22}$ and $\tilde{\alpha}_{33}$ are the non-zero elements of the linear expansivity tensor.

Equations (14.1)–(14.4) establish relationships between the effective properties of a laminated composite and its local properties, and hold for various types of distributions of local properties, the best known examples being a piecewise-continuous and a piecewise-constant distribution, of which the latter is found most frequently in laminates. To cover all the possibilities, it proves advisable to introduce a set of functions defined by

$$W = \{f(t) \in L_\infty[0,1]: (f(t))^{-1} \in L_\infty[0,1]\},$$

where for any $f(t)$ there exists a positive number $\xi(f)$ such that $f(t) > \xi(f)$ for almost all $t \in [0,1]$. We will assume, from this point on, that the functions $c_v(t)$, $\lambda(t)$, $E(t)$ and $\alpha(t)$ belong to the set W, and we will treat equations (14.1)–(14.4) as the mapping

$$I : (c_v(t),\ \lambda(t),\ E(t),\ \alpha(t)) \in W^4 \to R^{16}$$

of the set of distributions of local properties into the set of values of its effective properties. Denoting by $I(W^4)$ the image of the set W^4 upon the mapping I, we can express the problem of designing a laminated composite with prescribed effective properties as follows:

first, obtain a composite material possessing the prescribed effective properties ($\tilde{c}_v^{(0)}$, $\tilde{\lambda}_{ij}^{(0)}$, $\tilde{h}_{\alpha\beta}^{(0)}$, $\tilde{\alpha}_{ij}^{(0)}) \in I(W^4)$, *and second, give a method for producing such a composite.*

The problem is thus solved by first ascertaining whether or not a material with the prescribed properties exists at all (at which stage a description of the set $I(W^4)$ is all the information needed) and then by finding a way to construct a set of functions ($c_v(t)$, $\lambda(t)$, $E(t)$, $\alpha(t)$) such that $I(c_v, \lambda, E, \alpha) = \boldsymbol{x}$ for $\boldsymbol{x} \in R^{16}$. The second of the above subproblems is thus that of synthesis for the image I (see Pontryagin *et al.* 1976, Alekseev *et al.* 1979). It should also be noted that equations (14.1)–(14.4) are mutually independent and the effective properties they determine are in one-to-one correspondence with the functionals of the form

$$\int_0^1 u_1(t)\,\mathrm{d}t, \quad \int_0^1 \frac{\mathrm{d}t}{u_1(t)}, \quad \int_0^1 u_2(t)\,\mathrm{d}t, \quad \int_0^1 u_1(t)u_2(t)\,\mathrm{d}t, \tag{14.5}$$

where, for equations (14.3) and (14.4), $u_1(t) = E(t)$, $u_2(t) = \alpha(t)$, and in (14.1) and (14.2) the functionals of the first and second kinds are involved. It suffices, therefore, to solve the problem for the mapping (14.5) of the set W^2 onto R^4. We reproduce here only the final results of the analysis, referring the reader to the original paper, Kolpakov and Rakin (1986), for the treatment in detail.

It is stated, first of all, that, to within the boundary of the sets involved, laminated structures composed of isotropic materials with equal Poisson ratios may have only the following effective properties.

- Heat capacity:

$$\tilde{c}_v = X \quad (X > 0); \tag{14.6}$$

- thermal conductivity tensor:

$$\tilde{\lambda}_{11} = Y, \qquad \tilde{\lambda}_{22} = \tilde{\lambda}_{33} = Z \quad (Z > 0,\ Y > 1/Z); \tag{14.7}$$

- Young's moduli, Poisson's ratios and shear moduli, respectively:

$$\begin{aligned} \tilde{E}_1 &= \frac{(1-\nu)x}{(1+\nu)(1-2\nu)xy + 2\nu^2}, \qquad \tilde{E}_2 = \tilde{E}_3 = x, \\ \tilde{\nu}_{12} &= \tilde{\nu}_{13} = \frac{\nu(1-\nu)}{(1+\nu)(1-2\nu)xy + 2\nu^2}, \qquad \tilde{\nu}_{23} = \nu, \\ \tilde{G}_{12} &= \tilde{G}_{13} = \frac{2}{(1+\nu)y}, \qquad \tilde{G}_{23} = \frac{2x}{1+\nu}; \end{aligned} \tag{14.8}$$

and, finally,

- the linear expansivity tensor:

$$\tilde{\alpha}_{11} = \frac{1+\nu}{1-\nu} z - \frac{2\nu}{1-\nu}\frac{t}{x}, \qquad \tilde{\alpha}_{22} = \tilde{\alpha}_{33} = \frac{t}{x} \tag{14.9}$$

$$(x > 0,\ y > 1/x,\ z > 0,\ t > 0).$$

The variables X, Y, Z, x, y, z and t involved in (14.6)–(14.9) assume independent values in their respective domains, and any one of the allowable properties determined by these equations may characterize a laminated composite composed of no more than three dissimilar materials.

We can thus suggest the following procedure for designing a laminated composite with prescribed effective properties:

(a) Equate the left-hand sides of equations (14.6)–(14.9) to the (prescribed) values $\tilde{c}_v^{(0)}$, $\tilde{\lambda}_{ij}^{(0)}$, $\tilde{E}_i^{(0)}$, $\tilde{\nu}_{ij}^{(0)}$, $\tilde{G}_{ij}^{(0)}$ and $\tilde{\alpha}_{ij}^{(0)}$ and solve the resulting algebraic equations for X, Y, Z, x, y, z and t.

(b) If the system of these equations turns out to be unsolvable or if its solutions invalidate the inequalities indicated in (14.6)–(14.9), a material with the required properties cannot exist in the given class of composites.

(c) If the system is solvable and the inequalities are satisfied, a material with the required properties does exist in the given class of composites and will be found by constructing the functions $(c_v(t), \lambda(t), E(t), \alpha(t)) \in W^4$ such that $\langle c_v \rangle = X$, $\langle 1/\lambda \rangle = Y$, $\langle \lambda \rangle = Z$, $\langle E \rangle = x$, $\langle 1/E \rangle = y$, $\langle \alpha \rangle = z$ and $\langle E\alpha \rangle = t$.

As an interesting example of the application of the above principles, it can be shown that a laminate with a negative expansivity can in principle be composed of components with positive expansivities; in particular, the effective property $\tilde{\alpha}_{11}$ can be made negative by the proper choice of local properties. To see this, we repeat here the first of equations (14.9):

$$\tilde{\alpha}_{11} = \frac{1+\nu}{1-\nu} z - \frac{2\nu}{1+\nu}\frac{t}{x} \quad (x > 0,\ z > 0,\ t > 0,\ y > 1/x), \tag{14.10}$$

Table 14.1 *The properties of designed laminated composites with negative values of effective thermal expansivity* $\tilde{\alpha}_{11}$.

Ingredients of composite	i	$E_i \times 10^{-10}$ (Pa)	$\alpha_i \times 10^6$ (K^{-1})	v_i	$\tilde{\alpha}_{11} \times 10^6$ (K^{-1})
Iridium (Ir)	1	52.8	6.5	0.1	−3.79
Invar	2	13.5	0.2	0.9	
Teflon (PTFE)	1	9.8	220.0	0.1	−49.0
SRB paper laminate	2	1.2	20.0	0.9	
Iridium (Ir)	1	52.8	6.5	0.05	
Tungsten (W)	2	39.0	4.5	0.05	−2.79
Invar	3	13.5	0.2	0.9	

which shows that the set of all possible values of $\tilde{\alpha}_{11}$ is $(-\infty, \infty)$ and hence it is possible for a composite with a negative value of $\tilde{\alpha}_{11}$ to exist. We restrict attention to a class of composites having no more than three constituents, and denote by E_i, α_i and v_i, respectively, the Young's modulus, linear expansivity and volume fraction of the ith material ($i = 1,2,3$). All the possible values of $\tilde{\alpha}_{11}$ will then be given by equation (14.10) with

$$
\begin{gathered}
x = \sum_{i=1}^{3} E_i v_i, \qquad y = \sum_{i=1}^{3} v_i/E_i, \\
z = \sum_{i=1}^{3} \alpha_i v_i, \qquad t = \sum_{i=1}^{3} E_i \alpha_i v_i, \\
0 \leqslant v_i \leqslant 1, \qquad \sum_{i=1}^{3} v_i = 1, \qquad E_i, \alpha_i > 0.
\end{gathered}
\tag{14.11}
$$

The synthesis problem that arises is a finite-dimensional one which, when stated in a discrete formulation, leads to the problem of an exhaustive search, among a finite number of possibilities, for the sets E_i, α_i and v_i ($i = 1,2,3$) that ensure the prescribed $\tilde{\alpha}_{11}$ value. The results obtained on a computer for a number of practical materials are summarized in Table 14.1, where binary compositions were formally treated as ternary ones with two components considered to have identical properties.

5 General Homogenization Models for Composite Shells and Plates with Rapidly Varying Thickness

Currently, the preponderance of uses for composite materials is in the form of plate and shell structural members the strength and reliability of which, combined with reduced weight and concomitant material savings, offer the designer very impressive possibilities in some commercial applications. A fact which is of interest for us here is whether the reinforcing effect comes from 'real' elements (say, fibres) or 'formal' ones (such as voids, holes and similar design features), it often happens that these elements form a regular structure with a period much smaller than the characteristic dimension of the structural member; consequently the asymptotic homogenization analysis becomes applicable. The homogenized models of plates with periodic non-homogeneities in tangential coordinate(s) have been developed in this way by Duvaut (1976, 1977), Duvaut and Metellus (1976) and Artola and Duvaut (1977) and, more recently, by Andrianov and Manevich (1983), Andrianov *et al.* (1985) and other workers (see Kalamkarov *et al.* 1987a for a review). It should be noted, however, that the asymptotic homogenization method cannot be applied to a two-dimensional plate and shell theory if the space non-homogeneities of the material vary on a scale comparable with the small thickness of the three-dimensional body under consideration. A refined approach developed by Caillerie (1981a, b) in his heat conduction studies consists of applying the two-scale formalism directly to the three-dimensional problem of a thin non-homogeneous layer. Accordingly, Caillerie introduces two sets of 'rapid' coordinates. One of these, in the tangential directions, is associated with rapid periodic oscillations in the composite properties or layer thickness; in the transverse (thickness) direction, there is no periodicity, and the corresponding 'rapid' variable is associated with the small thickness of the layer. The two small parameters that arise from this approach, δ and ε, are determined by, respectively, the period of the coefficients of the pertinent equations and the layer thickness, and may or may not be of the same order of magnitude. Kohn and Vogelius (1984, 1985, 1986) adopted this approach in their study of the pure bending of a linearly elastic, thin homogeneous plate whose thickness was taken to be described by a rapidly oscillating function with a period $\varepsilon = \delta^a$, $0 < a < \infty$, δ denoting the mean plate thickness; see also Lewinski (1991).

Of special practical interest, however, is the case of proportionality between the two small parameters,

$$\varepsilon = h\delta, \quad \delta \ll 1, \quad h \sim 1,$$

which has been treated by Caillerie (1982, 1984, 1987) in his linearly elastic asymptotic analysis of a periodically non-homogeneous plate. For a still more special case of $\varepsilon = \delta$, Panasenko and Reztsov (1987) have been able to provide necessary justifications for the three-dimensional linear elasticity solutions they obtained for a thin periodically non-homogeneous plate.

In the present chapter the modified $\varepsilon = h\delta$ homogenization method is applied to the study of a curved thin composite plate with a regular structure and wavy surfaces in the contexts of, successively, linear and geometrically non-linear elasticity theory (Kalamkarov *et al.* 1987a, b, Kalamkarov 1988a, b, Parton and Kalamkarov 1992, Parton *et al.* 1989); heat conduction theory (Kalamkarov 1987b); and thermoelasticity (Kalamkarov *et al.* 1987c, d, Parton and Kalamkarov 1988a, Kalamkarov 1989). The starting point in each particular case is the exact three-dimensional formulation of the corresponding problem, without resource to the Kirchhoff–Love hypotheses or any similar simplifying assumptions. Owing to the presence of the small parameter δ, each original three-dimensional problem then proves to be amenable to a rigorous asymptotic analysis unifying an asymptotic three-to-two dimensions process and a homogenization composite material–homogeneous material process. The general homogenized models so obtained have many practical applications in the design of stiffened thin-walled elements of construction possessing a regular structure.

15. ELASTICITY PROBLEM FOR A SHELL OF REGULARLY NON-HOMOGENEOUS MATERIAL WITH WAVY SURFACES

15.1. Formulation in an orthogonal coordinate system

Let $(\alpha_1, \alpha_2, \gamma)$ be an orthogonal coordinate system such that the coordinate lines α_1 and α_2 coincide with the main curvature lines of the mid-surface of the shell and the γ-axis is normal to the mid-surface (Fig. 15.1). All three coordinates are assumed to have been made dimensionless by dividing by a certain characteristic dimension of the body, L.

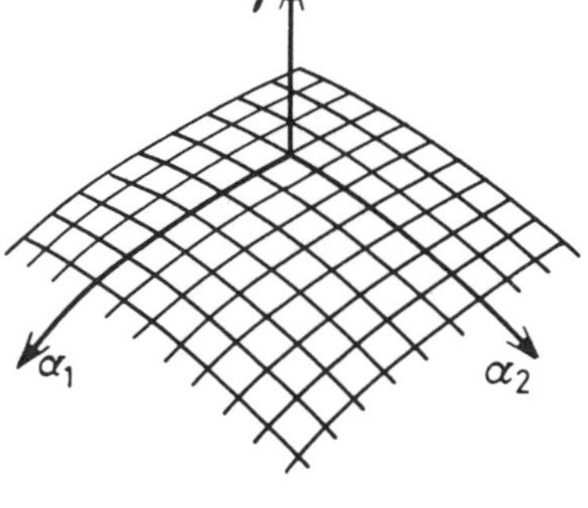

Fig. 15.1 *Orthogonal coordinate system $\alpha_1, \alpha_2, \gamma$.*

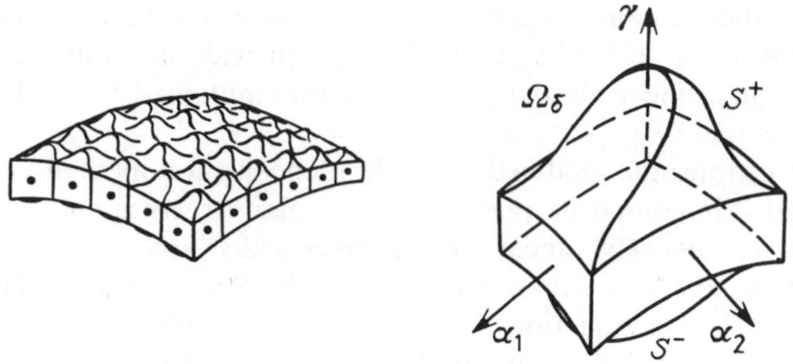

Fig. 15.2 *Curvilinear thin regularly nonhomogeneous (composite) layer with wavy surfaces; unit cell Ω_δ.*

The metric tensor of this coordinate system is

$$g_{ij} = \begin{pmatrix} H_1^2 & 0 & 0 \\ 0 & H_2^2 & 0 \\ 0 & 0 & 1 \end{pmatrix}$$

and the Lamé coefficients H_1 and H_2 are given by

$$H_1 = A_1(1 + k_1\gamma), \qquad H_2 = A_2(1 + k_2\gamma),$$

where $A_1(\alpha_1, \alpha_2)$ and $A_2(\alpha_1, \alpha_2)$ are the coefficients of the first quadratic form and $k_1(\alpha_1, \alpha_2)$ and $k_2(\alpha_1, \alpha_2)$ are the main curvatures of the mid-surface. The unit cell of the problem (see Fig. 15.2) is defined by the inequalities

$$\left\{ -\frac{\delta h_1}{2} < \alpha_1 < \frac{\delta h_1}{2}, -\frac{\delta h_2}{2} < \alpha_2 < \frac{\delta h_2}{2}, \gamma^- < \gamma < \gamma^+ \right\},$$

$$\gamma^\pm = \pm\frac{\delta}{2} \pm \delta F^\pm\left(\frac{\alpha_1}{\delta h_1}, \frac{\alpha_2}{\delta h_2}\right),$$

in which the dimensionless small parameter δ determines the thickness of the shell; the parameters h_1 and h_2 characterize the ratio of the tangential to thickness dimensions of the unit cell; and the functions F^+ and F^- describe the profiles of the upper and lower surfaces of the shell (respectively, S^+ and S^-) and are generally 1-periodic in $\alpha_1/\delta h_1$ and $\alpha_2/\delta h_2$.

With the above expressions for H_1 and H_2, the Cauchy relations for the physical strain and displacement components are (cf.(1.36))

$$e_{11} = \frac{1}{H_1}\frac{\partial u_1}{\partial \alpha_1} + \frac{1}{H_1 A_2}\frac{\partial A_1}{\partial \alpha_2}u_2 + \frac{A_1 k_1}{H_1}u_3,$$

$$e_{22} = \frac{1}{H_2}\frac{\partial u_2}{\partial \alpha_2} + \frac{1}{H_2 A_1}\frac{\partial A_2}{\partial \alpha_1}u_1 + \frac{A_2 k_2}{H_2}u_3,$$

$$e_{33} = \frac{\partial u_3}{\partial \gamma},$$

$$2e_{12}=\frac{1}{H_2}\frac{\partial u_1}{\partial \alpha_2}-\frac{1}{H_2A_1}\frac{\partial A_2}{\partial \alpha_1}u_2+\frac{1}{H_1}\frac{\partial u_2}{\partial \alpha_1}-\frac{1}{H_1A_2}\frac{\partial A_1}{\partial \alpha_2}u_1,$$

$$2e_{13}=\frac{\partial u_1}{\partial \gamma}+\frac{1}{H_1}\frac{\partial u_3}{\partial \alpha_1}-\frac{A_1k_1}{H_1}u_1,$$

$$2e_{23}=\frac{\partial u_2}{\partial \gamma}+\frac{1}{H_2}\frac{\partial u_3}{\partial \alpha_2}-\frac{A_2k_2}{H_2}u_2, \tag{15.1}$$

where (and throughout this chapter) the subscripts 1, 2, and 3 refer, respectively, to α_1, α_2 and γ and where the Gauss–Codazzi relations

$$\frac{1}{H_1}\frac{\partial H_2}{\partial \alpha_1}=\frac{1}{A_1}\frac{\partial A_2}{\partial \alpha_1}, \qquad \frac{1}{H_2}\frac{\partial H_1}{\partial \alpha_2}=\frac{1}{A_2}\frac{\partial A_1}{\partial \alpha_2} \tag{15.2}$$

have been used.

The equilibrium equations are given by (cf. (1.62))

$$\frac{\partial(H_2\sigma_{11})}{\partial \alpha_1}+\frac{\partial(H_1\sigma_{12})}{\partial \alpha_2}+\frac{\partial(H_1H_2\sigma_{13})}{\partial \gamma}-\frac{H_1}{A_1}\frac{\partial A_2}{\partial \alpha_1}\sigma_{22}$$
$$+\frac{H_2}{A_2}\frac{\partial A_1}{\partial \alpha_2}\sigma_{21}+H_2A_1k_1\sigma_{31}+H_1H_2P_1=0,$$

$$\frac{\partial(H_2\sigma_{21})}{\partial \alpha_1}+\frac{\partial(H_1\sigma_{22})}{\partial \alpha_2}+\frac{\partial(H_1H_2\sigma_{23})}{\partial \gamma}-\frac{H_2}{A_2}\frac{\partial A_1}{\partial \alpha_2}\sigma_{11}$$
$$+\frac{H_1}{A_1}\frac{\partial A_2}{\partial \alpha_1}\sigma_{12}+H_1A_2k_2\sigma_{32}+H_1H_2P_2=0, \tag{15.3}$$

$$\frac{\partial(H_2\sigma_{31})}{\partial \alpha_1})+\frac{\partial(H_1\sigma_{32})}{\partial \alpha_2}+\frac{\partial(H_1H_2\sigma_{33})}{\partial \gamma}-H_2A_1k_1\sigma_{11}$$
$$-H_1A_2k_2\sigma_{22}+H_1H_2P_3=0;$$

where $P_i (i=1,2,3)$ are the body force components.

The physical stresses and strains are related by the generalized Hooke's law (cf. (1.80)):

$$\sigma_{ij}=c_{ijmn}e_{mn}. \tag{15.4}$$

(We shall always assume, unless otherwise stated, that Latin indices range from 1 to 3 and Greek ones from 1 to 2.)

The surfaces of the shell are subject to the conditions

$$\sigma^{ij}n_j=p^i,$$

where σ^{ij} is the stress tensor, n_j the unit vector in the outward direction to the surface and p^i the vector of the outer surface force.

Now the physical components of interest are given by (cf. (1.61))

$$\sigma_{11}=\sigma_{\alpha_1\alpha_1}=\sigma^{11}g_{11}=\sigma^{11}H_1^2,$$
$$\sigma_{12}=\sigma_{\alpha_1\alpha_2}=\sigma^{12}\sqrt{g_{11}}\sqrt{g_{22}}=\sigma^{12}H_1H_2,$$
$$\sigma_{22}=\sigma_{\alpha_2\alpha_2}=\sigma^{22}g_{22}=\sigma^{22}H_2^2,$$
$$\sigma_{31}=\sigma_{\gamma\alpha_1}=\sigma^{31}\sqrt{g_{11}}=\sigma^{31}H_1,$$
$$\sigma_{32}=\sigma_{\gamma\alpha_2}=\sigma^{32}\sqrt{g_{22}}=\sigma^{32}H_2,$$
$$\sigma_{33}=\sigma_{\gamma\gamma}=\sigma^{33}\sqrt{g_{33}}=\sigma^{33},$$
$$p_1=p_{\alpha_1}=p^1\sqrt{g_{11}}=p^1H_1,\qquad p_2=p_{\alpha_2}=p^2\sqrt{g_{22}}=p^2H_2,$$
$$p_3=p_\gamma=p^3\sqrt{g_{33}}=p^3,$$

and substituting these into the surface boundary conditions we obtain:

$$\frac{\sigma_{11}}{H_1}n_1+\frac{\sigma_{12}}{H_2}n_2+\sigma_{13}n_3=p_1,$$
$$\frac{\sigma_{12}}{H_1}n_1+\frac{\sigma_{22}}{H_2}n_2+\sigma_{23}n_3=p_2,$$
$$\frac{\sigma_{13}}{H_1}n_1+\frac{\sigma_{23}}{H_2}n_2+\sigma_{33}n_3=p_3,$$

which, after multiplication by H_1H_2, leads to the result:

$$H_2\sigma_{i1}n_1^{\pm}+H_1\sigma_{i2}n_2^{\pm}+H_1H_2\sigma_{i3}n_3^{\pm}=\pm H_1H_2p_i^{\pm}\qquad(i=1,2,3),\tag{15.5}$$

for $\gamma=\gamma^{\pm}$, where $\boldsymbol{n}^+$ ($\boldsymbol{n}^-$) is the outward (inward) unit normal vector (hence the minus sign on the right-hand side of (15.5) for $\gamma=\gamma^-$).

The normal vector components are defined in a curvilinear coordinate system by (see, for example, Berdichevskii 1983)

$$n_i=\frac{1}{\sqrt{a}}\varepsilon_{ijm}r_1^j r_2^m$$

where ε_{ijm}, the Levi-Cività symbols, have the property that

$$\varepsilon_{ijm}=\begin{cases}\sqrt{g}, & \text{if all three indices are different and obtained by an even number of interchanges from 123,}\\ -\sqrt{g}, & \text{if all three indices are different and obtained by an odd number of interchanges from 123,}\\ 0, & \text{if any two of the three indices are the same.}\end{cases}$$

For the surfaces $\gamma=\gamma^{\pm}$ we find

$$\boldsymbol{n}^{\pm}=\left\{-\frac{\partial\gamma^{\pm}}{\partial\alpha_1},-\frac{\partial\gamma^{\pm}}{\partial\alpha_2},1\right\}\left[1+\frac{1}{H_1^2}\left(\frac{\partial\gamma^{\pm}}{\partial\alpha_1}\right)^2+\frac{1}{H_2^2}\left(\frac{\partial\gamma^{\pm}}{\partial\alpha_2}\right)^2\right]^{-1/2}.\tag{15.6}$$

The static (quasi-static) formulation of the above elasticity problem will be complete by adjoining the proper specification of boundary conditions at the lateral surfaces of the shell. The usual practice is to make some requirements on the stresses or displacement fields on these surfaces or, in the case of closed shells, to impose periodicity conditions on the corresponding coordinates.

15.2. Asymptotic analysis

We begin by introducing the 'rapid' coordinates of the problem

$$y_1 = \frac{\alpha_1}{\delta h_1}, \qquad y_2 = \frac{\alpha_2}{\delta h_2}, \qquad z = \frac{\gamma}{\delta}, \qquad \boldsymbol{y} = (y_1, y_2)$$

in terms of which the unit cell Ω is defined by the inequalities

$$\{-\tfrac{1}{2} < y_1, y_2 < \tfrac{1}{2}, z^- < z < z^+\}, z^\pm = \pm\tfrac{1}{2} \pm F^\pm(\boldsymbol{y}),$$

and the normal vector expression (15.6) is rewritten as

$$\boldsymbol{n}^\pm = \left\{ \mp \frac{1}{h_1}\frac{\partial F^\pm}{\partial y_1}, \mp \frac{1}{h_2}\frac{\partial F^\pm}{\partial y_2}, 1 \right\} \left[1 + \frac{1}{H_1^2 h_1^2}\left(\frac{\partial F^\pm}{\partial y_1}\right)^2 + \frac{1}{H_2^2 h_2^2}\left(\frac{\partial F^\pm}{\partial y_2}\right)^2 \right]^{-1/2}. \tag{15.7}$$

The regular non-homogeneity of the material is mathematically modelled by requiring that the functions $c_{ijmn}(\boldsymbol{y}, z)$ and $\boldsymbol{y} = (y_1, y_2)$ be periodic with the unit cell Ω in the coordinates y_1 and y_2.

The solution of the problem will be represented as the asymptotic expansion

$$u_i = u_i^{(0)}(\boldsymbol{\alpha}) + \delta u_i^{(1)}(\boldsymbol{\alpha}, \boldsymbol{y}, z) + \delta^2 u_i^{(2)}(\boldsymbol{\alpha}, \boldsymbol{y}, z) + \cdots, \tag{15.8}$$

where the functions $u_i^{(l)}(\boldsymbol{\alpha}, \boldsymbol{y}, z)$, $l = 1, 2, \ldots$, are periodic in y_1 and y_2 with the unit cell Ω; $\boldsymbol{\alpha} = (\alpha_1, \alpha_2)$.

Now let the radii of curvature of the shell mid-surface ($\gamma = 0$) be much larger than the shell thickness. We are then justified in assuming the asymptotic forms

$$k_1 = \delta K_2(\boldsymbol{\alpha}), \qquad k_2 = \delta K_1(\boldsymbol{\alpha}), \qquad k_1 + k_2 = \delta K_3(\boldsymbol{\alpha}) \tag{15.9}$$

for the main curvatures of the shell. For the external forces we write

$$p_\nu^\pm = \delta^2 r_\nu^\pm(\boldsymbol{\alpha}, \boldsymbol{y}), \qquad p_3^\pm = \delta^3 q_3^\pm(\boldsymbol{\alpha}, \boldsymbol{y}), \tag{15.10}$$

$$P_\nu = \delta f_\nu(\boldsymbol{\alpha}, \boldsymbol{y}, z), \qquad P_3 = \delta^2 g_3(\boldsymbol{\alpha}, \boldsymbol{y}, z), \tag{15.11}$$

where all functions involved are again periodic in y_1 and y_2 with the unit cell Ω.

To simplify the solution we introduce the notation

$$\xi_1 = A_1 y_1, \qquad \xi_2 = A_2 y_2$$

and define the following differential operators to be applied to the φ_{nm} components:

$$\mathscr{B}_1(\varphi_{n\mu}) = \frac{1}{A_1 A_2}\left[\frac{\partial(A_2\varphi_{11})}{\partial\alpha_1} + \frac{\partial(A_1\varphi_{21})}{\partial\alpha_2} + \frac{\partial A_1}{\partial\alpha_2}\varphi_{12} - \frac{\partial A_2}{\partial\alpha_1}\varphi_{22}\right],$$

$$\mathscr{B}_2(\varphi_{n\mu}) = \frac{1}{A_1 A_2}\left[\frac{\partial(A_2\varphi_{12})}{\partial\alpha_1} + \frac{\partial(A_1\varphi_{22})}{\partial\alpha_2} + \frac{\partial A_2}{\partial\alpha_1}\varphi_{21} - \frac{\partial A_1}{\partial\alpha_2}\varphi_{11}\right],$$

$$\mathscr{B}_3(\varphi_{n\mu}) = \frac{1}{A_1 A_2}\left[\frac{\partial(A_2\varphi_{31})}{\partial\alpha_1} + \frac{\partial(A_1\varphi_{32})}{\partial\alpha_2}\right], \tag{15.12}$$

$$\mathscr{K}_1(\varphi_{n\mu}) = K_2\varphi_{31}, \mathscr{K}_2(\varphi_{n\mu}) = K_1\varphi_{32},$$

$$\mathscr{K}_3(\varphi_{n\mu}) = -K_2\varphi_{11} - K_1\varphi_{22}.$$

It is essential in the asymptotic homogenization method to distinguish between 'rapid' and 'slow' variables when differentiating, that is,

$$\frac{\partial}{\partial\alpha_\mu} \to \frac{\partial}{\partial\alpha_\mu} + \frac{1}{\delta h_\mu}\frac{\partial}{\partial y_\mu}, \qquad \frac{\partial}{\partial\gamma} = \frac{1}{\delta}\frac{\partial}{\partial z}.$$

where μ assumes the values 1 and 2 and is not summed.

Using (15.8) and (15.9) in (15.1) we obtain:

$$e_{ij} = e_{ij}^{(0)} + \delta e_{ij}^{(1)} + \delta^2 e_{ij}^{(2)} + \ldots, \tag{15.13}$$

where

$$e_{11}^{(0)} = \frac{1}{A_1}\frac{\partial u_1^{(0)}}{\partial\alpha_1} + \frac{1}{h_1 A_1}\frac{\partial u_1^{(1)}}{\partial y_1} + \frac{1}{A_1 A_2}\frac{\partial A_1}{\partial\alpha_2}u_2^{(0)},$$

$$e_{22}^{(0)} = \frac{1}{A_2}\frac{\partial u_2^{(0)}}{\partial\alpha_2} + \frac{1}{h_2 A_2}\frac{\partial u_2^{(1)}}{\partial y_2} + \frac{1}{A_1 A_2}\frac{\partial A_2}{\partial\alpha_1}u_1^{(0)},$$

$$e_{33}^{(0)} = \frac{\partial u_3^{(1)}}{\partial z},$$

$$2e_{12}^{(0)} = \left[\left(\frac{1}{A_2}\frac{\partial u_2^{(0)}}{\partial\alpha_2} - \frac{1}{A_1 A_2}\frac{\partial A_2}{\partial\alpha_1}u_2^{(0)}\right) + \left(\frac{1}{A_1}\frac{\partial u_2^{(0)}}{\partial\alpha_1} - \frac{1}{A_1 A_2}\frac{\partial A_1}{\partial\alpha_2}u_1^{(0)}\right)\right. \tag{15.14}$$
$$\left. + \left(\frac{1}{h_2 A_2}\frac{\partial u_1^{(1)}}{\partial y_2} + \frac{1}{h_1 A_1}\frac{\partial u_2^{(1)}}{\partial y_1}\right)\right],$$

$$2e_{13}^{(0)} = \frac{\partial u_1^{(1)}}{\partial z} + \frac{1}{A_1}\frac{\partial u_3^{(0)}}{\partial\alpha_1} + \frac{1}{h_1 A_1}\frac{\partial u_3^{(1)}}{\partial y_1},$$

$$2e_{23}^{(0)} = \frac{\partial u_2^{(1)}}{\partial z} + \frac{1}{A_2}\frac{\partial u_3^{(0)}}{\partial\alpha_2} + \frac{1}{h_2 A_2}\frac{\partial u_3^{(1)}}{\partial y_2},$$

$$e_{11}^{(1)} = \frac{1}{A_1}\frac{\partial u_1^{(1)}}{\partial\alpha_1} + \frac{1}{A_1 A_2}\frac{\partial A_1}{\partial\alpha_2}u_2^{(1)} + K_2 u_3^{(0)} + \frac{1}{h_1 A_1}\frac{\partial u_1^{(2)}}{\partial y_1},$$

$$e_{22}^{(1)} = \frac{1}{A_2}\frac{\partial u_2^{(1)}}{\partial \alpha_2} + \frac{1}{A_1 A_2}\frac{\partial A_2}{\partial \alpha_1} u_1^{(1)} + K_1 u_3^{(0)} + \frac{1}{h_2 A_2}\frac{\partial u_2^{(2)}}{\partial y_2},$$

$$2e_{12}^{(1)} = \left[\left(\frac{1}{A_2}\frac{\partial u_1^{(1)}}{\partial \alpha_2} - \frac{1}{A_1 A_2}\frac{\partial A_2}{\partial \alpha_1} u_2^{(1)}\right) + \left(\frac{1}{A_1}\frac{\partial u_2^{(1)}}{\partial \alpha_1} - \frac{1}{A_1 A_2}\frac{\partial A_1}{\partial \alpha_2} u_1^{(1)}\right)\right.$$
$$\left. + \left(\frac{1}{h_2 A_2}\frac{\partial u_1^{(2)}}{\partial y_2} + \frac{1}{h_1 A_1}\frac{\partial u_2^{(2)}}{\partial y_1}\right)\right], \tag{15.15}$$

$$e_{33}^{(1)} = \frac{\partial u_3^{(2)}}{\partial z},$$

$$2e_{13}^{(1)} = \frac{\partial u_1^{(2)}}{\partial z} + \frac{1}{A_1}\frac{\partial u_3^{(1)}}{\partial \alpha_1} - K_2 u_1^{(0)} + \frac{1}{h_1 A_1}\frac{\partial u_3^{(2)}}{\partial y_1},$$

$$2e_{23}^{(1)} = \frac{\partial u_2^{(2)}}{\partial z} + \frac{1}{A_2}\frac{\partial u_3^{(1)}}{\partial \alpha_2} - K_1 u_2^{(0)} + \frac{1}{h_2 A_2}\frac{\partial u_3^{(2)}}{\partial y_2}.$$

Substituting (15.13) into (15.4) results in

$$\sigma_{ij} = \sigma_{ij}^{(0)} + \delta\sigma_{ij}^{(1)} + \delta^2\sigma_{ij}^{(2)} + \cdots,$$
$$\sigma_{ij}^{(l)} = c_{ijmn}(y, z)e_{mn}^{(l)} \quad (l = 0, 1, 2, \cdots). \tag{15.16}$$

From (15.9) and (15.11) we substitute into the equilibrium equations (15.3) and then expand in powers of $\delta^l (l = -1, 0, 1, 2)$ to get

$$\frac{1}{h_\mu}\frac{\partial \sigma_{i\mu}^{(0)}}{\partial \xi_\mu} + \frac{\partial \sigma_{i3}^{(0)}}{\partial z} = 0, \tag{15.17}$$

$$\frac{1}{h_\mu}\frac{\partial \sigma_{i\mu}^{(1)}}{\partial \xi_\mu} + \frac{\partial \sigma_{i3}^{(1)}}{\partial z} + \mathscr{B}_i(\sigma_{n\mu}^{(0)}) = 0, \tag{15.18}$$

$$\frac{1}{h_\mu}\frac{\partial}{\partial \xi_\mu}\left(\sigma_{i\mu}^{(2)} + zK_\mu \sigma_{i\mu}^{(0)}\right) + \frac{\partial}{\partial z}\left(\sigma_{i3}^{(2)} + zK_3\sigma_{i3}^{(0)}\right)$$
$$+ \mathscr{B}_i(\sigma_{n\mu}^{(1)}) + \mathscr{K}_i(\sigma_{n\mu}^{(0)}) + f_i = 0, \tag{15.19}$$

$$\frac{1}{h_\mu}\frac{\partial}{\partial \xi_\mu}\left(\sigma_{i\mu}^{(3)} + zK_\mu \sigma_{i\mu}^{(1)}\right) + \frac{\partial}{\partial z}\left(\sigma_{i3}^{(3)} + zK_3\sigma_{i3}^{(1)}\right)$$
$$+ \mathscr{B}_i(\sigma_{n\mu}^{(2)} + zKn\sigma_{n\mu}^{(0)}) + \mathscr{K}_i(\sigma_{n\mu}^{(1)}) + g_i = 0, \tag{15.20}$$

where

$$\frac{\partial}{\partial \xi_1} = \frac{1}{A_1}\frac{\partial}{\partial y_1}, \qquad \frac{\partial}{\partial \xi_2} = \frac{1}{A_2}\frac{\partial}{\partial y_2},$$

and where relations (15.2) have been used, which may be rewritten in the form

$$A_2\frac{\partial K_1}{\partial \alpha_1} = \left(K_2 - K_1\right)\frac{\partial A_2}{\partial \alpha_1}, \qquad A_1\frac{\partial K_2}{\partial \alpha_2} = \left(K_1 - K_2\right)\frac{\partial A_1}{\partial \alpha_2}$$

using (15.9). Note that $f_3 = 0$ and $g_\mu = 0$ in accordance with (15.11) and that the index n is not summed in the third term in (15.20).

Our next aim is to expand the boundary conditions (15.5) in powers of δ. Consider first the components of the unit normal vector (15.7). From (15.9) we see that they can be expanded as

$$
\begin{aligned}
n_i^{\pm} &= n_i^{\pm(0)} + \delta^2 n_i^{\pm(2)} + O(\delta^4),\\
\boldsymbol{n}^{\pm(0)} &= \left\{ \mp\frac{1}{h_1}\frac{\partial F^{\pm}}{\partial y_1}, \mp\frac{1}{h_2}\frac{\partial F^{\pm}}{\partial y_2}, 1 \right\}\left[1 + \frac{1}{A_1^2h_1^2}\left(\frac{\partial F^{\pm}}{\partial y_1}\right)^2 + \frac{1}{A_2^2h_2^2}\left(\frac{\partial F^{\pm}}{\partial y_2}\right)^2\right]^{-1/2},\\
n_i^{\pm(2)} &= n_i^{\pm(0)}\zeta^{\pm},\\
\zeta^{\pm} &= z^{\pm}\left[\frac{K_2}{A_1^2h_1^2}\left(\frac{\partial F^{\pm}}{\partial y_1}\right)^2 + \frac{K_1}{A_2^2h_2^2}\left(\frac{\partial F^{\pm}}{\partial y_2}\right)^2\right]\left[1 + \frac{1}{A_1^2h_1^2}\left(\frac{\partial F^{\pm}}{\partial y_1}\right)^2 + \frac{1}{A_2^2h_2^2}\left(\frac{\partial F^{\pm}}{\partial y_2}\right)^2\right]^{-1},
\end{aligned}
\tag{15.21}
$$

and using this along with (15.9) and (15.10), conditions (15.5) are expanded as follows in terms of $\delta^l\,(l = 0, 1, 2, 3)$:

$$\sigma_{ij}^{(l)} N_j^{\pm} = 0 \quad (z = z^{\pm}, l = 0, 1), \tag{15.22}$$

$$(\sigma_{ij}^{(2)} + z^{\pm}K_j\sigma_{ij}^{(0)})N_j^{\pm} = \pm\,\omega^{\pm} r_i^{\pm} \quad (z = z^{\pm}), \tag{15.23}$$

$$(\sigma_{ij}^{(3)} + z^{\pm}K_j\sigma_{ij}^{(1)})N_j^{\pm} = \pm\,\omega^{\pm} q_i^{\pm} \quad (z = z^{\pm}), \tag{15.24}$$

where we have defined

$$\boldsymbol{N}^{\pm} = \left\{ \mp\frac{1}{h_1}\frac{\partial F^{\pm}}{\partial \xi_1}, \mp\frac{1}{h_2}\frac{\partial F^{\pm}}{\partial \xi_2}, 1 \right\}, \tag{15.25}$$

$$\omega^{\pm} = \left[1 + \frac{1}{h_1^2}\left(\frac{\partial F^{\pm}}{\partial \xi_1}\right)^2 + \frac{1}{h_2^2}\left(\frac{\partial F^{\pm}}{\partial \xi_2}\right)^2\right]^{1/2}. \tag{15.26}$$

Note that, in view of (15.10),

$$r_3^{\pm} = 0, \qquad q_\mu^{\pm} = 0 \quad (\mu = 1, 2)$$

in (15.23) and (15.24).

Introducing the averaging procedure

$$\langle \varphi \rangle = \int_\Omega \varphi \,\mathrm{d}y_1 \mathrm{d}y_2 \mathrm{d}z \tag{15.27}$$

with respect to the coordinates y_1, y_2 and z over the unit cell Ω, and noting that this operation is interchangeable with differentiation with respect to α_1 and α_2, let us prove that if a function Q_1 is periodic with the unit cell Ω in y_1 and y_2, the following formula holds:

$$\left\langle \frac{1}{h_\mu}\frac{\partial Q_\mu}{\partial \xi_\mu} + \frac{\partial Q_3}{\partial z} \right\rangle = \int_{-1/2}^{1/2}\int_{-1/2}^{1/2} (Q_i^{+}N_i^{+} - Q_i^{-}N_i^{-})\,\mathrm{d}y_1\mathrm{d}y_2, \tag{15.28}$$

where $N_i^{\pm}$ are the components of the vectors (15.25), and $Q_i^{\pm}$ are values of the function

Q_i assumed at the surfaces $S^{\pm}$ defined by $z = z^{\pm} \equiv \pm 1/2 \pm F^{\pm}(\boldsymbol{y})$. Using the Green–Gauss theorem, we find first that

$$\left\langle \frac{1}{h_\mu}\frac{\partial Q_\mu}{\partial \xi_\mu} + \frac{\partial Q_3}{\partial z} \right\rangle = \left\langle \frac{\partial}{\partial y_\mu}\left(\frac{Q_\mu}{h_\mu A_\mu}\right) + \frac{\partial Q_3}{\partial z} \right\rangle$$

$$= \int_{S^+} \left(\frac{Q_\mu^+}{h_\mu A_\mu} n_{y\mu}^+ + Q_3^+ n_{y3}^+ \right) \mathrm{d}S_\Omega^+ - \int_{S^-} \left(\frac{Q_\mu^-}{h_\mu A_\mu} n_{y\mu}^- + Q_3^- n_{y3}^- \right) \mathrm{d}S_\Omega^- . \tag{15.29}$$

Here the integrals over the lateral surfaces of Ω cancel because of the periodicity of the functions Q_1 in y_1 and y_2, and by the definition of the surfaces $S^{\pm}$ we have

$$\mathrm{d}S_\Omega^{\pm} = \left[1 + \left(\frac{\partial F^{\pm}}{\partial y_1}\right)^2 + \left(\frac{\partial F^{\pm}}{\partial y_2}\right)^2 \right]^{1/2} \mathrm{d}y_1\,\mathrm{d}y_2,$$

$$\boldsymbol{n}_y^{\pm} = \left\{ \mp \frac{\partial F^{\pm}}{\partial y_1}, \mp \frac{\partial F^{\pm}}{\partial y_2}, 1 \right\} \left[1 + \left(\frac{\partial F^{\pm}}{\partial y_1}\right)^2 + \left(\frac{\partial F^{\pm}}{\partial y_2}\right)^2 \right]^{-1/2}, \tag{15.30}$$

where the minus sign before the second integral in (15.29) is due to the fact that the normal vector, $\boldsymbol{n}_y^-$, is an inner with respect to the volume Ω. The proof is completed by inserting (15.30) into (15.29) and recalling the notation (15.25).

Using (15.14) in (15.16), the following equation is obtained:

$$\sigma_{ij}^{(0)} = c_{ijn\mu} \frac{1}{h_\mu} \frac{\partial u_n^{(1)}}{\partial \xi_\mu} + c_{ijn3} \frac{\partial u_n^{(1)}}{\partial z} + c_{ijn\mu} \varepsilon_{n\mu}^{(0)}, \tag{15.31}$$

in which

$$\varepsilon_{11}^{(0)} = \frac{1}{A_1}\frac{\partial u_1^{(0)}}{\partial \alpha_1} + \frac{1}{A_1 A_2}\frac{\partial A_1}{\partial \alpha_2} u_2^{(0)},$$

$$\varepsilon_{22}^{(0)} = \frac{1}{A_2}\frac{\partial u_2^{(0)}}{\partial \alpha_2} + \frac{1}{A_1 A_2}\frac{\partial A_2}{\partial \alpha_1} u_1^{(0)},$$

$$\varepsilon_{12}^{(0)} = \varepsilon_{21}^{(0)} = \frac{1}{2}\left[\frac{A_1}{A_2}\frac{\partial}{\partial \alpha_2}\left(\frac{u_1^{(0)}}{A_1}\right) + \frac{A_1}{A_2}\frac{\partial}{\partial \alpha_1}\left(\frac{u_2^{(0)}}{A_2}\right) \right], \tag{15.32}$$

$$\varepsilon_{31}^{(0)} = \frac{1}{A_1}\frac{\partial u_3^{(0)}}{\partial \alpha_1}, \qquad \varepsilon_{32}^{(0)} = \frac{1}{\mathrm{A}_2}\frac{\partial u_3^{(0)}}{\partial \alpha_2},$$

and we introduce the notation:

$$\boldsymbol{L}_{ijn} = c_{ijn\mu} \frac{1}{h_\mu} \frac{\partial}{\partial \xi_\mu} + c_{ijn3} \frac{\partial}{\partial z}, \tag{15.33}$$

$$\boldsymbol{D}_{in} = \frac{1}{h_\beta} \frac{\partial}{\partial \xi_\beta} \boldsymbol{L}_{i\beta n} + \frac{\partial}{\partial z} \boldsymbol{L}_{i3n}, \tag{15.34}$$

$$C_{in\mu}(\xi, z) = \frac{1}{h_\beta} \frac{\partial c_{i\beta n\mu}}{\partial \xi_\beta} + \frac{\partial c_{i3n\mu}}{\partial z}, \tag{15.35}$$

where $\boldsymbol{\xi} = (\xi_1, \xi_2)$, and substituting (15.31) into (15.17), one has to consider

$$\boldsymbol{D}_{im} u_m^{(1)} = -\boldsymbol{C}_{in\mu} \varepsilon_{n\mu}^{(0)}. \tag{15.36}$$

The solution of (15.36) must be periodic in ξ_1 and ξ_2 (with periods A_1 and A_2, respectively) as well as satisfying the boundary conditions (15.22) for $l = 0$, which can be rewritten as

$$(\boldsymbol{L}_{ijm} u_m^{(1)} + c_{ijn\mu} \varepsilon_{n\mu}^{(0)}) N_j^{\pm} = 0, \quad (z = z^{\pm}), \tag{15.37}$$

following the notation just introduced. To meet these requirements we write

$$u_m^{(1)} = U_m^{n\mu}(\boldsymbol{\xi}, z) \varepsilon_{n\mu}^{(0)}(\boldsymbol{\alpha}) + v_m^{(1)}(\boldsymbol{\alpha}), \tag{15.38}$$

where the functions $U_m^{n\mu}$ are periodic in ξ_1 and ξ_2 and solve the problem

$$\boldsymbol{D}_{im} U_m^{n\mu} = -\boldsymbol{C}_{in\mu},$$

$$(\boldsymbol{L}_{ijm} U_m^{n\mu} + c_{ijn\mu}) N_j^{\pm} = 0 \quad (z = z^{\pm}). \tag{15.39}$$

For $n\mu = 31, 32$, the problem (15.39) is solved exactly and we find that

$$\begin{aligned} U_1^{31} &= -z, & U_2^{31} &= U_3^{31} = 0, \\ U_2^{32} &= -z, & U_1^{32} &= U_3^{32} = 0. \end{aligned} \tag{15.40}$$

This is easily verified by writing equations (15.39) in the explicit form

$$\frac{1}{h_\beta} \frac{\partial}{\partial \xi_\beta} \left(\frac{c_{i\beta m\nu}}{h_\nu} \frac{\partial U_m^{n\mu}}{\partial \xi_\nu} + c_{i\beta m3} \frac{\partial U_m^{n\mu}}{\partial z} \right) + \frac{\partial}{\partial z} \left(\frac{c_{i3m\nu}}{h_\nu} \frac{\partial U_m^{n\mu}}{\partial \xi_\nu} + c_{i3m3} \frac{\partial U_m^{n\mu}}{\partial z} \right)$$

$$= -\frac{1}{h_\nu} \frac{\partial c_{i\beta n\mu}}{\partial \xi_\beta} - \frac{\partial c_{i3n\mu}}{\partial z}, \quad \left(\frac{c_{ijm\nu}}{h_\nu} \frac{\partial U_m^{n\mu}}{\partial \xi_\nu} + c_{ijm3} \frac{\partial U_m^{n\mu}}{\partial z} + c_{ijn\mu} \right) N_j^{\pm} = 0 \quad (z = z^{\pm}),$$

and observing that they turn into identities after substituting (15.40).

To proceed further, define

$$b_{ij}^{n\mu} = \boldsymbol{L}_{ijm} U_m^{n\mu} + c_{ijn\mu} \tag{15.41}$$

and note that, by (15.40)

$$b_{ij}^{3\mu} \equiv 0. \tag{15.42}$$

Upon substituting (15.38) into (15.31) we then find

$$\sigma_{ij}^{(0)} = b_{ij}^{\beta\mu} \varepsilon_{\beta\mu}^{(0)}. \tag{15.43}$$

Now averaging (15.18), using (15.28) and conditions (15.22) for $l = 1$, yields

$$\mathscr{B}_i(\langle \sigma_{n\mu}^{(0)} \rangle) = 0. \tag{15.44}$$

Using (15.43), it is readily seen that the homogeneous equations (15.44) subjected to conditions (15.22) at $l=0$ have the zero solution

$$\varepsilon_{11}^{(0)}=\varepsilon_{22}^{(0)}=\varepsilon_{12}^{(0)}=0.$$

From (15.32), (15.38) and (15.40) it is found that

$$\begin{aligned} &u_1^{(0)}=u_2^{(0)}=0, && u_3^{(0)}=w(\boldsymbol{\alpha}),\\ &u_\mu^{(1)}=v_\mu^{(1)}(\boldsymbol{\alpha})-z\frac{1}{A_\mu}\frac{\partial w}{\partial\alpha_\mu}, && u_3^{(1)}=v_3^{(1)}(\boldsymbol{\alpha}), \end{aligned} \tag{15.45}$$

where μ is not summed.

Now since

$$\sigma_{ij}^{(0)}=0 \tag{15.46}$$

from (15.43), we find, using (15.15) and (15.16), that

$$\sigma_{ij}^{(1)}=\boldsymbol{L}_{ijm}u_m^{(2)}+c_{ijmv}\varepsilon_{mv}^{(1)}+zc_{ij\mu v}\tau_{\mu v}, \tag{15.47}$$

where

$$\begin{aligned} \varepsilon_{11}^{(1)}&=\frac{1}{A_1}\frac{\partial v_1^{(1)}}{\partial\alpha_1}+\frac{1}{A_1A_2}\frac{\partial A_1}{\partial\alpha_2}v_2^{(1)}+K_2w,\\ \varepsilon_{22}^{(1)}&=\frac{1}{A_2}\frac{\partial v_2^{(1)}}{\partial\alpha_2}+\frac{1}{A_1A_2}\frac{\partial A_2}{\partial\alpha_1}v_1^{(1)}+K_1w,\\ \varepsilon_{12}^{(1)}=\varepsilon_{21}^{(1)}&=\frac{1}{2}\left[\frac{A_1}{A_2}\frac{\partial}{\partial\alpha_2}\left(\frac{v_1^{(1)}}{A_1}\right)+\frac{A_2}{A_1}\frac{\partial}{\partial\alpha_1}\left(\frac{v_2^{(1)}}{A_2}\right)\right],\\ \varepsilon_{31}^{(1)}&=\frac{1}{A_1}\frac{\partial v_3^{(1)}}{\partial\alpha_1},\qquad \varepsilon_{32}^{(1)}=\frac{1}{A_2}\frac{\partial v_3^{(1)}}{\partial\alpha_2}, \end{aligned} \tag{15.48}$$

$$\begin{aligned} \tau_{11}&=-\frac{1}{A_1}\frac{\partial}{\partial\alpha_1}\left(\frac{1}{A_1}\frac{\partial w}{\partial\alpha_1}\right)-\frac{1}{A_1A_2^2}\frac{\partial A_1}{\partial\alpha_2}\frac{\partial w}{\partial\alpha_2},\\ \tau_{22}&=-\frac{1}{A_2}\frac{\partial}{\partial\alpha_2}\left(\frac{1}{A_2}\frac{\partial w}{\partial\alpha_2}\right)-\frac{1}{A_1^2A_2}\frac{\partial A_2}{\partial\alpha_1}\frac{\partial w}{\partial\alpha_1},\\ \tau_{12}=\tau_{21}&=-\frac{1}{A_1A_2}\left(\frac{\partial^2 w}{\partial\alpha_1\partial\alpha_2}-\frac{1}{A_1}\frac{\partial A_1}{\partial\alpha_2}\frac{\partial w}{\partial\alpha_1}-\frac{1}{A_2}\frac{\partial A_2}{\partial\alpha_1}\frac{\partial w}{\partial\alpha_2}\right). \end{aligned} \tag{15.49}$$

Substituting (15.46) and (15.47) into (15.18) and (15.22) at $l=1$ now yields

$$\begin{aligned} &\boldsymbol{D}_{im}u_m^{(2)}=-C_{imv}\varepsilon_{mv}^{(1)}-(c_{i3\mu v}+zC_{i\mu v})\tau_{\mu v},\\ &(\boldsymbol{L}_{ijm}u_m^{(2)}+c_{ijmv}\varepsilon_{mv}^{(1)}+z^{\pm}c_{ij\mu v}\tau_{\mu v})N_j^{\pm}=0\quad(z=z^{\pm}). \end{aligned} \tag{15.50}$$

The solution of (15.50) is assumed to be periodic in ξ_1 and ξ_2 (with respective periods A_1 and A_2) and can be expressed in the form

$$u_m^{(2)}=U_m^{lv}\varepsilon_{lv}^{(1)}+V_m^{\mu v}\tau_{\mu v}, \tag{15.51}$$

where the functions $U_m^{l\nu}(\boldsymbol{\xi}, z)$ are the solutions of the problem posed by (15.39), and the functions $V_m^{\mu\nu}(\boldsymbol{\xi}, z)$ are periodic in ξ_1 and ξ_2 and must satisfy the following equations:

$$\begin{aligned} \boldsymbol{D}_{im} V_m^{\mu\nu} &= -c_{i3\mu\nu} - zC_{i\mu\nu}, \\ (\boldsymbol{L}_{ijm} V_m^{\mu\nu} + z^{\pm} c_{ij\mu\nu}) N_j^{\pm} &= 0 \quad (z = z^{\pm}). \end{aligned} \tag{15.52}$$

Let

$$b_{ij}^{*\mu\nu} = \boldsymbol{L}_{ijm} V_m^{\mu\nu} + z c_{ij\mu\nu}. \tag{15.53}$$

Using (15.41) and (15.42) along with (15.51) then reduces (15.47) to the form

$$\sigma_{ij}^{(1)} = b_{ij}^{\mu\nu} \varepsilon_{\mu\nu}^{(1)} + b_{ij}^{*\mu\nu} \tau_{\mu\nu}. \tag{15.54}$$

Using (15.42) in (15.39) and (15.52), we obtain:

$$\left\{ \begin{aligned} & \frac{1}{h_\beta} \frac{\partial}{\partial \xi_\beta} b_{i\beta}^{\mu\nu} + \frac{\partial}{\partial z} b_{i3}^{\mu\nu} = 0, \\ & b_{ij}^{\mu\nu} N_j^{\pm} = 0 \quad (z = z^{\pm}), \end{aligned} \right. \tag{15.55}$$

$$\left\{ \begin{aligned} & \frac{1}{h_\beta} \frac{\partial}{\partial \xi_\beta} b_{i\beta}^{*\mu\nu} + \frac{\partial}{\partial z} b_{i3}^{*\mu\nu} = 0, \\ & b_{ij}^{*\mu\nu} N_j^{\pm} = 0 \quad (z = z^{\pm}), \end{aligned} \right. \tag{15.56}$$

according to the previous notation.

Together with (15.41) and (15.53), the problems expressed by (15.55) and (15.56) (and referred to as the unit cell local problems from now on) provide the functions $U_m^{\mu\nu}(\boldsymbol{\xi}, z)$ and $V_m^{\mu\nu}(\boldsymbol{\xi}, z)$, periodic in ξ_1 (with period A_1) and ξ_2 (with period A_2) and determine in turn, the coefficients $b_{ij}^{\mu\nu}$ and $b_{ij}^{*\mu\nu}$ in (15.54). It should be noted that, unlike the unit cell problems of 'classical' homogenization schemes (Sanchez-Palencia 1980, Bakhvalov and Panasenko 1984), those set by (15.55) and (15.56) depend on the boundary conditions at $z = z^{\pm}$ rather than on periodicity in the z direction. Clearly, in this case also the solutions $U_m^{\mu\nu}$ and $V_m^{\mu\nu}$ are unique up to constant terms, an ambiguity which may be removed by imposing the conditions

$$\langle U_m^{\mu\nu} \rangle_\xi = 0, \qquad \langle V_m^{\mu\nu} \rangle_\xi = 0 \quad \text{when } z = 0, \tag{15.57}$$

where $\langle \cdots \rangle_\xi$ indicates the average with respect to ξ_1 and ξ_2 only. Combined with (15.8), (15.45) and (15.51), these conditions make clear the mechanical meaning of the functions $v_1^{(1)}(\boldsymbol{\alpha})$, $v_2^{(1)}(\boldsymbol{\alpha})$ and $w(\boldsymbol{\alpha})$ characterizing the displacements of the mid-surface of the shell, $z = 0$.

To actually solve the local problems (15.55) and (15.56), a more convenient form of the boundary conditions at $z = z^{\pm}$ may be derived. Recalling (15.25) and using the expressions for the ξ_1-, ξ_2- and z-components for the normal $n_i^{\pm}$ to these surfaces, we find

$$\frac{1}{h_\beta} n_\beta^{\pm(\xi)} b_{i\beta}^{\mu\nu} + n_3^{\pm(\xi)} b_{i3}^{\mu\nu} = 0 \quad (z = z^{\pm}), \tag{15.58}$$

$$\frac{1}{h_\beta} n_\beta^{\pm(\xi)} b_{i\beta}^{*\mu\nu} + n_3^{\pm(\xi)} b_{i3}^{*\mu\nu} = 0 \quad (z = z^{\pm}). \tag{15.59}$$

An important point to be made, in conclusion, is that although the functions $c_{ijmn}(y,z)$ have been assumed to be smooth in the above discussion, we can readily generalize the local problems (15.55) and (15.56) to the case of piecewise-smooth elastic constants c_{ijmn}, with discontinuities of the first kind on a finite number of non-intersecting contact surfaces between dissimilar constituents (such as matrix and fibres, binder and inclusions, etc.). This is effected by adding the continuity conditions on the contact surfaces,

$$[\![U_m^{\mu\nu}]\!] = 0, \qquad \left[\!\!\left[\frac{1}{h_\beta} n_\beta^{(k)} b_{i\beta}^{\mu\nu} + n_3^{(k)} b_{i3}^{\mu\nu} \right]\!\!\right] = 0, \tag{15.60}$$

$$[\![V_m^{\mu\nu}]\!] = 0, \qquad \left[\!\!\left[\frac{1}{h_\beta} n_\beta^{(k)} b_{i\beta}^{*\mu\nu} + n_3^{(k)} b_{i3}^{*\mu\nu} \right]\!\!\right] = 0, \tag{15.61}$$

where $[\![\cdot]\!]$ denotes a jump in a function on the contact surface, and $n_i^{(k)}$, the components of the vector normal to this surface, are related to the coordinate system (ξ_1, ξ_2, z). Conditions (15.60) and (15.61) express the continuity of the displacements (15.51) and stresses (15.54) across the contact surface or, in other words, describe an ideal contact (or bonding) between dissimilar components of the composite. Note also that if we rewrite these conditions with respect to the functions $Q_m(b_{im}^{\mu\nu} \to Q_m,\ b_{im}^{*\mu\nu} \to Q_m)$, a generalization of formula (15.28) to the case of piecewise-smooth functions Q_m becomes possible.

15.3. Governing equations of the homogenized shell

To derive these equations we transform (15.55) and (15.56) by multiplying by z and z^2 and averaging. Using (15.28) and the boundary conditions on the surfaces $z = z^{\pm}$ gives

$$\langle b_{i3}^{\mu\nu} \rangle = \langle z b_{i3}^{\mu\nu} \rangle = \langle b_{i3}^{*\mu\nu} \rangle = \langle z b_{i3}^{*\mu\nu} \rangle = 0, \tag{15.62}$$

and averaging (15.54) we then have

$$\langle \sigma_{i3}^{(1)} \rangle = \langle z \sigma_{i3}^{(1)} \rangle = 0. \tag{15.63}$$

The averaging procedure is next applied to equations (15.19) and (15.20). Using again (15.28) and employing conditions (15.23) and (15.24) along with relations (15.46) and (15.63), we find

$$\mathscr{B}_\lambda(\langle \sigma_{\nu\mu}^{(1)} \rangle) + r_\lambda + \langle f_\lambda \rangle = 0, \tag{15.64}$$

$$\mathscr{B}_\lambda(\langle \sigma_{\nu\mu}^{(2)} \rangle) = 0, \tag{15.65}$$

$$\mathscr{B}_3(\langle \sigma_{3\mu}^{(2)} \rangle) + \mathscr{K}_3(\langle \sigma_{\nu\mu}^{(1)} \rangle) + q_3 + \langle g_3 \rangle = 0, \tag{15.66}$$

where

$$r_\lambda(\boldsymbol{\alpha}) = \int_{-1/2}^{1/2} \int_{-1/2}^{1/2} (\omega^+ r_\lambda^+ + \omega^- r_\lambda^-) \mathrm{d}y_1 \, \mathrm{d}y_2,$$

$$q_3(\boldsymbol{\alpha}) = \int_{-1/2}^{1/2} \int_{-1/2}^{1/2} (\omega^+ q_3^+ + \omega^- q_3^-) \mathrm{d}y_1 \, \mathrm{d}y_2. \tag{15.67}$$

Now if (15.19) is averaged after multiplying by z, then, by (15.23), we have for $i = 1, 2$:

$$\mathscr{B}_\mu(\langle z\sigma_{\nu\beta}^{(1)}\rangle) - \langle \sigma_{3\mu}^{(2)}\rangle + \rho_\mu + \langle zf_\mu\rangle = 0, \tag{15.68}$$

where

$$\rho_\mu(\boldsymbol{\alpha}) = \int_{-1/2}^{1/2}\int_{-1/2}^{1/2}(z^+\omega^+ r_\mu^+ + z^-\omega^- r_\mu^-)\,dy_1\,dy_2. \tag{15.69}$$

Equations (15.66) and (15.68) may be combined to eliminate $\langle \sigma_{3\mu}^{(2)}\rangle$, giving (cf. (15.12))

$$\frac{1}{A_1A_2}\left\{\frac{\partial}{\partial\alpha_1}[A_2(\mathscr{B}_1\langle z\sigma_{\nu\beta}^{(1)}\rangle + \rho_1 + \langle zf_1\rangle)] + \frac{\partial}{\partial\alpha_2}[A_1(\mathscr{B}_2\langle z\sigma_{\nu\beta}^{(1)}\rangle + \rho_2 + \langle zf_2\rangle)]\right\}$$
$$- K_2\langle\sigma_{11}^{(1)}\rangle - K_1\langle\sigma_{22}^{(1)}\rangle + q_3 + \langle g_3\rangle = 0. \tag{15.70}$$

Note that as far as the stresses are concerned, equations (15.64) and (15.70) only contain the averaged quantities $\langle\sigma_{\mu\nu}^{(1)}\rangle$ and $\langle z\sigma_{\mu\nu}^{(1)}\rangle$, which may be represented in their respective forms

$$\langle\sigma_{\mu\nu}^{(1)}\rangle = \langle b_{\mu\nu}^{\beta\lambda}\rangle\varepsilon_{\beta\lambda}^{(1)} + \langle b_{\mu\nu}^{*\beta\lambda}\rangle\tau_{\beta\lambda}, \tag{15.71}$$

$$\langle z\sigma_{\mu\nu}^{(1)}\rangle = \langle zb_{\mu\nu}^{\beta\lambda}\rangle\varepsilon_{\beta\lambda}^{(1)} + \langle zb_{\mu\nu}^{*\beta\lambda}\rangle\tau_{\beta\lambda}. \tag{15.72}$$

The system of three governing equations for the unknown functions of the problem can now be written down by substituting (15.71) and (15.72) into (15.64) and (15.70) and noting that $\varepsilon_{\beta\lambda}^{(1)}$ and $\tau_{\beta\lambda}$ are expressed in terms of $v_1^{(1)}(\boldsymbol{\alpha})$, $v_2^{(1)}(\boldsymbol{\alpha})$ and $w(\boldsymbol{\alpha})$; see equations (15.48) and (15.49).

We conclude that the original problem for the regularly non-homogeneous (composite material) shell with rapidly oscillating thickness reduces to two simpler types of problem. One type involves a pair of local problems, the first of which is set by equations (15.55), (15.57) and (15.60) and yields the functions $U_m^{\mu\nu}(\xi_1,\xi_2,z)$, periodic in ξ_1 and ξ_2, with periods A_1 and A_2, respectively; once these are known, the functions $b_{ij}^{\mu\nu}(\xi_1,\xi_2,z)$ can be determined by means of (15.41). The second local problem is posed by equations (15.56), (15.57) and (15.61) and produces the functions $V_m^{\mu\nu}(\xi_1,\xi_2,z)$, also periodic in ξ_1 and ξ_2, in terms of which the functions $b_{ij}^{*\mu\nu}(\xi_1,\xi_2,z)$ are calculated from (15.53). Both local problems are in fact elasticity theory problems formulated for a generally non-homogeneous (unit cell Ω) body subjected to certain artificial forces (surface and bulk), which are expressed in terms of the elastic coefficeints and their derivatives, as clearly seen from (15.39) and (15.52). With regard to the boundary conditions, there are periodicity requirements along the ξ_1 and ξ_2 coordinates, and on the surfaces $z = z^\pm$ the conditions expressed by the problems (15.55) and (15.56) or relations (15.58) and (15.59) must be satisfied.

The local problems having been solved, the functions $b_{ij}^{\mu\nu}$ and $b_{ij}^{*\mu\nu}$ are averaged by application of (15.27), giving the effective stiffness moduli of the homogenized shell, $\langle b_{\mu\nu}^{\beta\lambda}\rangle$, $\langle zb_{\mu\nu}^{\beta\lambda}\rangle$, $\langle b_{\mu\nu}^{*\beta\lambda}\rangle$ and $\langle zb_{\mu\nu}^{*\beta\lambda}\rangle$, entering the elastic relations (15.71) and (15.72) as coefficients. One can then proceed to the second of the two above-mentioned types of problems, namely the boundary value problem for the homogenized shell, expressed by relations (15.64), (15.70)–(15.72), (15.48) and (15.49) and yielding the functions $v_1^{(1)}(\boldsymbol{\alpha})$, $v_2^{(1)}(\boldsymbol{\alpha})$ and $w(\boldsymbol{\alpha})$. It will be noted that the solution of the homogenized problem corresponds to the application of the effective modulus method.

An important feature of the method discussed is that the solutions of the local problems and of the homogenized problem enable us to make very accurate predictions concerning the three-dimensional local structure of the displacement and stress fields. In particular, equations (15.8), (15.45) and (15.51) yield

$$
\begin{aligned}
u_1 &= \delta v_1^{(1)}(\boldsymbol{\alpha}) - \gamma \frac{1}{A_1}\frac{\partial w}{\partial \alpha_1} + \delta U_1^{\mu\nu}\delta\varepsilon_{\mu\nu}^{(1)} + \delta^2 V_1^{\mu\nu}\tau_{\mu\nu} + \cdots, \\
u_2 &= \delta v_2^{(1)}(\boldsymbol{\alpha}) - \gamma \frac{1}{A_2}\frac{\partial w}{\partial \alpha_2} + \delta U_2^{\mu\nu}\delta\varepsilon_{\mu\nu}^{(1)} + \delta^2 V_2^{\mu\nu}\tau_{\mu\nu} + \cdots, \\
u_3 &= w(\boldsymbol{\alpha}) + \delta U_3^{\mu\nu}\delta\varepsilon_{\mu\nu}^{(1)} + \delta^2 V_3^{\mu\nu}\tau_{\mu\nu} + \cdots,
\end{aligned}
\qquad (15.73)
$$

for the displacement components. Note that the function $v_3^{(1)}(\boldsymbol{\alpha})$ involved in (15.38) and (15.45) can be set equal to zero without loss of generality because it has been proved earlier that this particular function does not enter into the system of governing equations of the problem.

Similarly, it is found from (15.16) and (15.46) that the stresses are given by

$$
\sigma_{ij} = b_{ij}^{\mu\nu}\delta\varepsilon_{\mu\nu}^{(1)} + \delta b_{ij}^{*\mu\nu}\tau_{\mu\nu} + \ldots. \qquad (15.74)
$$

From the way they are stated, it is seen that the local problems are completely determined by the structure of the unit cell of the composite and are totally independent of the formulation of the global boundary value problem. It follows that the solutions of these problems, and particularly the effective elastic moduli of the homogenized shell, are universal in their nature and, once found, may therefore be utilized in studying very different types of boundary value problems associated with a given composite structure.

It should be noted, finally, that the coordinates ξ_1 and ξ_2 involved in the local problems are defined in terms of the quantities $A_1(\boldsymbol{\alpha})$ and $A_2(\boldsymbol{\alpha})$, so that if these latter are not constant (and constant they may be in the case of developable surfaces), the effective stiffness moduli will also depend on the 'slow' coordinates α_1 and α_2. This means that even in the case of an originally homogeneous material we may come up with a structural non-homogeneity after the homogenization procedure.

15.4. The symmetry of effective properties

When the properties of a material are periodic in all three coordinates, it can be shown (Bakhvalov and Panasenko 1984) that the symmetry properties of the coefficients involved remain the same after the homogenization process. In the case we treat here, there is no periodicity along the thickness coordinate z of the shell and the question of symmetry is therefore given special consideration.

Let us first return to definitions (15.41) and (15.53) for the quantities b_{ij}^{mn} and b_{ij}^{*mn}. Modifying them by allowing all the indices to range from 1 to 3 and using relations

(15.33)–(15.35), (15.39) and (15.52) we find that

$$b_{ij}^{mn} = b_{ji}^{mn} = b_{ij}^{nm}, \qquad b_{ij}^{*mn} = b_{ji}^{*mn} = b_{ij}^{*nm}, \tag{15.75}$$

where the well-known symmetry properties of the elastic coefficients c_{ijmn} have also been taken into account.

We wish now to prove the following symmetry properties:

$$\langle b_{ij}^{mn} \rangle = \langle b_{mn}^{ij} \rangle, \qquad \langle z b_{ij}^{mn} \rangle = \langle b_{mn}^{*ij} \rangle,$$
$$\langle z b_{ij}^{*mn} \rangle = \langle z b_{mn}^{*ij} \rangle. \tag{15.76}$$

To this end, define

$$\zeta_1 = h_1 \xi_1, \qquad \zeta_2 = h_2 \xi_2, \qquad \zeta_3 = z.$$

Equations (15.33) then reduce to

$$\boldsymbol{L}_{ijl} = c_{ijlr} \frac{\partial}{\partial \zeta_r}, \qquad \boldsymbol{D}_{il} = \frac{\partial}{\partial \zeta_p} c_{iplr} \frac{\partial}{\partial \zeta_r}, \qquad \boldsymbol{C}_{imn} = \frac{\partial c_{iqmn}}{\partial \zeta_q}.$$

Equations (15.41) and (15.53) become

$$b_{ij}^{mn} = c_{ijlr} \left(\frac{\partial U_l^{mn}}{\partial \zeta_r} + \delta_{lm} \delta_{rn} \right), \tag{15.77}$$

$$b_{ij}^{*mn} = c_{ijlr} \left(\frac{\partial V_l^{mn}}{\partial \zeta_r} + z \delta_{lm} \delta_{rn} \right), \tag{15.78}$$

and the local problems (15.55) and (15.56) can be rewritten as

$$\frac{\partial b_{ij}^{mn}}{\partial \zeta_j} = 0, \qquad N_j^{\pm} b_{ij}^{mn} = 0 \quad (z = z^{\pm}), \tag{15.79}$$

$$\frac{\partial b_{ij}^{*mn}}{\partial \zeta_j} = 0, \qquad N_j^{\pm} b_{ij}^{*mn} = 0 \quad (z = z^{\pm}). \tag{15.80}$$

Now if the function $\varphi(\zeta_1, \zeta_2, \zeta_3)$ is periodic in ζ_1 and ζ_2 with the unit cell Ω, then the following equations hold:

$$\left\langle b_{qp}^{mn} \frac{\partial \varphi}{\partial \zeta_p} \right\rangle = 0, \qquad \left\langle b_{qp}^{*mn} \frac{\partial \varphi}{\partial \zeta_p} \right\rangle = 0. \tag{15.81}$$

To show this, employ (15.79) to obtain:

$$\frac{\partial}{\partial \zeta_p} (b_{qp}^{mn} \varphi) = \varphi \frac{\partial b_{qp}^{mn}}{\partial \zeta_p} + b_{qp}^{mn} \frac{\partial \varphi}{\partial \zeta_p} = b_{qp}^{mn} \frac{\partial \varphi}{\partial \zeta_p}.$$

Applying (15.28) and making use of the conditions at the $z = z^{\pm}$ surfaces, we then find

$$\left\langle \frac{\partial}{\partial \zeta_p} (b_{qp}^{mn} \varphi) \right\rangle = \int_{-1/2}^{1/2} \int_{-1/2}^{1/2} [(\varphi b_{qp}^{mn})^+ N_p^+ - (\varphi b_{qp}^{mn})^- N_p^-] \mathrm{d}y_1 \mathrm{d}y_2 = 0,$$

thereby proving the first of equations (15.81). The second is proved in exactly the same way by taking (15.80) as the starting point.

Next, we observe that adding the identity

$$\langle b_{ij}^{mn} \rangle = \langle b_{qp}^{mn} \delta_{qi} \delta_{pj} \rangle$$

and the equation

$$\langle b_{qp}^{mn} \partial U_q^{ij} / \partial \zeta_p \rangle = 0,$$

which is simply the first of (15.81) for $\varphi = U_q^{ij}$, we obtain:

$$\begin{aligned} \langle b_{ij}^{mn} \rangle &= \left\langle b_{qp}^{mn} \left(\frac{\partial U_q^{ij}}{\partial \zeta_p} + \delta_{qi} \delta_{pj} \right) \right\rangle \\ &= \left\langle c_{qplr} \left(\frac{\partial U_l^{mn}}{\partial \zeta_r} + \delta_{lm} \delta_{rn} \right) \left(\frac{\partial U_q^{ij}}{\partial \zeta_p} + \delta_{qi} \delta_{pj} \right) \right\rangle, \end{aligned} \tag{15.82}$$

which actually proves the first of equations (15.76) if the symmetry properties of the elastic coefficients are properly taken into account.

Likewise, setting φ successively equal to U_q^{ij} and V_q^{ij}, it follows from (15.77), (15.78) and (15.81) that

$$\langle z b_{ij}^{*mn} \rangle = \left\langle c_{qplr} \left(\frac{\partial V_l^{mn}}{\partial \zeta_r} + z \delta_{lm} \delta_{rn} \right) \left(\frac{\partial V_q^{ij}}{\partial \zeta_p} + z \delta_{qi} \delta_{pj} \right) \right\rangle, \tag{15.83}$$

$$\begin{aligned} \langle b_{ij}^{*mn} \rangle &= \left\langle c_{qplr} \left(\frac{\partial V_l^{mn}}{\partial \zeta_r} + z \delta_{lm} \delta_{rn} \right) \left(\frac{\partial U_q^{ij}}{\partial \zeta_p} + \delta_{qi} \delta_{pj} \right) \right\rangle, \\ \langle z b_{ij}^{mn} \rangle &= \left\langle c_{qplr} \left(\frac{\partial U_l^{mn}}{\partial \zeta_r} + \delta_{lm} \delta_{rn} \right) \left(\frac{\partial V_q^{ij}}{\partial \zeta_p} + z \delta_{qi} \delta_{pj} \right) \right\rangle. \end{aligned} \tag{15.84}$$

Using again the symmetry of the elastic coefficients, the third of equations (15.76) is derived from (15.83), and the second from (15.84).

We note that if one or more of the indices is 3, (15.76) produces an identical zero by virtue of (15.42) and (15.62); otherwise, these formulas are non-trivial and secure the symmetry of the coefficient matrix of the elastic relations (15.71) and (15.72).

15.5. *Homogenized shell model versus thin-shell theory*

To carry out this comparison, the basic results we have obtained by homogenizing the shell equations will be rewritten here in terms of the standard theory of thin shells as described, for example, in Novozhilov (1962) or Ambartsumyan (1974).

Using (15.16), (15.46) and (15.63), the leading-order expressions for the stress resultants N_1, N_2 and N_{12}, moment resultants M_1, M_2 and M_{12}, and the shearing forces Q_1 and Q_2 are found to be given by

$$\begin{aligned} N_1 &= \delta^2 \langle \sigma_{11}^{(1)} \rangle, & M_1 &= \delta^3 \langle z \sigma_{11}^{(1)} \rangle \quad (1 \leftrightarrow 2), \\ N_{12} &= \delta^2 \langle \sigma_{12}^{(1)} \rangle, & M_{12} &= \delta^3 \langle z \sigma_{12}^{(1)} \rangle, \\ Q_1 &= \delta^3 \langle \sigma_{31}^{(2)} \rangle, & Q_2 &= \delta^3 \langle \sigma_{32}^{(2)} \rangle. \end{aligned} \tag{15.85}$$

To determine the components of the displacement vector, again at leading order, formulas (15.73) must be used. Denoting

$$v_1(\boldsymbol{\alpha}) = \delta v_1^{(1)}(\boldsymbol{\alpha}), \qquad v_2(\boldsymbol{\alpha}) = \delta v_2^{(1)}(\boldsymbol{\alpha}), \qquad \varepsilon_{\mu\nu} = \delta \varepsilon_{\mu\nu}^{(1)},$$

and using (15.9), equations (15.48) take the form

$$\begin{aligned}
\varepsilon_{11} &= \frac{1}{A_1}\frac{\partial v_1}{\partial \alpha_1} + \frac{1}{A_1 A_2}\frac{\partial A_1}{\partial \alpha_2} v_2 + k_1 w, \\
\varepsilon_{22} &= \frac{1}{A_2}\frac{\partial v_2}{\partial \alpha_2} + \frac{1}{A_1 A_2}\frac{\partial A_2}{\partial \alpha_1} v_1 + k_2 w, \\
\varepsilon_{12} = \varepsilon_{21} &= \frac{1}{2}\left[\frac{A_1}{A_2}\frac{\partial}{\partial \alpha_2}\left(\frac{v_1}{A_1}\right) + \frac{A_2}{A_1}\frac{\partial}{\partial \alpha_1}\left(\frac{v_2}{A_2}\right)\right],
\end{aligned} \tag{15.86}$$

and (15.73) may be rewritten as

$$\begin{aligned}
u_1 &= v_1(\boldsymbol{\alpha}) - \frac{\gamma}{A_1}\frac{\partial w}{\partial \alpha_1} + \delta U_1^{\mu\nu}\varepsilon_{\mu\nu} + \delta^2 V_1^{\mu\nu}\tau_{\mu\nu} + \cdots, \\
u_2 &= v_2(\boldsymbol{\alpha}) - \frac{\gamma}{A_2}\frac{\partial w}{\partial \alpha_2} + \delta U_2^{\mu\nu}\varepsilon_{\mu\nu} + \delta^2 V_2^{\mu\nu}\tau_{\mu\nu} + \cdots, \\
u_3 &= w(\boldsymbol{\alpha}) + \delta U_3^{\mu\nu}\varepsilon_{\mu\nu} + \delta^2 V_3^{\mu\nu}\tau_{\mu\nu} + \cdots.
\end{aligned} \tag{15.87}$$

A comparison of relations (15.86), (15.49) and (15.87) with the corresponding thin-shell theory results (Novozhilov 1962, Ambartsumyan 1974) shows that the functions $v_1(\boldsymbol{\alpha})$, $v_2(\boldsymbol{\alpha})$ and $w(\boldsymbol{\alpha})$ are the mid-surface displacements of the shell; the quantities $\varepsilon_{11} = \varepsilon_1$, $\varepsilon_{22} = \varepsilon_2$ and $\varepsilon_{12} = \varepsilon_{21} = \omega/2$ are the elongations and shears; and, finally, $\tau_{11} = \kappa_1, \tau_{22} = \kappa_2$ and $\tau_{12} = \tau_{21} = \tau$ are the torsional and flexural mid-surface strains as calculated in the framework of the Donnell–Mushtari–Vlasov model (see Novozhilov 1962). It will be clear that the correspondence between the two sets of results is due to the asymptotic form adopted above for the main curvatures of the shell mid-surfaces, i.e. equations (15.9).

The elastic relations of the homogenized shell, i.e. those between the stress and moment resultants on the one hand and the mid-surface strains on the other, are found from equations (15.71), (15.72) and (15.85). We have

$$\begin{aligned}
N_\beta &= \delta\langle b_{\beta\beta}^{\mu\nu}\rangle\varepsilon_{\mu\nu} + \delta^2\langle b_{\beta\beta}^{*\mu\nu}\rangle\tau_{\mu\nu}, \\
N_{12} &= \delta\langle b_{12}^{\mu\nu}\rangle\varepsilon_{\mu\nu} + \delta^2\langle b_{12}^{*\mu\nu}\rangle\tau_{\mu\nu}, \\
M_\beta &= \delta^2\langle z b_{\beta\beta}^{\mu\nu}\rangle\varepsilon_{\mu\nu} + \delta^3\langle z b_{\beta\beta}^{*\mu\nu}\rangle\tau_{\mu\nu}, \\
M_{12} &= \delta^2\langle z b_{12}^{\mu\nu}\rangle\varepsilon_{\mu\nu} + \delta^3\langle z b_{12}^{*\mu\nu}\rangle\tau_{\mu\nu},
\end{aligned} \tag{15.88}$$

where β takes the values 1 and 2, and is not summed. We should remark here that the symmetry of the 6×6 coefficient matrix involved in this equation is ensured by the symmetry properties of the effective elastic moduli, i.e. equations (15.76) above.

Using (15.67), (15.69), (15.10), (15.11) and (15.85) and following the notation of (15.9) and (15.12), equations (15.64)–(15.66) and (15.68) can be written as

$$\frac{\partial(A_2N_1)}{\partial\alpha_1}-\frac{\partial A_2}{\partial\alpha_1}N_2+\frac{\partial(A_1N_{12})}{\partial\alpha_2}+\frac{\partial A_1}{\partial\alpha_2}N_{12}=-A_1A_2G_1,$$

$$\frac{\partial(A_1N_2)}{\partial\alpha_2}-\frac{\partial A_1}{\partial\alpha_2}N_1+\frac{\partial(A_2N_{12})}{\partial\alpha_1}+\frac{\partial A_2}{\partial\alpha_1}N_{12}=-A_1A_2G_2,$$

$$k_1N_1+k_2N_2-\frac{1}{A_1A_2}\left[\frac{\partial(A_2Q_1)}{\partial\alpha_1}+\frac{\partial(A_1Q_2)}{\partial\alpha_2}\right]=G_3, \tag{15.89}$$

$$Q_1=\frac{1}{A_1A_2}\left[\frac{\partial(A_2M_1)}{\partial\alpha_1}-\frac{\partial A_2}{\partial\alpha_1}M_2+\frac{\partial(A_1M_{12})}{\partial\alpha_2}+\frac{\partial A_1}{\partial\alpha_2}M_{12}\right]+m_1,$$

$$Q_2=\frac{1}{A_1A_2}\left[\frac{\partial(A_1M_2)}{\partial\alpha_2}-\frac{\partial A_1}{\partial\alpha_2}M_1+\frac{\partial(A_2M_{12})}{\partial\alpha_1}+\frac{\partial A_2}{\partial\alpha_1}M_{12}\right]+m_2.$$

In the above, the external loads are given by

$$G_i=\int_{-1/2}^{1/2}\int_{-1/2}^{1/2}(\omega^+p_i^++\omega^-p_i^-)\mathrm{d}y_1\mathrm{d}y_2+\delta\langle P_i\rangle,$$

$$m_\beta=\int_{-1/2}^{1/2}\int_{-1/2}^{1/2}(\gamma^+\omega^+p_\beta^++\gamma^-\omega^-p_\beta^-)\mathrm{d}y_1\mathrm{d}y_2+\delta\langle\gamma P_\beta\rangle, \tag{15.90}$$

where the functions $\omega^\pm$, defined by (15.26), are determined by the profiles of the top and bottom shell surfaces, $\gamma=\gamma^\pm$, and where the presence of the coefficient δ is due to the fact that the averaging operation (15.27) involves integration over the coordinate z rather than over $\gamma=\delta z$. We thus see that equations (15.89) are identical to the equilibrium equations of the engineering formulation of thin shell theory (Ambartsumyan 1974).

If we take as an example the limiting case of a homogeneous isotropic shell of constant thickness, we have the equations $F^\pm\equiv 0$, $N_i^\pm=\{0,0,1\}$, $\omega^\pm\equiv 1$, $c_{ijmn}=\text{const.}$ and

$$c_{ijmn}=\frac{E}{2(1+\nu)}\left(\frac{2\nu}{1-2\nu}\delta_{ij}\delta_{mn}+\delta_{im}\delta_{jn}+\delta_{in}\delta_{jm}\right) \tag{15.91}$$

as the initial formulation of the problem, where E and ν denote, as usual, the Young's modulus and the Poisson's ratio of the material. Since, in the case, none of the quantities of interest depends on y_1 or y_2, the problem is solvable exactly to give

$$U_3^{11}=U_3^{22}=-\frac{\nu z}{1-\nu},\qquad V_3^{11}=V_3^{22}=-\frac{\nu z^2}{2(1-\nu)} \tag{15.92}$$

for the non-zero solutions of the local problems (15.55)–(15.57). The set of non-zero elastic moduli then follows from (15.41) and (15.53) as

$$\langle b_{11}^{11} \rangle = \langle b_{22}^{22} \rangle = \frac{E}{1-\nu^2}, \qquad \langle b_{11}^{22} \rangle = \langle b_{22}^{11} \rangle = \frac{E\nu}{1-\nu^2},$$

$$\langle b_{12}^{12} \rangle = \frac{E}{2(1+\nu)}, \qquad \langle z b_{\mu\theta}^{\beta\lambda} \rangle = 0, \qquad \langle b_{\mu\theta}^{*\beta\lambda} \rangle = 0, \tag{15.93}$$

$$\langle z b_{\mu\theta}^{*\beta\lambda} \rangle = \tfrac{1}{12} \langle b_{\mu\theta}^{\beta\lambda} \rangle,$$

which when substituted into (15.88) leads—a point worthy of emphasis—to the elastic relations of the theory of thin isotropic shells.

An important conclusion to be drawn from the above comparison is that we may utilize the well-studied formalism of the theory of anisotropic shells for the solution of the homogenized problem, using the elastic relations (15.88) to describe the particular type of anisotropy of the problem under consideration.

With regard to the edge conditions necessary for the solution of the homogenized boundary value problem, we may take these in the form known from ordinary shell theory but, obviously, the edge effects remain unaccounted for in this approach. The remedies are to solve the problem in its exact three-dimensional formulation or, alternatively, to employ the method of boundary layer solutions that has been developed in the framework of the asymptotic homogenization method by a number of investigators (Panasenko 1979, Bakhvalov and Panasenko 1984, Sanchez-Palencia 1987).

15.6. *Cylindrical shell*

In this special case, of well-known importance for engineering and other purposes, the mid-surface of the shell is a cylindrical surface. If we introduce a coordinate system (α_1, α_2) such that α_1 is measured along the generator and α_2 along the directrix of the cylinder, we have

$$A_1 \equiv A_2 \equiv 1, \qquad k_1 = 0, \qquad k_2 = \frac{1}{r(\alpha_2)},$$

where $r(\alpha_2)$ is constant in the special case of a circular cylinder. Clearly, α_1 and α_2 can be made dimensionless by dividing by a certain characteristic dimension L, which we take to be equal to unity for the sake of simplicity.

From (15.49) and (15.86):

$$\varepsilon_{11} = \frac{\partial v_1}{\partial \alpha_1}, \qquad \varepsilon_{22} = \frac{\partial v_2}{\partial \alpha_2} + \frac{w}{r},$$

$$\varepsilon_{12} = \frac{1}{2}\left(\frac{\partial v_1}{\partial \alpha_2} + \frac{\partial v_2}{\partial \alpha_1}\right), \qquad \tau_{\lambda\mu} = -\frac{\partial^2 w}{\partial \alpha_\lambda \partial \alpha_\mu}, \tag{15.94}$$

and equations (15.89) become

$$\frac{\partial N_1}{\partial \alpha_1} + \frac{\partial N_{12}}{\partial \alpha_2} = -G_1,$$
$$\frac{\partial N_2}{\partial \alpha_2} + \frac{\partial N_{12}}{\partial \alpha_1} = -G_2, \qquad (15.95)$$
$$\frac{\partial^2 M_\beta}{\partial \alpha_\beta^2} + 2\frac{\partial^2 M_{12}}{\partial \alpha_1 \partial \alpha_2} - \frac{N_2}{r} = -G_3 - \frac{\partial m_\mu}{\partial \alpha_\mu}.$$

The stress and moment resultants are expressed in terms of the mid-surface strains, equations (15.94), by means of the elastic relations (15.88). Since the quantities A_1 and A_2 are constant in this case, so too are the effective stiffness moduli, because the functions $b_{\mu\nu}^{\beta\theta}$ and $b_{\mu\nu}^{*\beta\theta}$ are independent of the coordinates α_1 and α_2. The values of the moduli are determined by the functions $c_{ijmn}(y, z)$ and $F^{\pm}(y)$ and are found from the local problems of the type we have discussed above.

15.7. *Plane shell*

This also is a geometry of great practical interest. Since the mid-surface obviously is a plane in this case, the coordinate system α_1, α_2, γ is naturally chosen to be a Cartesian one, so that $A_1 \equiv A_2 \equiv 1$, $k_1 = k_2 = 0$ and the governing relations of the problem follow from (15.94) and (15.95) in the limit as $r \to \infty$. The effective moduli are again constants, and we obtain the following system of three equations to determine the functions $v_1(\boldsymbol{\alpha})$, $v_2(\boldsymbol{\alpha})$ and $w(\boldsymbol{\alpha})$:

$$\delta \langle b_{\theta\mu}^{\beta\lambda} \rangle \frac{\partial^2 v_\beta}{\partial \alpha_\lambda \partial \alpha_\mu} - \delta^2 \langle b_{\theta\mu}^{*\beta\lambda} \rangle \frac{\partial^3 w}{\partial \alpha_\beta \partial \alpha_\lambda \partial \alpha_\mu} = -\mathrm{G}_\theta \quad (\theta = 1,2),$$
$$(15.96)$$
$$\delta^2 \langle z b_{\mu\nu}^{\beta\lambda} \rangle \frac{\partial^3 v_\beta}{\partial \alpha_\lambda \partial \alpha_\mu \partial \alpha_\nu} - \delta^3 \langle z b_{\mu\nu}^{*\beta\lambda} \rangle \frac{\partial^4 w}{\partial \alpha_\beta \partial \alpha_\lambda \partial \alpha_\mu \partial \alpha_\nu} = -G_3 - \frac{\partial m_\mu}{\partial \alpha_\mu},$$

where the elastic relations (15.88) and the properties (15.75) of the stiffness moduli have been used.

Probably the most important thing to note in this system is that, unless there are some special cases, it does not separate into an equation for the deflection $w(\boldsymbol{\alpha})$ and equations for the plate surface displacements $v_1(\boldsymbol{\alpha})$ and $v_2(\boldsymbol{\alpha})$. It is obvious, however, that separation does occur when either $\langle z b_{\mu\nu}^{\beta\lambda} \rangle = 0$ or $\langle b_{\mu\nu}^{*\beta\lambda} \rangle = 0$, two conditions that are equivalent in view of (15.76) and may be satisfied if certain types of 'good symmetry' happen to occur in the material properties and/or the unit cell geometry. This is illustrated by means of examples below.

In the limiting case of a homogeneous, isotropic plane plate of constant thickness, the effective elastic moduli are determined by equations (15.93); the system (15.96) separates and well-known equations of ordinary plates theory (cf. Ambartsumyan 1987) are obtained.

15.8. A laminated shell composed of homogeneous isotropic layers

The mathematical apparatus we have developed above can be successfully applied to the calculation of the effective moduli of a laminated shell composed of homogeneous isotropic layers. If we assume that the layers are perfectly bonded and are all parallel to the mid-surface of the shell, all the variables involved will be independent of the coordinates y_1 and y_2 (or, equivalently, of ξ_1 and ξ_2) and the local problems of relevance will be solved in an elementary way.

The unit cell of the problem, as shown in Fig. 15.3, is referred to the coordinate system (ξ_1, ξ_2, z) and is completely determined by the set of parameters $\delta_1, \delta_2, \ldots, \delta_M$, where M is the number of layers. The thickness of the mth layer is, in these coordinates, $\delta_m - \delta_{m-1}$, δ_0 and δ_M being zero and unity, respectively. The real thickness of the mth layer, referred to the original coordinate system $\alpha_1, \alpha_2, \gamma$, is $\delta(\delta_m - \delta_{m-1})$, and the total thickness of the packet is δ.

Since

$$z^{\pm} = \pm\tfrac{1}{2}, \qquad N_i^{\pm} = n_i^{(k)} = \{0, 0, 1\}, \qquad L_{ijn} = c_{ijn3}\,\mathrm{d}/\mathrm{d}z,$$

in this case the first local problem, equations (15.41), (15.55), (15.57) and (15.61), is transformed to

$$b_{ij}^{\lambda\nu} = c_{ijn3}\frac{\mathrm{d}U_n^{\lambda\mu}}{\mathrm{d}z} + c_{ij\lambda\mu}, \tag{15.97}$$

$$\frac{\mathrm{d}b_{i3}^{\lambda\mu}}{\mathrm{d}z} = 0, \qquad b_{i3}^{\lambda\mu} = 0 \quad (z = \pm\tfrac{1}{2}),$$

$$[\![U_i^{\lambda\mu}]\!] = 0, \qquad [\![b_{i3}^{\lambda\mu}]\!] = 0 \quad (z = \delta_m - \tfrac{1}{2};\ m = 1, 2, \ldots, M-1), \tag{15.98}$$

$$U_i^{\lambda\mu} = 0, \quad (z = 0),$$

and the corresponding second local problem, equations (15.53), (15.55), (15.57) and

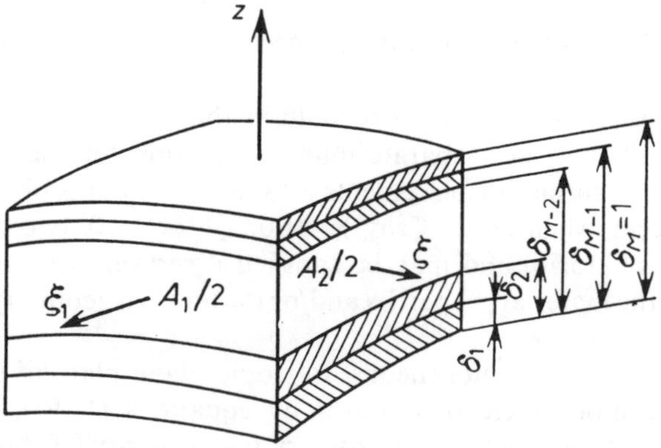

Fig. 15.3 *Laminated shell composed of M homogeneous layers.*

(15.61), differs from the above by having

$$b_{ij}^{*\lambda\mu} = c_{ijn3}\frac{\mathrm{d}V_n^{\lambda\mu}}{\mathrm{d}z} + zc_{ij\lambda\mu} \tag{15.99}$$

instead of (15.97), and by replacing

$$b_{i3}^{\lambda\mu} \to b_{i3}^{*\lambda\mu}, \qquad U_i^{\lambda\mu} \to V_i^{\lambda\mu}$$

everywhere in (15.98). Because of the assumed isotropy, the elastic coefficients in the above equations are determined by means of (15.91) in which the mth layer of the composite shell will be referenced by employing an index m on the quantities E_m and ν_m.

From (15.97) and (15.98), it follows that $b_{i3}^{\lambda\mu} = 0$ so that, under the isotropy assumption,

$$\begin{gathered}\frac{\mathrm{d}U_1^{\lambda\mu(m)}}{\mathrm{d}z} = \frac{\mathrm{d}U_2^{\lambda\mu(m)}}{\mathrm{d}z} = \frac{\mathrm{d}U_3^{12(m)}}{\mathrm{d}z} = 0,\\ \frac{\mathrm{d}U_3^{11(m)}}{\mathrm{d}z} = \frac{\mathrm{d}U_3^{22(m)}}{\mathrm{d}z} = -\frac{\nu_m}{1-\nu_m} \quad (m = 1, 2, \ldots, M).\end{gathered} \tag{15.100}$$

It should be understood that the condition at $z = 0$ and the interface conditions (15.98) with respect to the functions $U_i^{\lambda\mu}$ are satisfied by properly choosing the values of the constants arising from the integration of relations (15.100). According to the total number of layers, there are M such constants for each set i, λ, μ, and there are precisely M algebraic equations for their determination, of which $M-1$ express the interface conditions, and one expresses the condition at $z = 0$. Note, however, that evaluation of the elastic moduli does not require knowledge of these constants, because the quantities b, as given by equation (15.97), only contain the derivatives of the functions $U_i^{\lambda\mu}$, and may therefore be determined by the use of (15.100).

Substituting (15.100) into (15.97), we have, after averaging over the thickness,

$$\begin{aligned}\langle b_{11}^{11}\rangle = \langle b_{22}^{22}\rangle &= \sum_{m=1}^{M}\frac{(\delta_m - \delta_{m-1})E_m}{1-\nu_m^2},\\ \langle b_{11}^{22}\rangle &= \sum_{m=1}^{M}\frac{(\delta_m - \delta_{m-1})\nu_m E_m}{1-\nu_m^2},\\ \langle b_{12}^{12}\rangle &= \sum_{m=1}^{M}\frac{(\delta_m - \delta_{m-1})E_m}{2(1+\nu_m)},\end{aligned} \tag{15.101}$$

$$\begin{aligned}\langle zb_{11}^{11}\rangle = \langle zb_{22}^{22}\rangle &= \tfrac{1}{2}\sum_{m=1}^{M}\frac{E_m}{1-\nu_m^2}[\delta_m^2 - \delta_{m-1}^2 - (\delta_m - \delta_{m-1})],\\ \langle zb_{11}^{22}\rangle = \langle zb_{22}^{11}\rangle &= \tfrac{1}{2}\sum_{m=1}^{M}\frac{\nu_m E_m}{1-\nu_m^2}[\delta_m^2 - \delta_{m-1}^2 - (\delta_m - \delta_{m-1})],\\ \langle zb_{12}^{12}\rangle &= \tfrac{1}{2}\sum_{m=1}^{M}\frac{E_m}{2(1+\nu_m)}[\delta_m^2 - \delta_{m-1}^2 - (\delta_m - \delta_{m-1})].\end{aligned} \tag{15.102}$$

For the second local problem, we find in an identical manner that

$$\frac{\mathrm{d}V_1^{\lambda\mu(m)}}{\mathrm{d}z}=\frac{\mathrm{d}V_2^{\lambda\mu(m)}}{\mathrm{d}z}=\frac{\mathrm{d}V_3^{\lambda\mu(m)}}{\mathrm{d}z}=0,$$

$$\frac{\mathrm{d}V_3^{11(m)}}{\mathrm{d}z}=\frac{\mathrm{d}V_3^{22(m)}}{\mathrm{d}z}=-\frac{\nu_m z}{1-\nu_m}\quad (m=1,2,\ldots,M), \tag{15.103}$$

which when substituted into (15.99) give

$$\langle zb_{11}^{*11}\rangle=\langle zb_{22}^{*22}\rangle=\tfrac{1}{3}\sum_{m=1}^{M}\frac{E_m}{1-\nu_m^2}[\delta_m^3-\delta_{m-1}^3-\tfrac{3}{2}(\delta_m^2-\delta_{m-1}^2)+\tfrac{3}{4}(\delta_m-\delta_{m-1})],$$

$$\langle zb_{11}^{*22}\rangle=\tfrac{1}{3}\sum_{m=1}^{M}\frac{\nu_m E_m}{1-\nu_m^2}[\delta_m^3-\delta_{m-1}^3-\tfrac{3}{2}(\delta_m^2-\delta_{m-1}^2)+\tfrac{3}{4}(\delta_m-\delta_{m-1})], \tag{15.104}$$

$$\langle zb_{12}^{*12}\rangle=\tfrac{1}{3}\sum_{m=1}^{M}\frac{E_m}{2(1+\nu_m)}[\delta_m^3-\delta_{m-1}^3-\tfrac{3}{2}(\delta_m^2-\delta_{m-1}^2)+\tfrac{3}{4}(\delta_m-\delta_{m-1})],$$

after averaging over the thickness. Equations (15.101), (15.102), and (15.104), together with the very general relations (15.75) and (15.76) proved above, enable us to determine all the effective elastic moduli of the laminated shell under study.

16. THERMAL CONDUCTIVITY OF A CURVED THIN SHELL OF A REGULARLY NON-HOMOGENEOUS MATERIAL WITH CORRUGATED SURFACES

16.1. Orthogonal coordinate formulation

We consider here the problem of heat conductance for a thin, curved, regularly non-homogeneous (i.e. composite) shell described in Section 15.

We assume that the physical components of the heat flow vector, q_i, and the temperature, θ, are related by Fourier's law, which in terms of the orthogonal coordinates α_1, α_2 and γ takes the form (cf. (12.4))

$$q_i=-\lambda_{i\mu}\frac{1}{H_\mu}\frac{\partial\theta}{\partial\alpha_\mu}-\lambda_{i3}\frac{\partial\theta}{\partial\gamma}, \tag{16.1}$$

where λ_{ij} are the heat conductivity coefficients, and where Latin indices range from 1 to 3, and Greek ones from 1 to 2.

In the approximation where the strain-related energy dissipation may be ignored, the heat balance equation can be written as (cf. (12.5))

$$-f+c_v\frac{\partial\theta}{\partial t}=-\frac{1}{H_1H_2}\left[\frac{\partial(H_2q_1)}{\partial\alpha_1}+\frac{\partial(H_1q_2)}{\partial\alpha_2}+\frac{\partial(H_1H_2q_3)}{\partial\gamma}\right], \tag{16.2}$$

where f denotes the density of the internal heat sources and c_v is the bulk heat capacity.

The heat transfer conditions to be satisfied on the top and bottom faces of the shell, $s^{\pm}$, are taken in the form (cf. (12.7))

$$n_\mu^{\pm}\frac{1}{H_\mu}q_\mu + n_3^{\pm}q_3 = \pm\alpha_S^{\pm}\theta \mp g_S^{\pm} \quad (\gamma=\gamma^{\pm}), \tag{16.3}$$

where $\alpha_S^{\pm}$ and $g_S^{\pm}$ are, respectively, the heat transfer coefficient and external heat flows on the surfaces $S^{\pm}$, and $n_i^{\pm}$, the components of the unit vectors normal to these surfaces, are given by equation (15.6). In the special case of convective heat transfer on the surfaces $S^{\pm}$, heat transfer conditions of the third kind (cf. (12.8))

$$g_S^{\pm} = \alpha_S^{\pm}\theta_S^{\pm}, \tag{16.4}$$

where $\theta_S^{\pm}$ is the value the ambient temperature takes on shell faces. If $\alpha_S^{\pm}=0$, (16.3) expresses the heat transfer conditions of the second kind.

The edge surface of the shell, which we denote by Σ, is a ruled surface whose directrix is the contour Γ bounding the shell mid-surface, and whose generatrix is normal to the mid-surface. The heat transfer conditions on this surface may be given in the form

$$q_\mu n_\mu^{\Gamma} = \alpha_\Sigma\theta - g_\Sigma(\boldsymbol{\alpha},\gamma,t) \quad (\alpha\in\Gamma), \tag{16.5}$$

which embraces the conditions of the second and third kinds and in which α_Σ is the heat transfer coefficient, g_Σ the external heat flow and n_μ^{Γ} the components of the unit vector normal to Σ. Alternatively, a condition of the first kind may be imposed:

$$\theta|_\Sigma = \theta_\Sigma(\boldsymbol{\alpha},\gamma,t) \quad (\boldsymbol{\alpha}\in\Gamma). \tag{16.6}$$

Specification of the initial temperature distribution,

$$\theta|_{t=0} = \theta^*, \tag{16.7}$$

completes the formulation of the heat conduction problem for the regularly non-homogeneous (composite) shell with a rapidly oscillating thickness, where we have assumed that all the thermal properties of the material, i.e. the quantities $\lambda_{ij}(y_1, y_2, z)$, $c_v(y_1, y_2, z)$ and $\alpha_S(y_1, y_2)$, are periodic functions of y_1 and y_2 with the unit cell Ω. The external factors $f(\boldsymbol{\alpha}, \boldsymbol{y}, z, t)$, $g_S^{\pm}(\boldsymbol{\alpha}, \boldsymbol{y}, t)$ and $\theta^*(\boldsymbol{\alpha}, \boldsymbol{y}, z)$ generally depend both on the 'slow' coordinates α_1 and α_2 and on $\boldsymbol{y}=(y_1, y_2)$, with the same periodicity cell.

16.2. Asymptotic analysis of the problem

We solve the heat conduction problem (16.1)–(16.7) by expressing the function θ in the form of a two-scale asymptotic expansion in powers of the small parameters, δ:

$$\begin{aligned}&\theta = \theta_1 + z\theta_2,\\ &\theta_\nu = \theta_\nu^{(0)}(\boldsymbol{\alpha},t) + \delta\theta_\nu^{(1)}(\boldsymbol{\alpha},t,\boldsymbol{y},z) + \delta^2\theta_\nu^{(2)}(\boldsymbol{\alpha},t,\boldsymbol{y},z) + \cdots,\end{aligned} \tag{16.8}$$

where the functions $\theta_\nu^{(l)}(\boldsymbol{\alpha},t,\boldsymbol{y},z)$ $(\nu=1,2;\ l=1,2,\ldots)$ are periodic in y_1 and y_2 with the unit cell Ω. The leading term of this expansion,

$$\theta^{(0)} = \theta_1^{(0)}(\boldsymbol{\alpha},t) + z\theta_2^{(0)}(\boldsymbol{\alpha},t),$$

corresponds to the linear thickness coordinate dependence of the temperature, a commonly adopted assumption in the treatment of the heat conduction of plate and shell structures (Podstrigach and Shvets 1978, Podstrigach *et al.* 1984).

The asymptotic behaviour of the mid-surface curvatures of the shell, see equations (15.9), may be conveniently rewritten in the form

$$k_\nu = \delta k'_\nu(\boldsymbol{\alpha}) \quad (\nu = 1, 2), \tag{16.9}$$

for the purposes of the present section, and in conditions (16.3) we put

$$\alpha_S^\pm = \delta\alpha^\pm(\boldsymbol{y}), \qquad g_S^\pm = \delta g^\pm(\boldsymbol{\alpha}, \boldsymbol{y}, t). \tag{16.10}$$

It is useful to note—and it will be demonstrated later on in our discussion—that the asymptotic forms (16.9) and (16.10) are equivalent to neglecting terms in k_1 and k_2 in the derivation of thin shell heat conduction equations and in the determination of reduced thermal characteristics of a thin-walled structure (see again Podstrigach and Shvets 1978, or Podstrigach *et al.* 1984).

In a manner similar to that discussed in Section 15, we now separate the 'rapid' and 'slow' coordinates for the purposes of differentiating, and we combine (16.1) and (16.8) to obtain:

$$q_i = \delta^{-1} q_i^{(-1)} + q_i^{(0)} + \delta q_i^{(1)} + \ldots, \tag{16.11}$$

where

$$\begin{aligned} q_i^{(-1)} &= -\lambda_{3i}\theta_2^{(0)}, \\ q_i^{(l)} &= -\lambda_{i\mu}\frac{1}{h_\mu A_\mu}\frac{\partial}{\partial y_\mu}(\theta_1^{(l+1)} + z\theta_2^{(l+1)}) \\ &\quad - \lambda_{i3}\frac{\partial}{\partial z}(\theta_1^{(l+1)} + z\theta_2^{(l+1)}) - \lambda_{i\mu}\frac{1}{A_\mu}\frac{\partial}{\partial \alpha_\mu}(\theta_1^{(l)} + z\theta_2^{(l)}) \quad (l = 0, 1). \end{aligned} \tag{16.12}$$

We now define the differential operators

$$\partial_1\varphi = \frac{1}{A_1 A_2}\frac{\partial(A_2\varphi)}{\partial\alpha_1}, \qquad \partial_2\varphi = \frac{1}{A_1 A_2}\frac{\partial(A_1\varphi)}{\partial\alpha_2} \tag{16.13}$$

and proceed to consistently expand (16.2) in powers of δ. In doing so, the expressions for the Lamé coefficients H_1 and H_2, the Gauss–Codazzi relations (15.2), and equations (16.8), (16.9) and (16.11)–(16.13) will be employed. For the right-hand side of (16.2), the expansion in terms of δ^m ($m \leqslant 0$) can be written as

$$\begin{aligned} &\frac{1}{H_1 H_2}\left[\frac{\partial(H_2 q_1)}{\partial\alpha_1} + \frac{\partial(H_1 q_2)}{\partial\alpha_2} + \frac{\partial(H_1 H_2 q_3)}{\partial\gamma}\right] \\ &= -\delta^{-1}\lambda_{3\mu}\partial_\mu\theta_2^{(0)} + \partial_\mu q_\mu^{(0)} + \delta^{-2}\frac{1}{h_\mu}\theta_2^{(0)}\frac{\partial\lambda_{3\mu}}{\partial\xi_\mu} + \delta^{-1}\frac{1}{h_\mu}\frac{\partial q_\mu^{(0)}}{\partial\xi_\mu} + \frac{1}{h_\mu}\frac{\partial q_\mu^{(1)}}{\partial\xi_\mu} \\ &\quad + z\frac{k'_\mu\theta_2^{(0)}}{h_\mu}\frac{\partial\lambda_{3\mu}}{\partial\xi_\mu} - \delta^{-2}\theta_2^{(0)}\frac{\partial\lambda_{33}}{\partial z} + \delta^{-1}\frac{\partial q_3^{(0)}}{\partial z} + \frac{\partial q_3^{(1)}}{\partial z} - (k'_1 + k'_2)\lambda_{33}\theta_2^{(0)}. \end{aligned}$$

Substituting this into (16.2) and using (16.8), we obtain the following expansion of (16.2) in powers of δ^m for $m \leqslant 0$:

$$-f + c_v\left(\frac{\partial\theta_1^{(0)}}{\partial t} + z\frac{\partial\theta_2^{(0)}}{\partial t}\right) = \delta^{-1}\lambda_{3\mu}\partial_\mu\theta_2^{(0)} - \partial_\mu q_\mu^{(0)} + (k_1' + k_2')\lambda_{33}\theta_2^{(0)}$$
$$-\frac{1}{h_\mu}\frac{\partial}{\partial\xi_\mu}\left(\delta^{-1}q_\mu^{(0)} + q_\mu^{(1)} + \lambda_{3\mu}k_\mu' z\theta_2^{(0)} - \delta^{-2}\lambda_{3\mu}\theta_2^{(0)}\right)$$
$$-\frac{\partial}{\partial z}(\delta^{-1}q_3^{(0)} + q_3^{(1)} - \delta^{-2}\lambda_{33}\theta_2^{(0)}). \tag{16.14}$$

To expand conditions (16.3) we use equations (15.21), (15.25) and (15.26) and relations (16.8)–(16.12). Neglecting terms of second and higher orders in δ, we find

$$n_\mu^{\pm}\frac{1}{H_\mu}q_\mu + n_3^{\pm}q_3 = \frac{N_i^{\pm}}{\omega^{\pm}}(-\delta^{-1}\lambda_{3i}\theta_2^{(0)} + q_i^{(0)} + \delta q_i^{(1)})$$
$$+\frac{N_\mu^{\pm}}{\omega^{\pm}}\delta\lambda_{3\mu}k_\mu' z^{\pm}\theta_2^{(0)} - \frac{N_i^{\pm}}{\omega^{\pm}}\delta\zeta^{\pm}\lambda_{3i}\theta_2^{(0)}.$$

If we substitute this into (16.3) and use equation (16.8) and (16.10), the expansion of the conditions at $z = z^{\pm}$ in powers of δ^m takes the form, for $m \leqslant 1$,

$$N_\mu^{\pm}(q_\mu^{(1)} + \delta^{-1}q_\mu^{(0)} + \lambda_{3\mu}k_\mu' z^{\pm}\theta_2^{(0)} - \delta^{-2}\lambda_{3\mu}\theta_2^{(0)})$$
$$+ N_3^{\pm}(q_3^{(1)} + \delta^{-1}q_3^{(0)} - \delta^{-2}\lambda_{33}\theta_2^{(0)})$$
$$= \pm\omega^{\pm}[\alpha^{\pm}(\theta_1^{(0)} + z^{\pm}\theta_2^{(0)}) - g^{\pm}] + N_i^{\pm}\zeta^{\pm}\lambda_{3i}\theta_2^{(0)} \quad (z = z^{\pm}), \tag{16.15}$$

where the functions $\zeta^{\pm}$, $N_i^{\pm}$ and $\omega^{\pm}$ are defined by formulae (15.21) (15.25) and (15.26), respectively.

Now equations (16.14) and (16.15) may be reduced to

$$\frac{1}{h_\mu}\frac{\partial q_\mu^{(l)}}{\partial\xi_\mu} + \frac{\partial q_3^{(l)}}{\partial z} = 0,$$
$$N_i^{\pm}q_i^{(l)} = 0, \quad (z = z^{\pm},\ l = 0, 1), \tag{16.16}$$

$$-f + c_v\left(\frac{\partial\theta_1^{(0)}}{\partial t} + z\frac{\partial\theta_2^{(0)}}{\partial t}\right) = \delta^{-1}\lambda_{3\mu}\partial_\mu\theta_2^{(0)} - \partial_\mu q_\mu^{(0)} + (k_1' + k_2')\lambda_{33}\theta_2^{(0)}$$
$$+\frac{1}{h_\mu}\frac{\partial}{\partial\xi_\mu}(\delta^{-2}\lambda_{3\mu}\theta_2^{(0)} - \lambda_{3\mu}k_\mu' z\theta_2^{(0)}) + \frac{\partial}{\partial z}(\delta^{-2}\lambda_{33}\theta_2^{(0)}), \tag{16.17}$$

$$N_\mu^{\pm}(\delta^{-2}\lambda_{3\mu}\theta_2^{(0)} - \lambda_{3\mu}k_\mu' z^{\pm}\theta_2^{(0)}) + N_3^{\pm}\delta^{-2}\lambda_{33}\theta_2^{(0)}$$
$$= \mp\omega^{\pm}[\alpha^{\pm}(\theta_1^{(0)} + z^{\pm}\theta_2^{(0)}) - g^{\pm}] - N_i^{\pm}\lambda_{3i}\zeta^{\pm}\theta_2^{(0)} \quad (z = z^{\pm}).$$

We now introduce the following notation:

$$L_i = \frac{\lambda_{i\nu}}{h_\nu}\frac{\partial}{\partial \xi_\nu} + \lambda_{i3}\frac{\partial}{\partial z}, \qquad D = \frac{1}{h_\mu}\frac{\partial}{\partial \xi_\mu} L_\mu + \frac{\partial}{\partial z} L_3, \tag{16.18}$$

$$\Lambda_i = \frac{1}{h_\mu}\frac{\partial \lambda_{i\mu}}{\partial \xi_\mu} + \frac{\partial \lambda_{i3}}{\partial z}. \tag{16.19}$$

Taking $l = 0$ in (16.16) and using $q_i^{(0)}$ from (16.12), we then obtain:

$$\begin{gathered} D(\theta_1^{(1)} + z\theta_2^{(1)}) = -\Lambda_\mu \frac{1}{A_\mu}\frac{\partial \theta_1^{(0)}}{\partial \alpha_\mu} - (\lambda_{3\mu} + z\Lambda_\mu)\frac{1}{A_\mu}\frac{\partial \theta_2^{(0)}}{\partial \alpha_\mu}, \\ N_i^\pm\left[L_i(\theta_1^{(1)} + z^\pm\theta_2^{(1)}) + \frac{\lambda_{i\mu}}{A_\mu}\left(\frac{\partial \theta_1^{(0)}}{\partial \alpha_\mu} + z^\pm \frac{\partial \theta_2^{(0)}}{\partial \alpha_\mu}\right)\right] = 0 \quad (z = z^\pm). \end{gathered} \tag{16.20}$$

We may give a solution of (16.20) in the form

$$\theta_1^{(1)} = W_\mu(\xi, z)\frac{1}{A_\mu}\frac{\partial \theta_1^{(0)}}{\partial \alpha_\mu}, \qquad z\theta_2^{(1)} = W_\mu^*(\xi, z)\frac{1}{A_\mu}\frac{\partial \theta_2^{(0)}}{\partial \alpha_\mu}, \tag{16.21}$$

provided that the functions $W_\mu(\xi, z)$ and $W_\mu^*(\xi, z)$ are periodic in ξ_1 and ξ_2 (with respective periods A_1 and A_2) and solve the problems

$$\begin{gathered} DW_\mu = -\Lambda_\mu, \\ N_i^\pm(L_i W_\mu + \lambda_{i\mu}) = 0 \quad (z = z^\pm), \end{gathered} \tag{16.22}$$

$$\begin{gathered} DW_\mu^* = -(\lambda_{3\mu} + z\Lambda_\mu), \\ N_i^\pm(L_i W_\mu^* + z^\pm \lambda_{i\mu}) = 0 \quad (z = z^\pm). \end{gathered} \tag{16.23}$$

Let

$$l_{i\mu}(\xi, z) = L_i W_\mu + \lambda_{i\mu}, \qquad l_{i\mu}^*(\xi, z) = L_i W_\mu^* + z\lambda_{i\mu}. \tag{16.24}$$

Then, from (16.12) with $l = 0$, and using (16.21), we find that

$$q_i^{(0)} = -l_{i\mu}\frac{1}{A_\mu}\frac{\partial \theta_1^{(0)}}{\partial \alpha_\mu} - l_{i\mu}^*\frac{1}{A_\mu}\frac{\partial \theta_2^{(0)}}{\partial \alpha_\mu}. \tag{16.25}$$

Using the notation introduced in (16.18), (16.19) and (16.24), the problems (16.22) and (16.23) may be rewritten as

$$\begin{cases} \dfrac{1}{h_\beta}\dfrac{\partial}{\partial \xi_\beta} l_{\beta\mu} + \dfrac{\partial}{\partial z} l_{3\mu} = 0, \\ N_i^\pm l_{i\mu} = 0 \quad (z = z^\pm), \end{cases} \tag{16.26}$$

$$\begin{cases} \dfrac{1}{h_\beta}\dfrac{\partial}{\partial \xi_\beta} l_{\beta\mu}^* + \dfrac{\partial}{\partial z} l_{3\mu}^* = 0, \\ N_i^\pm l_{i\mu}^* = 0 \quad (z = z^\pm). \end{cases} \tag{16.27}$$

Equations (16.26) and (16.27) are the local problems of heat conduction theory. They are very similar in form to the local problems of the theory of elasticity, i.e. equations (15.55) and (15.56), and their solutions are unique up to a constant term, an ambiguity which is removed by imposing the conditions

$$\langle W_\mu \rangle_\xi = 0, \qquad \langle W_\mu^* \rangle_\xi = 0 \quad \text{when } z = 0 \ (\mu = 1, 2), \tag{16.28}$$

analogous to those given by (15.57). Together with (16.8) and (16.21), these conditions elucidate the meaning of the function $\theta_1^{(0)}$ which characterizes the distribution of temperature over the mid-surface of the shell, $z = 0$.

The conditions for the faces $z = z^\pm$ involved in the local problems (16.26) and (16.27) may be represented in the same form as equations (16.58) and (16.59), namely

$$\begin{gathered} \frac{1}{h_\beta} n_\beta^{\pm(\xi)} l_{\beta\mu} + n_3^{\pm(\xi)} l_{3\mu} = 0 \quad (z = z^\pm), \\ l_{i\mu} \leftrightarrow l_{i\mu}^*. \end{gathered} \tag{16.29}$$

As was the case in Section 15, the local heat conduction problems may also be generalized to the case of piecewise-smooth heat conduction coefficients, $\lambda_{ij}(y, z)$, with discontinuities of the first kind on the (non-intersecting) interfaces between dissimilar phases of the composite. The formulations of the local problems are then augmented by the conditions of continuity on the interfaces:

$$[\![W_\mu]\!] = 0, \qquad \left[\!\!\left[\frac{1}{h_\beta} n_\beta^{(k)} l_{\beta\mu} + n_3^{(k)} l_{3\mu} \right]\!\!\right] = 0, \tag{16.30}$$

$$[\![W_\mu^*]\!] = 0, \qquad \left[\!\!\left[\frac{1}{h_\beta} n_\beta^{(k)} l_{\beta\mu}^* + n_3^{(k)} l_{3\mu}^* \right]\!\!\right] = 0, \tag{16.31}$$

which are analogous to conditions (15.60) and (15.61) and correspond to an ideal thermal contact in the sense that both the temperature and the heat flow vector have no jumps at the surface of a contact.

16.3. Governing equations for the heat conduction of a homogenized shell

To derive homogenized heat conduction equations use will be made of equations (15.28). We begin by averaging equation (16.17), using conditions (16.17) at $z = z^\pm$. Multiplying by δ and employing the notation described in (16.9) and (16.10), we find the first of the governing heat conduction equations for a homogenized shell:

$$\begin{aligned} -\delta\langle f \rangle + \delta\langle c_v \rangle \frac{\partial \theta_1^{(0)}}{\partial t} + \delta\langle z c_v \rangle \frac{\partial \theta_2^{(0)}}{\partial t} \\ = \langle \lambda_{3\mu} \rangle \partial_\mu \theta_2^{(0)} - \delta \partial_\mu \langle q_\mu^{(0)} \rangle - \mathfrak{J}_0 \theta_1^{(0)} \\ - [\mathfrak{J}_1 - (k_1 + k_2)\langle \lambda_{33} \rangle + \delta Z_0] \theta_2^{(0)} + R_0, \end{aligned} \tag{16.32}$$

where

$$\mathfrak{I}_m = \int_{-1/2}^{1/2}\int_{-1/2}^{1/2} [(z^+)^m\omega^+\alpha_S^+ + (z^-)^m\omega^-\alpha_S^-]\mathrm{d}y_1\mathrm{d}y_2 \quad (m=0,1,2),$$

$$R_m = \int_{-1/2}^{1/2}\int_{-1/2}^{1/2} [(z^+)^m\omega^+ g_S^+ + (z^-)^m\omega^- g_S^-]\mathrm{d}y_1\mathrm{d}y_2 \quad (m=0,1), \tag{16.33}$$

$$Z_m = \int_{-1/2}^{1/2}\int_{-1/2}^{1/2} [(z^+)^m N_i^+\lambda_{3i}(y,z^+)\zeta^+ - (z^-)^m N_i^-\lambda_{3i}(y,z^-)\zeta^-]\,\mathrm{d}y_1\mathrm{d}y_2$$

The second equation, for the functions $\theta_1^{(0)}(\boldsymbol{\alpha},t)$ and $\theta_2^{(0)}(\boldsymbol{\alpha},t)$, may be derived by averaging (16.17) after first multiplying it by z. Multiplying by z the conditions (16.17) at $z=z^\pm$ we obtain, using (15.28):

$$\begin{aligned} -\delta\langle zf\rangle + \delta\langle zc_v\rangle\frac{\partial\theta_1^{(0)}}{\partial t} + \delta\langle z^2c_v\rangle\frac{\partial\theta_2^{(0)}}{\partial t} &= \langle z\lambda_{3\mu}\rangle\partial_\mu\theta_2^{(0)} - \delta\partial_\mu\langle zq_\mu^{(0)}\rangle - \mathfrak{I}_1\theta_1^{(0)} \\ &\quad - [\mathfrak{I}_2 - (k_1+k_2)\langle z\lambda_{33}\rangle + \delta^{-1}\langle\lambda_{33}\rangle + \delta Z_1]\theta_2^{(0)} + R_1. \end{aligned} \tag{16.34}$$

The components $\langle q_\mu^{(0)}\rangle$ and $\langle zq_\mu^{(0)}\rangle$ involved in (16.32) and (16.34) can be evaluated by averaging relations (16.35) to give

$$\begin{aligned} \langle q_\mu^{(0)}\rangle &= -\langle l_{\mu\nu}\rangle\frac{1}{A_\nu}\frac{\partial\theta_1^{(0)}}{\partial\alpha_\nu} - \langle l^*_{\mu\nu}\rangle\frac{1}{A_\nu}\frac{\partial\theta_2^{(0)}}{\partial\alpha_\nu}, \\ \langle zq_\mu^{(0)}\rangle &= -\langle zl_{\mu\nu}\rangle\frac{1}{A_\nu}\frac{\partial\theta_1^{(0)}}{\partial\alpha_\nu} - \langle zl^*_{\mu\nu}\rangle\frac{1}{A_\nu}\frac{\partial\theta_2^{(0)}}{\partial\alpha_\nu}. \end{aligned} \tag{16.35}$$

Using (16.26) and (16.27) it is readily proved that

$$\langle l_{3\mu}\rangle = \langle zl_{3\mu}\rangle = \langle l^*_{3\mu}\rangle = \langle zl^*_{3\mu}\rangle = 0, \tag{16.36}$$

and therefore, by (16.25),

$$\langle q_3^{(0)}\rangle = \langle zq_3^{(0)}\rangle = 0.$$

Substitution of (16.35) into (16.32) and (16.34) now gives a system of two governing equations for $\theta_2^{(0)}(\boldsymbol{\alpha},t)$ and $\theta_2^{(0)}(\boldsymbol{\alpha},t)$ which, together with the solutions of the local problems (16.26)–(16.31), enable us to estimate closely the three-dimensional local structure of the temperature and heat flow fields of the problem. The necessary formulae follow from (16.8), (16.11), (16.12), (16.21) and (16.25) as

$$\begin{aligned} \theta &= \theta_1^{(0)} + z\theta_2^{(0)} + \delta\left(W_\mu\frac{1}{A_\mu}\frac{\partial\theta_1^{(0)}}{\partial\alpha_\mu} + W^*_\mu\frac{1}{A_\mu}\frac{\partial\theta_2^{(0)}}{\partial\alpha_\mu}\right) + \ldots, \\ q_i &= -\delta^{-1}\lambda_{i3}\theta_2^{(0)} - l_{i\mu}\frac{1}{A_\mu}\frac{\partial\theta_1^{(0)}}{\partial\alpha_\mu} - l^*_{i\mu}\frac{1}{A_\mu}\frac{\partial\theta_2^{(0)}}{\partial\alpha_\mu} + \ldots. \end{aligned} \tag{16.37}$$

The functions $\theta_1^{(2)}$ and $\theta_2^{(2)}$ are determined from the higher order local problems which result from combining the problem (16.16) for $l=1$ with the expression for $q_i^{(1)}$ from (16.12). These problems will not be considered here, however, because the accuracy provided by equations (16.37) is quite enough for our purposes.

We next turn our attention to the intial and boundary conditions at the contour Γ. To derive them, we average equations (16.5)–(16.7) and retain only the leading terms in the expansions for the temperature and the heat flow components, i.e. equations (16.37). In the case of boundary equations of the second or third kind, we see from (16.5) that

$$(-\delta^{-1}\langle z^m\lambda_{\mu 3}\rangle\theta_2^{(0)}+\langle z^m q_\mu^{(0)}\rangle)n_\mu^\Gamma=\langle z^m\alpha_\Sigma\rangle\theta_1^{(0)}$$
$$+\langle z^{m+1}\alpha_\Sigma\rangle\theta_2^{(0)}-\langle z^m g_\Sigma\rangle,\quad \boldsymbol{\alpha}\in\Gamma\quad(m=0,1), \tag{16.38}$$

and for boundary conditions of the first kind we find

$$\langle z^m\rangle\theta_1^{(0)}+\langle z^{m+1}\rangle\theta_2^{(0)}=\langle z^m\theta_\Sigma\rangle,\quad \boldsymbol{\alpha}\in\Gamma\quad(m=0,1), \tag{16.39}$$

using (16.6). The initial conditions for the functions $\theta_1^{(0)}$ and $\theta_2^{(0)}$ follow from (16.7) as

$$(\langle z^m\rangle\theta_1^{(0)}+\langle z^{m+1}\rangle\theta_2^{(0)})|_{t=0}=\langle z^m\theta^*\rangle\quad(m=0,1). \tag{16.40}$$

16.4. Symmetry of the effective heat conduction properties

By the effective heat conduction properties we mean the quantities $\langle l_{\mu\nu}\rangle$, $\langle zl_{\mu\nu}\rangle$, $\langle l_{\mu\nu}^*\rangle$ and $\langle zl_{\mu\nu}^*\rangle$ $(\mu,\nu=1,2)$, which enter as coefficients in equations (16.35) and are determined from two local problems, one of which is posed by equations (16.24), (16.26), (16.28) and (16.30), and the other by equations (16.24), (16.27), (16.28) and (16.31). Note that, quite analogous to Section 15, these problems involve the coefficients $A_1(\boldsymbol{\alpha})$ and $A_2(\boldsymbol{\alpha})$ of the first quadratic form of the shell mid-surface and the effective properties may therefore depend on the slow coordiantes α_1 and α_2, even in the case of an originally homogeneous material. Our object here is to prove that

$$\langle l_{\mu\nu}\rangle=\langle l_{\nu\mu}\rangle,\qquad \langle zl_{\mu\nu}\rangle=\langle l_{\nu\mu}^*\rangle,\qquad \langle zl_{\mu\nu}^*\rangle=\langle zl_{\nu\mu}^*\rangle. \tag{16.41}$$

From (16.18) and (16.19), using the auxiliary coordinates ζ_i defined in Section 15, we find

$$\boldsymbol{L}_i=\lambda_{ir}\frac{\partial}{\partial\zeta_r},\qquad \boldsymbol{D}=\frac{\partial}{\partial\zeta_p}\lambda_{pr}\frac{\partial}{\partial\zeta_r},\qquad \Lambda_i=\frac{\partial\lambda_{ir}}{\partial\zeta_r}.$$

Now if we redefine $l_{i\mu}$ and $l_{i\mu}^*$ for the entire range of indices, that is, for 1, 2, and 3, we obtain:

$$l_{ij}=\lambda_{ir}\left(\frac{\partial W_j}{\partial\zeta_r}+\delta_{rj}\right),\qquad l_{ij}^*=\lambda_{ir}\left(\frac{\partial W_j^*}{\partial\zeta_r}+z\delta_{rj}\right) \tag{16.42}$$

from (16.24), and the local problems (16.26) and (16.27) can be rewritten as

$$\frac{\partial l_{pj}}{\partial \zeta_p} = 0, \qquad N_p^{\pm} l_{pj} = 0 \quad (z = z^{\pm}), \qquad l_{pj} \leftrightarrow l^*_{pj}, \tag{16.43}$$

which, combined with (15.28), leads to the results

$$\left\langle l_{pj} \frac{\partial \varphi}{\partial \zeta_p} \right\rangle = 0, \qquad \left\langle l^*_{pj} \frac{\partial \varphi}{\partial \zeta_p} \right\rangle = 0, \tag{16.44}$$

which hold for any function $\varphi(\zeta_1, \zeta_2, \zeta_3)$ periodic in ζ_1 and ζ_2 with the unit cell Ω.

Setting $\varphi = W_i$ and then $\varphi = W^*_i$ in (16.44) and using (16.43), it is readily verified that

$$\begin{aligned}
\langle l_{ij} \rangle &= \left\langle \lambda_{pr} \left(\frac{\partial W_i}{\partial \zeta_p} + \delta_{pi} \right) \left(\frac{\partial W_j}{\partial \zeta_r} + \delta_{rj} \right) \right\rangle, \\
\langle z l^*_{ij} \rangle &= \left\langle \lambda_{pr} \left(\frac{\partial W^*_i}{\partial \zeta_p} + z\delta_{pi} \right) \left(\frac{\partial W^*_j}{\partial \zeta_r} + z\delta_{rj} \right) \right\rangle \\
\langle z l_{ij} \rangle &= \left\langle \lambda_{pr} \left(\frac{\partial W_j}{\partial \zeta_r} + \delta_{rj} \right) \left(\frac{\partial W^*_i}{\partial \zeta_p} + z\delta_{pi} \right) \right\rangle, \\
\langle l^*_{ij} \rangle &= \left\langle \lambda_{pr} \left(\frac{\partial W_i}{\partial \zeta_p} + \delta_{pi} \right) \left(\frac{\partial W^*_j}{\partial \zeta_r} + z\delta_{rj} \right) \right\rangle,
\end{aligned} \tag{16.45}$$

and the symmetry properties (16.41) now immediately follow from (16.45) and the symmetry of the heat conductivity tensor, λ_{pr}. If one or more of the indices is equal to 3, it is seen that both sides of equations (16.45) turn to zero in view of (16.36).

16.5. Homogenized heat conduction problem versus results from a thin-walled model

In this section we apply the above analysis to the limiting case of a homogeneous shell of constant thickness, which implies that the quantities λ_{ij}, c_v and $\alpha_S^{\pm}$ are constant and $F^{\pm}$ are identically zero. Because there is no dependence on ξ_1 and ξ_2 in this case, the local problems are solved without difficulty to give

$$W_\mu = -z \frac{\lambda_{3\mu}}{\lambda_{33}}, \qquad W^*_\mu = -\frac{z^2}{2} \frac{\lambda_{3\mu}}{\lambda_{33}}. \tag{16.46}$$

Substitution into (16.24) results in

$$l_{\mu\nu} = \lambda_{\mu\nu} - \frac{\lambda_{3\mu}\lambda_{3\nu}}{\lambda_{33}}, \qquad l^*_{\mu\nu} = z l_{\mu\nu}. \tag{16.47}$$

Now since $z^{\pm} = \pm\frac{1}{2}$ in this particular case, it follows from (16.35) that

$$\langle q_{\mu}^{(0)} \rangle = -l_{\mu\nu} \frac{1}{A_{\nu}} \frac{\partial \theta_1^{(0)}}{\partial \alpha_{\nu}}, \qquad \langle z q_{\mu}^{(0)} \rangle = -\frac{l_{\mu\nu}}{12} \frac{1}{A_{\nu}} \frac{\partial \theta_2^{(0)}}{\partial \alpha_{\nu}}. \tag{16.48}$$

Using (16.33), (16.47) and (16.48), equations (16.32) and (16.34) can be written in the form

$$\begin{aligned} -\delta\langle f \rangle + \delta c_{\nu} \frac{\partial \theta_1^{(0)}}{\partial t} &= \lambda_{3\mu} \partial_{\mu} \theta_2^{(0)} + \delta\left(\lambda_{\mu\nu} - \frac{\lambda_{3\mu}\lambda_{3\nu}}{\lambda_{33}}\right) \partial_{\mu}\left(\frac{1}{A_{\nu}} \frac{\partial \theta_1^{(0)}}{\partial \alpha_{\nu}}\right) \\ &\quad - (\alpha_S^+ + \alpha_S^-)\theta_1^{(0)} - \left[\frac{\alpha_S^+ - \alpha_S^-}{2} - (k_1 + k_2)\lambda_{33}\right]\theta_2^{(0)} + g_S^+ + g_S^-, \\ -12\delta\langle z f \rangle + \delta c_{\nu} \frac{\partial \theta_2^{(0)}}{\partial t} &= \delta\left(\lambda_{\mu\nu} - \frac{\lambda_{3\mu}\lambda_{3\nu}}{\lambda_{33}}\right) \partial_{\mu}\left(\frac{1}{A_{\nu}} \frac{\partial \theta_2^{(0)}}{\partial \alpha_{\nu}}\right) \\ &\quad - 6(\alpha_S^+ - \alpha_S^-)\theta_1^{(0)} - 3(\alpha_S^+ + \alpha_S^- + 4\delta^{-1}\lambda_{33})\theta_2^{(0)} + \frac{g_S^+ - g_S^-}{2}, \end{aligned} \tag{16.49}$$

which is the system of equations for the heat conduction problem of an anisotropic homogeneous shell (or plate). The quantities $g_S^{\pm}$ involved in (16.49) are determined by (16.4) if the heat transfer proceeds by convection across the faces of the shell.

Returning to (16.37), we see that the leading term in the temperature equation,

$$\theta \approx \theta_1^{(0)}(\boldsymbol{\alpha}, t) + \frac{\gamma}{\delta}\theta_2^{(0)}(\boldsymbol{\alpha}, t), \tag{16.50}$$

describes a linear dependence of the temperature on the thickness coordinate, and we may use this dependence to calculate the integral temperature characteristics usually employed in the analysis of heat conduction in thin plates and shells. From Podstrigach and Shvets (1978) (see also Podstrigach *et al.* 1984) these characteristics are

$$T = \frac{1}{\delta}\int_{-\delta/2}^{\delta/2} \theta \, d\gamma = \theta_1^{(0)}, \qquad T^* = \frac{6}{\delta^2}\int_{-\delta/2}^{\delta/2} \gamma\theta \, d\gamma = \frac{\theta_2^{(0)}}{2}, \tag{16.51}$$

and their substitution into (16.50) gives the relation between the temperature and its integral characteristics,

$$\theta \approx \mathrm{T} + \frac{2\gamma}{\delta}\mathrm{T}^*,$$

which is well known from the heat conduction theory of thin shells. Referring to the same authors, similar integral expressions for the density of heat sources and for its first moment may be constructed:

$$\Phi = \int_{-\delta/2}^{\delta/2} f \, d\gamma = \delta\langle f \rangle, \qquad \Phi^* = \frac{6}{\delta}\int_{-\delta/2}^{\delta/2} \gamma f \, d\gamma = 6\delta\langle z f \rangle. \tag{16.52}$$

Equations (16.50)–(16.52) thus show that the system (16.49) coincides with the analogous heat conduction equations known for the cases when

(a) the shell is homogeneous and isotropic ($\lambda_{ij} = \lambda\delta_{ij}$) (see Podstrigach and Shvets 1978), or when

(b) the plane shell is anisotropic, but there exists a thermal symmetry plane at each point of the shell which is parallel to the plane of the shell ($\lambda_{31} = \lambda_{32} = 0$, $k_1 = k_2 = 0$, $A_1 = A_2 = 1$); see Podstrigach *et al.* (1984).

16.6. Laminated shell composed of homogeneous anisotropic layers

In this section the effective heat conduction coefficients of the laminated shell of Fig. 15.3 will be evaluated under the assumption that the layers (or laminae) of the shell are anisotropic while being homogeneous.

Of the two local problems that must be considered, the first, expressed by equations (16.24), (16.26), (16.28) and (16.30), in this case takes the following form:

$$\begin{aligned}
&l_{i\mu} = \lambda_{i3}\frac{\mathrm{d}W_\mu}{\mathrm{d}z} + \lambda_{i\mu} \\
&\frac{\mathrm{d}l_{3\mu}}{\mathrm{d}z} = 0, \qquad l_{3\mu} = 0 \quad (z = \pm\tfrac{1}{2}), \\
&[\![W_\mu]\!] = 0, \qquad [\![l_{3\mu}]\!] = 0 \quad (z = \delta_m - \tfrac{1}{2};\ m = 1, 2, \ldots, M-1), \\
&W_\mu = 0 \quad (z = 0)
\end{aligned} \tag{16.53}$$

The relevant form of the second local problem, equations (16.24), (16.27), (16.28) and (16.31), is obtained from the above equations by writing

$$l^*_{i\mu} = \lambda_{i3}\frac{\mathrm{d}W^*_\mu}{\mathrm{d}z} + z\lambda_{i\mu} \tag{16.55}$$

in place of (16.53) and making the replacements

$$l_{3\mu} \to l^*_{3\mu} \quad \text{and} \quad W_\mu \to W^*_\mu$$

in (16.54).

Since $l_{3\mu} = 0$ and $l^*_{3\mu} = 0$ from the solution of the local problems, it follows that

$$\frac{\mathrm{d}W^{(m)}_\mu}{\mathrm{d}z} = -\frac{\lambda^{(m)}_{3\mu}}{\lambda^{(m)}_{33}}, \qquad \frac{\mathrm{d}W^{*(m)}_\mu}{\mathrm{d}z} = -z\frac{\lambda^{(m)}_{3\mu}}{\lambda^{(m)}_{33}}, \tag{16.56}$$

where the superscript m indicates the layer. Analogous to Section 15, here again the interface conditions and $z = 0$ conditions are satisfied by the appropriate choice of the constants of integration in equations (16.56).

The effective heat conduction coefficients of the laminated shell are obtained by substituting (16.56) into (16.53) and (16.55) and then averaging over the shell thickness to give

$$\langle l_{\beta\mu}\rangle = \sum_{m=1}^{M}\left(\lambda_{\beta\mu}^{(m)} - \frac{\lambda_{3\beta}^{(m)}\lambda_{3\mu}^{(m)}}{\lambda_{33}^{(m)}}\right)(\delta_m - \delta_{m-1}),$$

$$\langle zl_{\beta\mu}\rangle = \langle l_{\beta\mu}^{*}\rangle = \tfrac{1}{2}\sum_{m-1}^{M}\left(\lambda_{\beta\mu}^{(m)} - \frac{\lambda_{3\beta}^{(m)}\lambda_{3\mu}^{(m)}}{\lambda_{33}^{(m)}}\right)[\delta_m^2 - \delta_{m-1}^2 - (\delta_m - \delta_{m-1})],$$

$$\langle zl_{\beta\mu}^{*}\rangle = \tfrac{1}{3}\sum_{m=1}^{M}\left(\lambda_{\beta\mu}^{(m)} - \frac{\lambda_{3\beta}^{(m)}\lambda_{3\mu}^{(m)}}{\lambda_{33}^{(m)}}\right)[\delta_m^3 - \delta_{m-1}^3 - \tfrac{3}{2}(\delta_m^2 - \delta_{m-1}^2) + \tfrac{3}{4}(\delta_m - \delta_{m-1})]. \tag{16.57}$$

If the layer material is isotropic, then $\lambda_{ij}^{(m)} = \lambda^{(m)}\delta_{ij}$, with resulting simplifications in (16.57) (in particular, $\lambda_{3\beta}^{(m)} = 0$).

17. THERMOELASTICITY OF A CURVED SHELL OF REGULARLY NON-HOMOGENEOUS MATERIAL WITH CORRUGATED SURFACES

17.1. Governing thermoelastic equations for the homogenized shell

In Sections 15 and 16 we derived the asymptotic formalism necessary to treat the elasticity and heat conduction problems associated with a thin, curved, regularly non-homogeneous (or composite) shell with rapidly oscillating thickness. With the results obtained, we are now in a position to consider the problem of the thermoelasticity for such a structure.

As is customary in the theory of thermoelasticity, the stresses and temperature changes will be taken to be related by the usual Duhamel–Neumann law (see (2.14) and (12.2)),

$$\sigma_{ij} = c_{ijmn}(e_{mn} - \alpha_{mn}^{\mathrm{T}}\theta), \tag{17.1}$$

where α_{mn}^{T} are the thermal elongation and shear coefficients. We assume that the thermal relaxation time is much greater than the attenuation time associated with mechanical oscillations, and we accordingly limit our consideration to the quasi-static formulation of the problem. This means that inertial effects may be neglected in the equations of motion, and these therefore reduce to the equilibrium equations (15.3), with time as a parameter.

If the thermoelastic problem is an uncoupled one (which is the case we consider), the heat conduction problem set by equations (16.1)–(16.7) is solved independently. Its solution, the temperature θ, then enters into (15.3) by means of (17.1) and determines the additional external forces caused by the effects of thermal deformation.

Because there are rapidly oscillating components in the coefficients in equation (17.1), clearly a homogenization technique must be applied here, along lines similar to those discussed in Section 15, but with equation (17.1) instead of Hooke's law (15.4).

As will be seen in the following discussion, the stress equations (15.64) and (15.70) are not affected by the thermal term in (17.1), but in the elastic relations (15.71) and (15.72) additional terms associated with thermal deformations will appear. These are expressible in terms of $\theta_1^{(0)}(\boldsymbol{\alpha}, t)$ and $\theta_2^{(0)}(\boldsymbol{\alpha}, t)$, for which functions the homogenized problem (16.32)–(16.35) and (16.38)–(16.40) was formulated in the previous section.

Assuming the asymptotic form

$$\alpha_{mn}^{\mathrm{T}} = \delta\alpha_{mn}(\boldsymbol{y}, z), \tag{17.2}$$

where the functions $\alpha_{mn}(\boldsymbol{y}, z)$ are periodic in y_1 and y_2 with the unit cell Ω, and using (15.13), (15.14) and (16.8) in combination with (17.1), we obtain for $\sigma_{ij}^{(0)}$ an expression identical in form to (15.31). Proceeding in a manner parallel to that described in Section 15, we also retrieve equations (15.45) for the leading order terms of the displacement components, and relations (15.46) for the stresses, $\sigma_{ij}^{(0)}$.

From (17.1), (17.2), (15.15) and (16.8) it is found that

$$\sigma_{ij}^{(1)} = \boldsymbol{L}_{ijm} u_m^{(2)} + c_{ijmv}\varepsilon_{mv}^{(1)} + zc_{ij\mu v}\tau_{\mu v} - c_{ijmn}\alpha_{mn}(\theta_1^{(0)} + z\theta_2^{(0)}), \tag{17.3}$$

which differs from (15.47) in having terms corresponding to thermal stresses. As before, the functions $\varepsilon_{mv}^{(1)}$ and $\tau_{\mu v}$ are determined by equations (15.48) and (15.49).

We now substitute (17.3) into equation (15.22) and into conditions (15.22) with $l = 1$. Using (15.46) and following the notation introduced in (15.33)–(15.35) we arrive at

$$\begin{aligned} \boldsymbol{D}_{im} u_m^{(2)} &= -C_{imv}\varepsilon_{mv}^{(1)} - (c_{i3\mu v} + zC_{i\mu v})\tau_{\mu v} + B_i\theta_1^{(0)} + (\beta_{i3} + zB_i)\theta_2^{(0)}, \\ &N_j^{\pm}[\boldsymbol{L}_{ijm} u_m^{(2)} + c_{ijmv}\varepsilon_{mv}^{(1)} + z^{\pm} c_{ij\mu v}\tau_{\mu v} \\ &- \beta_{ij}(\theta_1^{(0)} + z^{\pm}\theta_2^{(0)})] = 0 \quad (z = z^{\pm}), \end{aligned} \tag{17.4}$$

where we define

$$\beta_{ij} = c_{ijmn}\alpha_{mn}, \qquad B_i = \frac{1}{h_v}\frac{\partial \beta_{iv}}{\partial \xi_v} + \frac{\partial \beta_{i3}}{\partial z}. \tag{17.5}$$

We will have satisfied (17.4) and secured periodicity in ξ_1 and ξ_2 (with respective periods A_1 and A_2) by writing

$$u_m^{(2)} = U_m^{lv}\varepsilon_{lv}^{(1)} + V_m^{\mu v}\tau_{\mu v} + S_m\theta_1^{(0)} + S_m^*\theta_2^{(0)} \tag{17.6}$$

for the solution, where $U_m^{lv}(\boldsymbol{\xi}, z)$ and $V_m^{\mu v}(\boldsymbol{\xi}, z)$ are the solutions of the local elastic problems (15.39) and (15.52), and the functions $S_m(\boldsymbol{\xi}, z)$ and $S_m^*(\boldsymbol{\xi}, z)$ are periodic in ξ_1 and ξ_2 (with periods indicated) and solve the problems

$$\boldsymbol{D}_{im} S_m = B_i, \qquad (\boldsymbol{L}_{ijm} S_m - \beta_{ij})N_j^{\pm} = 0 \quad (z = z^{\pm}), \tag{17.7}$$

and

$$\begin{aligned} &\boldsymbol{D}_{im} S_m^* = \beta_{i3} + zB_i \\ &(\boldsymbol{L}_{ijm} S_m^* - z^{\pm}\beta_{ij})N_j^{\pm} = 0 \quad (z = z^{\pm}). \end{aligned} \tag{17.8}$$

Now from (17.3), denoting

$$s_{ij} = \beta_{ij} - \boldsymbol{L}_{ijm} S_m, \qquad s_{ij}^* = z\beta_{ij} - \boldsymbol{L}_{ijm} S_m^*, \tag{17.9}$$

and using (17.6) and (17.9) along with (15.41), (15.42) and (15.53), we obtain:

$$\sigma_{ij}^{(1)} = b_{ij}^{\mu\nu}\, \varepsilon_{\mu\nu}^{(1)} + b_{ij}^{*\mu\nu} \tau_{\mu\nu} - s_{ij}\theta_1^{(0)} - s_{ij}^{*}\theta_2^{(0)}, \tag{17.10}$$

which generalizes (15.54) to include the effects of thermal stresses on material behaviour.

Equations (17.7) and (17.8) represent the local problems of thermoelasticity and may be written in the form

$$\frac{1}{h_\mu}\frac{\partial}{\partial \xi_\mu} s_{i\mu} + \frac{\partial}{\partial z} s_{i3} = 0, \qquad s_{ij}\, N_j^{\pm} = 0, \quad (z = z^{\pm}), \quad (s_{ij} \leftrightarrow s_{ij}^{*}), \tag{17.11}$$

using (17.5) and (17.9). From this, multiplying by z and z^2 and averaging by means of (15.28), we find

$$\langle s_{i3} \rangle = \langle z s_{i3} \rangle = \langle s_{i3}^{*} \rangle = \langle z s_{i3}^{*} \rangle = 0, \tag{17.12}$$

and using this last equation together with (15.62), it follows from (17.10) that

$$\langle \sigma_{i3}^{(1)} \rangle = \langle z \sigma_{i3}^{(1)} \rangle = 0,$$

indicating that formulae (15.63) retain their truth in this case. Since, furthermore, relations (15.46) and (15.63) remain unchanged, so do equations (15.64) and (15.70), because these latter were derived by essentially using the former in Section 15. Instead of the elastic relations (15.71) and (15.72) for the homogenized equations, from (17.10) we find

$$\begin{aligned} \langle \sigma_{\alpha\kappa}^{(1)} \rangle &= \langle b_{\alpha\kappa}^{\mu\nu} \rangle \varepsilon_{\mu\nu}^{(1)} + \langle b_{\alpha\kappa}^{*\mu\nu} \rangle \tau_{\mu\nu} - \langle s_{\alpha\kappa} \rangle \theta_1^{(0)} - \langle s_{\alpha\kappa}^{*} \rangle \theta_2^{(0)}, \\ \langle z\sigma_{\alpha\kappa}^{(1)} \rangle &= \langle z b_{\alpha\kappa}^{\mu\nu} \rangle \varepsilon_{\mu\nu}^{(1)} + \langle z b_{\alpha\kappa}^{*\mu\nu} \rangle \tau_{\mu\nu} - \langle z s_{\alpha\kappa} \rangle \theta_1^{(0)} - \langle z s_{\alpha\kappa}^{*} \rangle \theta_2^{(0)}. \end{aligned} \tag{17.13}$$

Now if this is substituted into (15.64) and (15.70) we obtain, using (15.48) and (15.49), a system of three governing equations for the functions $v_1^{(1)}(\boldsymbol{\alpha}, t)$, $v_2^{(1)}(\boldsymbol{\alpha}, t)$ and $w(\boldsymbol{\alpha}, t)$ determining (at leading order) the components of the displacement vector (15.45). The functions $\theta_1^{(0)}(\boldsymbol{\alpha}, t)$ and $\theta_2^{(0)}(\boldsymbol{\alpha}, t)$, also involved in these equations, are found from the homogenized heat conduction problem.

17.2. Effective properties

The local thermoelastic problems posed by equations (17.9) and (17.12) are similar to the problems (15.41) and (15.55), and (15.53) and (15.56), and their solutions are unique only up to constant terms. This ambiguity is removed by imposing the conditions

$$\langle S_m \rangle_\xi = 0, \qquad \langle S_m^{*} \rangle_\xi = 0 \quad \text{when } z = 0, \tag{17.14}$$

in accordance with (17.6) and analogous to (15.57).

The $z = z^{\pm}$ conditions relevant to the local problems (17.9) and (17.11) may be written in a form analogous to (15.58) and (15.59). That is,

$$\frac{1}{h_\mu} n_\mu^{\pm(\xi)} s_{i\mu} + n_3^{\pm(\xi)} s_{i3} = 0 \quad (z = z^{\pm}),$$

$$s_{ij} \leftrightarrow s_{ij}^{*}. \tag{17.15}$$

If the functions $\alpha_{mn}(y,z)$ and $c_{ijmn}(y,z)$ are piecewise smooth and undergo discontinuities of the first kind at the contact surfaces between dissimilar components of the composite, the following interface continuity conditions should be added:

$$\begin{aligned} [\![S_m]\!] &= 0, \qquad \left[\!\!\left[\frac{1}{h_\mu} n_\mu^{(k)} s_{i\mu} + n_3^{(k)} s_{i3}\right]\!\!\right] = 0, \\ [\![S_m^*]\!] &= 0, \qquad \left[\!\!\left[\frac{1}{h_\mu} n_\mu^{(k)} s_{i\mu}^* + n_3^{(k)} s_{i3}^*\right]\!\!\right] = 0, \end{aligned} \tag{17.16}$$

which are readily recognizable as having the same form as (15.60) and (15.61).

Equations (17.9), (17.11) and (17.14)–(17.16) represent a complete formulation of the local thermoelastic problems, and their solutions enable us to reveal the local structure of the thermal stresses in (17.10) and to calculate the effective properties $\langle s_{\theta\kappa}\rangle$, $\langle s_{\theta\kappa}^*\rangle$, $\langle z s_{\theta\kappa}\rangle$ and $\langle z s_{\theta\kappa}^*\rangle$ occurring in (17.13).

By comparing the local problems of thermoelasticity with those of the theory of elasticity, it can be shown that the following equations hold:

$$\begin{aligned} \delta\langle s_{\theta\kappa}\rangle &= \langle \alpha_{ij}^{\mathrm{T}} b_{ij}^{\theta\kappa}\rangle, \qquad \delta\langle z s_{\theta\kappa}\rangle = \langle \alpha_{ij}^{\mathrm{T}} b_{ij}^{*\theta\kappa}\rangle, \\ \delta\langle s_{\theta\kappa}^*\rangle &= \langle z\alpha_{ij}^{\mathrm{T}} b_{ij}^{\theta\kappa}\rangle, \qquad \delta\langle z s_{\theta\kappa}^*\rangle = \langle z\alpha_{ij}^{\mathrm{T}} b_{ij}^{*\theta\kappa}\rangle, \end{aligned} \tag{17.17}$$

with the implication that it suffices to know the solutions of the local problems of ordinary elasticity in order to derive all the effective thermoelastic properies of the system.

In proving (17.17), the notation developed in Section 15 will be employed, Rewriting (17.11) in the form

$$\frac{\partial s_{ij}}{\partial \zeta_j} = 0, \qquad N_j^{\pm} s_{ij} = 0 \quad \text{when } z = z^{\pm}, \qquad s_{ij} \leftrightarrow s_{ij}^*, \tag{17.18}$$

and following the procedure similar to that outlined in Section 15, we find that for any function $\varphi(\zeta_1, \zeta_2, \zeta_3)$, periodic in ζ_1 and ζ_2 with the unit cell Ω.

$$\left\langle s_{qp} \frac{\partial \varphi}{\partial \zeta_p} \right\rangle = 0, \quad \left\langle s_{qp}^* \frac{\partial \varphi}{\partial \zeta_p} \right\rangle = 0. \tag{17.19}$$

Putting $\varphi = U_q^{\theta\kappa}$ and using (17.9), it follows from the first of (17.19) that

$$0 = \left\langle s_{qp} \frac{\partial U_q^{\theta\kappa}}{\partial \zeta_p} \right\rangle = \left\langle \beta_{qp} \frac{\partial U_q^{\theta\kappa}}{\partial \zeta_p} \right\rangle - \left\langle c_{qplr} \frac{\partial S_l}{\partial \zeta_r} \frac{\partial U_q^{\theta\kappa}}{\partial \zeta_p} \right\rangle,$$

giving

$$\left\langle \beta_{qp} \frac{\partial U_q^{\theta\kappa}}{\partial \zeta_p} \right\rangle = \left\langle c_{qplr} \frac{\partial S_l}{\partial \zeta_r} \frac{\partial U_q^{\theta\kappa}}{\partial \zeta_p} \right\rangle. \tag{17.20}$$

Similarly, by inserting $\varphi = S_q$ into the first of (15.81) we have

$$0 = \left\langle b_{qp}^{\theta\kappa} \frac{\partial S_q}{\partial \zeta_p} \right\rangle = \left\langle c_{qp\theta\kappa} \frac{\partial S_q}{\partial \zeta_p} \right\rangle + \left\langle c_{qplr} \frac{\partial U_l^{\theta\kappa}}{\partial \zeta_r} \frac{\partial S_q}{\partial \zeta_p} \right\rangle,$$

which yields

$$\left\langle c_{qp\theta\kappa} \frac{\partial S_q}{\partial \zeta_p} \right\rangle = -\left\langle c_{qplr} \frac{\partial U_l^{\theta\kappa}}{\partial \zeta_r} \frac{\partial S_q}{\partial \zeta_p} \right\rangle. \tag{17.21}$$

By comparing (17.20) and (17.21) we see that

$$\left\langle \beta_{qp} \frac{\partial U_q^{\theta\kappa}}{\partial \zeta_p} \right\rangle = -\left\langle c_{qp\theta\kappa} \frac{\partial S_q}{\partial \zeta_p} \right\rangle, \tag{17.22}$$

using the symmetry of the elastic coefficients c_{qplr}, and therefore, by (15.41) the first of (17.5) and (17.9):

$$\begin{aligned} \langle s_{\theta\kappa} \rangle &= \left\langle \beta_{\theta\kappa} - c_{\theta\kappa qp} \frac{\partial S_q}{\partial \zeta_p} \right\rangle = \langle \beta_{\theta\kappa} \rangle \\ &+ \left\langle \beta_{qp} \frac{\partial U_q^{\theta\kappa}}{\partial \zeta_p} \right\rangle = \left\langle \alpha_{ij} c_{ij\theta\kappa} + \alpha_{ij} c_{ijqp} \frac{\partial U_q^{\theta\kappa}}{\partial \zeta_p} \right\rangle \\ &= \langle \alpha_{ij} b_{ij}^{\theta\kappa} \rangle, \end{aligned}$$

which, in view of (17.2), proves the first of relations (17.17).

Now if we substitute $\varphi = S_q$ into the second of equations (15.81) and $\varphi = V_q^{\theta\kappa}$ into the first of (17.19), we obtain:

$$\left\langle \beta_{qp} \frac{\partial V_q^{\theta\kappa}}{\partial \zeta_p} \right\rangle = -\left\langle z c_{qp\theta\kappa} \frac{\partial S_q}{\partial \zeta_p} \right\rangle,$$

which leads to the second of relations (17.17) after using (15.53), (17.2), (17.5) and (17.9). The third of equations (17.17) is arrived at by setting $\varphi = S_q^*$ in the first of (15.81) and $\varphi = U_q^{\theta\kappa}$ in the second of (17.19). This gives

$$\left\langle z\beta_{qp} \frac{\partial U_q^{\theta\kappa}}{\partial \zeta_p} \right\rangle = -\left\langle c_{qp\theta\kappa} \frac{\partial S_q^*}{\partial \zeta_p} \right\rangle,$$

which must then be combined with (15.41), (17.2), (17.5) and (17.9). Finally substitute $\varphi = S_q^*$ into the second of (15.81) and $\varphi = V_q^{\theta\kappa}$ into the second of (17.19) to obtain:

$$\left\langle z\beta_{qp} \frac{\partial V_q^{\theta\kappa}}{\partial \zeta_p} \right\rangle = -\left\langle z c_{qp\theta\kappa} \frac{\partial S_q^*}{\partial \zeta_p} \right\rangle.$$

The last of relations (17.17) then follows with the help of (15.53), (17.2), (17.5) and (17.9).

17.3. Homogenized thermoelastic problem versus thin-shell thermoelasticity results

In Section 15 we were able to rewrite the elastic relations (15.71) and (15.72) in the form of (15.88) using the stress resultants N and moment resultants M defined in (15.85). Exactly analogous constitutive relations may be derived for the homogenized shell in the context of thermoelasticity theory. Using (15.85) again, equations (17.13) become

$$\begin{aligned} N_\beta &= \delta\langle b^{\mu\nu}_{\beta\beta}\rangle\varepsilon_{\mu\nu} + \delta^2\langle b^{*\mu\nu}_{\beta\beta}\rangle\tau_{\mu\nu} - \delta^2\langle s_{\beta\beta}\rangle\theta^{(0)}_1 - \delta^2\langle s^*_{\beta\beta}\rangle\theta^{(0)}_2,\\ N_{12} &= \delta\langle b^{\mu\nu}_{12}\rangle\varepsilon_{\mu\nu} + \delta^2\langle b^{*\mu\nu}_{12}\rangle\tau_{\mu\nu} - \delta^2\langle s_{12}\rangle\theta^{(0)}_1 - \delta^2\langle s^*_{12}\rangle\theta^{(0)}_2,\\ M_\beta &= \delta^2\langle zb^{\mu\nu}_{\beta\beta}\rangle\varepsilon_{\mu\nu} + \delta^3\langle zb^{*\mu\nu}_{\beta\beta}\rangle\tau_{\mu\nu} - \delta^3\langle zs_{\beta\beta}\rangle\theta^{(0)}_1 - \delta^3\langle zs^*_{\beta\beta}\rangle\theta^{(0)}_2,\\ M_{12} &= \delta^2\langle zb^{\mu\nu}_{12}\rangle\varepsilon_{\mu\nu} + \delta^3\langle zb^{*\mu\nu}_{12}\rangle\tau_{\mu\nu} - \delta^3\langle zs_{12}\rangle\theta^{(0)}_1 - \delta^3\langle zs^*_{12}\rangle\theta^{(0)}_2, \end{aligned} \tag{17.23}$$

where β takes the values 1 and 2 and is not summed.

Over the middle surface of the shell ($z = 0$), the strains $\varepsilon_{\mu\nu}$ and $\tau_{\mu\nu}$, and displacements $v_1 = \delta v^{(1)}_1$, $v_2 = \delta v^{(1)}_2$ and w, will again be related by equations (15.86) and (15.49).

For the displacement components using (15.8), (15.45) and (17.6), we obtain the relations

$$\begin{aligned} u_1 &= v_1(\boldsymbol{\alpha}, t) - \frac{\gamma}{A_1}\frac{\partial w}{\partial \alpha_1} + \delta U^{\mu\nu}_1\varepsilon_{\mu\nu} + \delta^2 V^{\mu\nu}_1\tau_{\mu\nu} + \delta^2 S_1\theta^{(0)}_1 + \delta^2 S^*_1\theta^{(0)}_2 + \cdots,\\ u_2 &= v_2(\boldsymbol{\alpha}, t) - \frac{\gamma}{A_2}\frac{\partial w}{\partial \alpha_2} + \delta U^{\mu\nu}_2\varepsilon_{\mu\nu} + \delta^2 V^{\mu\nu}_2\tau_{\mu\nu} + \delta^2 S_2\theta^{(0)}_1 + \delta^2 S^*_2\theta^{(0)}_2 + \cdots,\\ u_3 &= w(\boldsymbol{\alpha}, t) + \delta U^{\mu\nu}_3\varepsilon_{\mu\nu} + \delta^2 V^{\mu\nu}_3\tau_{\mu\nu} + \delta^2 S_3\theta^{(0)}_1 + \delta^2 S^*_3\theta^{(0)}_2 + \cdots, \end{aligned} \tag{17.24}$$

which complement equations (15.87) derived in the framework of the theory of elasticity.

The functions $\theta^{(0)}_1(\boldsymbol{\alpha}, t)$ and $\theta^{(0)}_2(\boldsymbol{\alpha}, t)$ are found from the homogenized heat conduction problem and determine the temperature and heat flow vector by means of equations (16.37). The integral temperature characteristics needed for the thin shell (thin plate) theory of thermoelasticity (Podstrigach and Shvets 1978, Podstrigach *et al.* 1984) follow from (16.37) as

$$\begin{aligned} T &= \langle\theta\rangle \approx \theta^{(0)}_1 + \langle z\rangle\theta^{(0)}_2,\\ T^* &= 6\langle z\theta\rangle \approx 6\langle z\rangle\theta^{(0)}_1 + 6\langle z^2\rangle\theta^{(0)}_2 \end{aligned} \tag{17.25}$$

at leading order. Note that if the shell is of constant thickness, i.e. $z^\pm = \pm(\frac{1}{2})$, then $\langle z\rangle = 0$, $\langle z^2\rangle = \frac{1}{2}$ and (16.51) follows in an obvious way from (17.25).

It should be re-emphasized that (15.64) and (15.70), the homogenized shell equations expressed in terms of stresses, and equations (15.89), written in terms of forces and moments, remain unchanged in the framework of thermoelasticity theory and are identical to the corresponding thin-shell equations.

Let us consider now the limiting case of a homogeneous isotropic shell of constant thickness, when $F^\pm \equiv 0$ and $\alpha^{\mathrm{T}}_{ij} = \alpha^{\mathrm{T}}\delta_{ij}$, α^{T} being the (linear) thermal expansion coefficient. It has been shown in Sections 15 and 16 that the elastic part of relations (17.23) and the homogenized heat conduction problem reduce in this case to the corresponding thin-shell relations known from the theories of elasticity and heat conduction. We show now that the temperature part of the constitutive equations (17.23) of such a shell is also reducible to the thin-shell thermoelasticity results.

The non-zero solutions of the local thermoelasticity problems (17.9), (17.11) and (17.14) are found to be given by

$$\delta S_3 = z\alpha^{\mathrm{T}}\frac{1+\nu}{1-\nu}, \qquad \delta S_3^* = \frac{z^2}{z}\alpha^{\mathrm{T}}\frac{1+\nu}{1-\nu}, \tag{17.26}$$

using (15.91) and (17.5). Substituting this into (17.9) and again making use of (17.1) and (17.5), we find

$$\begin{aligned}\delta\langle s_{11}\rangle &= \delta\langle s_{22}\rangle = \frac{\alpha^{\mathrm{T}}E}{1-\nu},\\ \delta\langle z s_{11}^*\rangle &= \delta\langle z s_{22}^*\rangle = \frac{\alpha^{\mathrm{T}}E}{12(1-\nu)},\end{aligned} \tag{17.27}$$

after averaging the expressions for all non-zero effective thermoelastic properties. (We note that the same equations may be derived from (17.17) using the solutions obtained in Section 15 for the local elastic problems (15.92).)

Now if we substitute (15.93) and (17.27) into (17.23) and use (16.51), the corresponding relations of the thermoelasticity theory of thin isotropic shells will result (Podstrigach and Shvets 1978).

17.4. *Laminated shell composed of homogeneous isotropic layers*

Referring once again to Fig. 15.3, we proceed to obtain the effective thermoelastic properties of a shell composed of homogeneous isotropic layers. The effective elastic and heat conduction properties of this structure have already been determined in Sections 15 and 16 from the corresponding local problems.

There are two principal ways of handling the problem of thermoelasticity in this case; (1) by employing the solutions of the local thermoelastic problems (17.9), (17.11) and (17.14)–(17.16); the (2) by combining equations (17.17) and relations (15.97) and (15.99) with the solutions of the laminated shell local elastic problems set by equations (15.100) and (15.103). Because of its relative simplicity, the second approach is preferred here.

Accordingly, we substitute (15.100) and (15.103) into, respectively, (15.97) and (15.99) to obtain, for the case of isotropic layer materials,

$$\begin{aligned}&b_{11}^{11(m)} = b_{22}^{22(m)} = \frac{E_m}{1-\nu_m^2}, \qquad b_{11}^{22(m)} = b_{22}^{11(m)} = \frac{\nu_m E_m}{1-\nu_m^2},\\ &b_{11}^{*11(m)} = b_{22}^{*22(m)} = \frac{zE_m}{1-\nu_m^2}, \qquad b_{11}^{*22(m)} = b_{22}^{*11(m)} = \frac{z\nu_m E_m}{1-\nu_m^2},\\ &b_{33}^{\theta\kappa(m)} = b_{33}^{*\theta\kappa(m)} = 0,\end{aligned} \tag{17.28}$$

with m labelling the layer ($m = 1, 2, \ldots, M$).

Using (17.28) in (17.17), averaging over the thickness, and noting that $\alpha_{ij}^{\mathrm{T}(m)} = \alpha_m^{\mathrm{T}} \delta_{ij}$ because of the assumed isotropy, we obtain the following set of non-zero effective thermoelastic properties:

$$\delta\langle s_{11}\rangle = \delta\langle s_{22}\rangle = \sum_{m=1}^{M} \frac{\alpha_m^{\mathrm{T}} E_m}{1-\nu_m}(\delta_m - \delta_{m-1}),$$

$$\delta\langle zs_{11}\rangle = \delta\langle zs_{22}\rangle = \delta\langle s_{11}^{*}\rangle = \delta\langle s_{22}^{*}\rangle$$

$$= \tfrac{1}{2}\sum_{m=1}^{M} \frac{\alpha_m^{\mathrm{T}} E_m}{1-\nu_m}[\delta_m^2 - \delta_{m-1}^2 - (\delta_m - \delta_{m-1})], \tag{17.29}$$

$$\delta\langle zs_{11}^{*}\rangle = \delta\langle zs_{22}^{*}\rangle = \tfrac{1}{3}\sum_{m=1}^{M} \frac{\alpha_m^{\mathrm{T}} E_m}{1-\nu_m}[\delta_m^3 - \delta_{m-1}^3 - \tfrac{3}{2}(\delta_m^2 - \delta_{m-1}^2) + \tfrac{3}{4}(\delta_m - \delta_{m-1})].$$

These, together with (15.101), (15.102), (15.104) and (16.57), give all the (non-zero) effective properties of the laminated shell which are involved in the homogenized equations (17.23) (for thermoelasticity) and (16.35) (for heat conduction).

18. GEOMETRICALLY NON-LINEAR PROBLEM FOR A THIN REGULARLY NON-HOMOGENEOUS SHELL WITH CORRUGATED FACES

In this section a three-dimensional asymptotic analysis of the problem given in the section headings is carried out, on the basis of which a non-linear homogenized shell model is developed (Kalamkarov 1988a) and local problems are formulated. Application of the homogenized model is limited to conditions under which shears and elongations are related by Hooke's law and relative fibre rotations are large compared with strains (Novozhilov 1948, Mushtari and Galimov 1957). It is shown that the effective coefficients of the leading order terms in the non-linear elastic relations coincide with the effective homogenized shell moduli evaluated in Section 15 in the framework of the linear theory. In the limiting case of a homogeneous material and constant thickness, it turns out that the homogenized shell model reduces to the relations usually adopted in the geometrically non-linear mean–flexure plate theory, the mid-surface strain expressions coinciding with the well-known von Karman's results (Novozhilov 1948, Ciarlet and Rabier 1980).

We thus apply the geometrically non-linear theory of elasticity to a thin, regularly non-homogeneous (composite material) layer with wavy (corrugated) faces, and we select a rectangular frame of reference x_1, x_2, x_3 such that the $x_3 = 0$ plane coincides with the middle surface of the shell. It is assumed that both the shape of the layer and the structure of the material are periodic in the coordinates x_1 and x_2, with the unit cell Ω_δ defined by

$$\left\{-\frac{\delta h_1}{2} < x_1 < \frac{\delta h_1}{2},\ -\frac{\delta h_2}{2} < x_2 < \frac{\delta h_2}{2},\ x_3^- < x_3 < x_3^+\right\},$$

where

$$x_3^{\pm} = \pm\frac{\delta}{2} \pm \sigma F^{\pm}\left(\frac{x_1}{\delta h_1}, \frac{x_2}{\delta h_2}\right), \quad \delta \ll 1.$$

It is assumed, futhermore, that elongations and strains are small and the non-linear equations of motion are therefore of the form (cf.(11.1) and (11.5))

$$t_{ij,i} + P_j^* = 0, \tag{18.1}$$

$$t_{ij} = \sigma_{ij} + \sigma_{li} u_{j,l}, \tag{18.2}$$

where

$$\frac{\partial u_j}{\partial x_l} \equiv u_{j,l}.$$

For stress and strain variables we have the usual Hooke's law relation (11.2), and the strains will be related to the displacements by means of equation (11.3), whose Cartesian coordinate representation is

$$2e_{kl} = u_{k,l} + u_{l.k} + 2u_{m,k} u_{m,l}. \tag{18.3}$$

Finally, the boundary conditions on the top and bottom faces of the shell, $S^\pm(x_3 = x_3^\pm)$, have to be specified. We write

$$t_{ij} n_i^\pm = \pm p_j^{*\pm}, \tag{18.4}$$

where $n_i^\pm$ are the components of the unit vector normal to $S^\pm$ prior to deformation, and P_j^* and $p_j^{*\pm}$ are the components of the bulk and surface forces acting in the already deformed shell.

18.1. Asymptotic analysis

As before, we introduce the 'rapid' coordinates $y_1 = x_1/(\delta h_1)$, $y_2 = x_2/(\delta h_2)$ and $z = x_3/\delta$ and distinguish between 'rapid' and 'slow' coordinates when performing differentiation. The solution of the problem is represented as an asymptotic series expansion in powers of the small parameter, namely

$$u_i = u_i^{(0)}(\boldsymbol{x}) + \delta u_i^{(1)}(\boldsymbol{x}, \boldsymbol{y}, z) + \delta^2 u_i^{(2)}(\boldsymbol{x}, \boldsymbol{y}, z) + \cdots, \tag{18.5}$$

where $\boldsymbol{x} = (x_1, x_2), \boldsymbol{y} = (y_1, y_2)$ and the functions $u_i^{(l)}(\boldsymbol{x}, \boldsymbol{y}, z)$ (for $l = 1, 2, \ldots$) are 1-periodic in y_1 and y_2.

As to the asymptotic behaviour of the external forces, we may write, as in (15.10) and (15.11),

$$\begin{aligned} P_\nu^* &= \delta f_\nu^*(\boldsymbol{x}, \boldsymbol{y}, z), & P_3^* &= \delta^2 f_3^*(\boldsymbol{x}, \boldsymbol{y}, z), \\ p_\nu^{*\pm} &= \delta^2 g_\nu^{*\pm}(\boldsymbol{x}, \boldsymbol{y}), & p_3^{*\pm} &= \delta^3 g_3^{*\pm}(\boldsymbol{x}, \boldsymbol{y}) \quad (\nu = 1, 2). \end{aligned} \tag{18.6}$$

where all functions involved are periodic in y_1 and y_2, with the unit cell Ω defined by

$$\{y_1, y_2 \in (-\tfrac{1}{2}, \tfrac{1}{2}), z \in (z^-, z^+)\}, \qquad z^\pm = \pm\tfrac{1}{2} \pm F^\pm(\boldsymbol{y}).$$

The same periodicity property has to be assumed in the elastic coefficients $c_{ijkl}(\boldsymbol{y}, z)$, which are visualized as piecewise-smooth functions undergoing discontinuities of the first kind at the (non-intersecting) contact surfaces between the dissimilar phases of the composite.

From (18.2), (18.3) and (18.5) it follows that

$$\sigma_{ij} = \sigma_{ij}^{(0)} + \delta\sigma_{ij}^{(1)} + \delta^2\sigma_{ij}^{(2)} + \dots,$$
$$t_{ij} = t_{ij}^{(0)} + \delta t_{ij}^{(1)} + \delta^2 t_{ij}^{(2)} + \dots. \tag{18.7}$$

Using these and (18.6) in (18.1) and (18.4) yields the following δ-expansions:

$$\delta^{-1}H_j^{(-1)} + H_j^{(0)} + \delta H_j^{(1)} + \delta^2 H_j^{(2)} + \dots = 0,$$
$$H_j^{(-1)} = t_{3j|3}^{(0)} + \frac{1}{h_\alpha} t_{\alpha j|\alpha}^{(0)},$$
$$\dot{H}_j^{(0)} = t_{\alpha j,\alpha}^{(0)} + t_{3j|3}^{(1)} + \frac{1}{h_\alpha} t_{\alpha j|\alpha}^{(1)}, \tag{18.8}$$
$$H_j^{(1)} = t_{\alpha j,\alpha}^{(1)} + t_{3j|3}^{(2)} + \frac{1}{h_\alpha} t_{\alpha j|\alpha}^{(2)} + f_j^*(\delta_{j1} + \delta_{j2}),$$
$$H_j^{(2)} = t_{\alpha j,\alpha}^{(2)} + t_{3j|3}^{(3)} + \frac{1}{h_\alpha} t_{\alpha j|\alpha}^{(3)} + f_j^*\delta_{j3},$$
$$(t_{ij}^{(0)} + \delta t_{ij}^{(1)} + \delta^2 t_{ij}^{(2)} + \delta^3 t_{ij}^{(3)} + \cdots)n_i^{\pm}$$
$$= \pm\delta^2 g_j^{*\pm}(\delta_{j1} + \delta_{j2}) \pm \delta^3 g_j^{*\pm}\delta_{j3} \quad (z = z^{\pm}), \tag{18.9}$$

where, as before, the range of Greek indices is 1 and 2, while Latin indices take on 1, 2 and 3; $\delta\varphi/\delta y_\alpha = \varphi_{|\alpha}$ and $\delta\varphi/\delta z = \varphi_{|3}$.

As in (15.33), let

$$\boldsymbol{L}_{ijn} = c_{ijn\mu}\frac{1}{h_\mu}\frac{\delta}{\delta y_\mu} + c_{ijn3}\frac{\delta}{\delta z}. \tag{18.10}$$

The leading terms in the expansions (18.7) may then be written as

$$\sigma_{ij}^{(0)} = \boldsymbol{L}_{ijk}u_k^{(1)} + c_{ijk\alpha}u_{k,\alpha}^{(0)} + \frac{1}{2h_\mu}u_{m|\mu}^{(1)}\boldsymbol{L}_{ij\mu}u_m^{(1)}$$
$$+ \tfrac{1}{2}u_{m|3}^{(1)}\boldsymbol{L}_{ij3}u_m^{(1)} + u_{m,\alpha}^{(0)}\boldsymbol{L}_{ij\alpha}u_m^{(1)} \tag{18.11}$$
$$+ \tfrac{1}{2}c_{ij\alpha\beta}u_{m,\alpha}^{(0)}u_{m,\beta}^{(0)},$$
$$t_{ij}^{(0)} = \sigma_{ij}^{(0)} + \sigma_{i\beta}^{(0)}u_{j,\beta}^{(0)} + \sigma_{i3}^{(0)}u_{j|3}^{(1)} + \frac{1}{h_\beta}\sigma_{i\beta}^{(0)}u_{j|\beta}^{(1)}.$$

The problem of determining $t_{ij}^{(0)}$ follows from (18.8) and (18.9) as

$$H_j^{(-1)} = 0, \qquad t_{ij}^{(0)}n_i^{\pm} = 0 \quad (z = z^{\pm}), \tag{18.12}$$

and substitution from (18.11) yields a problem for the functions $u_k^{(1)}$, in which, by the same kind of argument used in Section 11, we will everywhere ignore the terms containing products of three or more derivatives of displacement components with respect to the 'slow' coordinates $x_\alpha, \alpha = 1, 2$.

The solution of the problem (18.11) and (18.12) may be represented in the form

$$u_k^{(1)} = v_k^{(1)}(\boldsymbol{x}) + U_k^{n\mu}(\boldsymbol{y}, z)u_{n,\mu}^{(0)} + W_k^{mn\lambda\mu}(\boldsymbol{y}, z)u_{m,\lambda}^{(0)}u_{n,\mu}^{(0)}, \tag{18.13}$$

with the proviso that the functions $U_k^{n\mu}(\boldsymbol{y}, z)$ and $W_k^{mn\lambda\mu}(\boldsymbol{y}, z)$ are 1-periodic in y_1 and y_2 and solve the following local problems:

$$\begin{gathered}\frac{1}{h_\beta} b_{i\beta|\beta}^{n\mu} + b_{i3|3}^{n\mu} = 0, \qquad b_{ij}^{n\mu} = \boldsymbol{L}_{ijk} U_k^{n\mu} + c_{ijn\mu}, \\ b_{ij}^{n\mu} n_j^{\pm} = 0 \quad (z = z^{\pm}),\end{gathered} \tag{18.14}$$

$$\begin{aligned}&\frac{1}{h_\beta}\left(B_{i\beta}^{mn\lambda\mu} + b_{\alpha\beta}^{m\lambda}\frac{1}{h_\alpha}U_{i|\alpha}^{n\mu} + b_{3\beta}^{m\lambda}U_{i|3}^{n\mu}\right)_{|\beta} \\ &\quad + \left(B_{i3}^{mn\lambda\mu} + b_{\alpha 3}^{m\lambda}\frac{1}{h_\alpha}U_{i|\alpha}^{n\mu} + b_{33}^{m\lambda}U_{i|3}^{n\mu}\right)_{|3} = 0, \\ &B_{ij}^{mn\lambda\mu} n_j^{\pm} = 0 \quad (z = z^{\pm}), \\ &B_{ij}^{mn\lambda\mu} = \boldsymbol{L}_{ijk}W_k^{mn\lambda\mu} + \frac{1}{2h_\alpha}U_{k|\alpha}^{m\lambda}\boldsymbol{L}_{ij\alpha}U_k^{n\mu} + \tfrac{1}{2}U_{k|3}^{m\lambda}\boldsymbol{L}_{ij3}U_k^{n\mu} + \boldsymbol{L}_{ij\lambda}U_m^{n\mu} + \tfrac{1}{2}c_{ij\lambda\mu}\delta_{mn}.\end{aligned} \tag{18.15}$$

It is seen that problem (18.14) coincides with that posed by (15.39) and (15.41) for a thin shell in the framework of linear elasticity theory. The local problem (18.15) corresponds to problem (11.12) for three-dimensional periodic composites. It should be noted that at the surfaces where discontinuities in material properties occur, continuity conditions similar to those given by (15.60) must be added to the above two problems.

Now for $n\mu = 31$, 32, it was shown in Section 15 that problem (18.14) has an exact solution, given by equation (15.40) and securing relations (15.42), and it turns out that substituting (15.40) and (15.42) into (18.15) reduces this latter to a much simpler form for $mn = 33$. We have, that is,

$$\begin{aligned}&\frac{1}{h_\beta} B_{i\beta|\beta}^{33\lambda\mu} + B_{i3|3}^{33\lambda\mu} = 0, \\ &B_{ij}^{33\lambda\mu} n_j^{\pm} = 0 \quad (z = z^{\pm}), \\ &B_{ij}^{33\lambda\mu} = \boldsymbol{L}_{ijk} W_k^{33\lambda\mu} + \tfrac{1}{2}c_{ij33}\delta_{\lambda\mu} + \tfrac{1}{2}c_{ij\lambda\mu}.\end{aligned} \tag{18.16}$$

Comparing the local problems (18.14) for the functions $U_k^{\lambda\mu}$ and (18.16) for $W_k^{33\lambda\mu}$, it can be shown that

$$\begin{aligned}&W_\alpha^{33\lambda\mu} = \tfrac{1}{2}U_\alpha^{\lambda\mu} \quad (\alpha = 1, 2), \\ &W_3^{33\lambda\mu} = \tfrac{1}{2}(U_3^{\lambda\mu} - z\delta_{\lambda\mu}),\end{aligned} \tag{18.17}$$

and using this in (18.16) for $B_{ij}^{33\lambda\mu}$ yields

$$B_{ij}^{33\lambda\mu} = \tfrac{1}{2}b_{ij}^{\lambda\mu} \tag{18.18}$$

after comparing with (18.14) for $b_{ij}^{\lambda\mu}$.

Next we substitute (18.13) into (18.11) and use the notation of (18.14) and (18.15) to obtain:

$$\sigma_{ij}^{(0)} = b_{ij}^{\lambda\mu} u_{\lambda,\mu}^{(0)} + B_{ij}^{mn\lambda\mu} u_{m,\lambda}^{(0)} u_{n,\mu}^{(0)}. \tag{18.19}$$

Application of the method of homogenization and use of (18.12) now yields the problem from which the leading order terms in (18.8) and (18.12) can be determined:

$$H_j^{(0)} = \langle H_j^{(0)} \rangle, \qquad t_{ij}^{(1)} n_i^{\pm} = 0, \quad (z = z^{\pm}), \tag{18.20}$$

where the (volume) average is defined in (15.27) and where, using the periodicity in y_1, y_2 and conditions (18.20) at $z = z^{\pm}$, it is found from (18.8) that

$$\langle H_j^{(0)} \rangle = \langle t_{\alpha j}^{(0)} \rangle_{,\alpha}. \tag{18.21}$$

Referring once more to Section 15, where the solution of the linear version of the problem (18.20) and (18.21) was obtained, we write

$$u_1^{(0)} = u_2^{(0)} = 0, \qquad u_3^{(0)} = w(\boldsymbol{x}), \qquad v_3^{(1)}(\boldsymbol{x}) = 0. \tag{18.22}$$

This gives

$$\begin{aligned} u_\alpha^{(1)} &= v_\alpha^{(1)}(\boldsymbol{x}) - z w_{,\alpha} + \tfrac{1}{2} U_\alpha^{\lambda\mu} w_{,\lambda} w_{,\mu}, \\ u_3^{(1)} &= \tfrac{1}{2}(U_3^{\lambda\mu} - z\delta_{\lambda\mu}) w_{,\lambda} w_{,\mu} \end{aligned} \tag{18.23}$$

using (18.13) and (18.17), and hence by (18.19) and (18.18):

$$\sigma_{ij}^{(0)} = \tfrac{1}{2} b_{ij}^{\lambda\mu} w_{,\lambda} w_{,\mu}. \tag{18.24}$$

Using (18.2), (18.3), (18.22) and (18.23) along with (11.2) we find, within the accuracy of the calculation,

$$\begin{aligned} \sigma_{ij}^{(1)} &= L_{ijk} u_k^{(2)} + c_{ij\alpha\beta} \varepsilon_{\alpha\beta}^{(1)} + z c_{ij\alpha\beta} \tau_{\alpha\beta} \\ &\quad + w_{,\alpha}(L_{ij\alpha} u_3^{(2)} - L_{ij3} u_\alpha^{(2)}) - c_{ij3\beta} w_{,\alpha} \varepsilon_{\alpha\beta}^{(1)} \\ &\quad - c_{ijm\beta} U_m^{\alpha\mu} w_{,\alpha} \tau_{\mu\beta}, \end{aligned} \tag{18.25}$$

$$t_{ij}^{(1)} = \sigma_{ij}^{(1)} + \sigma_{i\beta}^{(1)} w_{,\beta} \delta_{j3} - \sigma_{i3}^{(1)} w_{,\beta} \delta_{j\beta}, \tag{18.26}$$

denoting

$$\varepsilon_{\alpha\beta}^{(1)} = v_{\alpha,\beta}^{(1)}, \qquad \tau_{\alpha\beta} = -w_{,\alpha\beta}, \tag{18.27}$$

in analogy with (15.48) and (15.49).

Now if we substitute (18.21) and (18.25)–(18.27) into (18.20) and make use of (18.11) and (18.24), a problem for determining the functions $u_k^{(2)}$ will be obtained, the solution of which may be represented to the same accuracy in the following form:

$$u_k^{(2)} = U_k^{\lambda\mu} \varepsilon_{\lambda\mu}^{(1)} + V_k^{\lambda\mu} \tau_{\lambda\mu} + Q_k^{\alpha\lambda\mu} w_{,\alpha} \varepsilon_{\lambda\mu}^{(1)} + R_k^{\alpha\lambda\mu} w_{,\alpha} \tau_{\lambda\mu}. \tag{18.28}$$

Here the functions $U_k^{\lambda\mu}(\boldsymbol{y}, z)$, $V_k^{\lambda\mu}(\boldsymbol{y}, z)$, $Q_k^{\alpha\lambda\mu}(\boldsymbol{y}, z)$ and $R_k^{\alpha\lambda\mu}(\boldsymbol{y}, z)$ are periodic in y_1 and y_2 with the unit cell Ω and solve the following local problems:

$$\frac{1}{h_\beta}b^{\lambda\mu}_{i\beta|\beta}+b^{\lambda\mu}_{i3|3}=0, \qquad b^{\lambda\mu}_{ij}n^{\pm}_j=0, \quad (z=z^{\pm}), \qquad (b^{\lambda\mu}_{ij}\leftrightarrow b^{*\lambda\mu}_{ij}\leftrightarrow q^{\alpha\lambda\mu}_{ij}), \tag{18.29}$$

$$\frac{1}{h_\beta}r^{\alpha\lambda\mu}_{i\beta|\beta}+r^{\alpha\lambda\mu}_{i3|3}=b^{\lambda\alpha}_{i\mu}-\langle b^{\lambda\alpha}_{i\mu}\rangle, \qquad r^{\alpha\lambda\mu}_{ij}n^{\pm}_j=0 \quad (z=z^{\pm}), \tag{18.30}$$

$$\begin{aligned} b^{\lambda\mu}_{ij}&=\boldsymbol{L}_{ijk}U^{\lambda\mu}_k+c_{ij\lambda\mu}, \qquad b^{*\lambda\mu}_{ij}=\boldsymbol{L}_{ijk}V^{\lambda\mu}_k+zc_{ij\lambda\mu},\\ q^{\alpha\lambda\mu}_{ij}&=\boldsymbol{L}_{ijk}Q^{\alpha\lambda\mu}_k+\boldsymbol{L}_{ij\alpha}U^{\lambda\mu}_3-\boldsymbol{L}_{ij3}U^{\lambda\mu}_\alpha-c_{ij3\mu}\delta_{\alpha\lambda},\\ r^{\alpha\lambda\mu}_{ij}&=\boldsymbol{L}_{ijk}R^{\alpha\lambda\mu}_k+\boldsymbol{L}_{ij\alpha}V^{\lambda\mu}_3-\boldsymbol{L}_{ij3}V^{\lambda\mu}_\alpha-c_{ijm\mu}U^{\alpha\lambda}_m. \end{aligned} \tag{18.31}$$

To the foregoing equations we have to adjoin the jump conditions at the material surfaces of discontinuity, which we give in the following form (cf.(15.60)):

$$\begin{aligned} &[\![U^{\lambda\mu}_k]\!]=0 \quad (U^{\lambda\mu}_k\leftrightarrow V^{\lambda\mu}_k\leftrightarrow Q^{\alpha\lambda\mu}_k\leftrightarrow R^{\alpha\lambda\mu}_k),\\ &\left[\!\!\left[\frac{1}{h_\beta}n^{(k)}_\beta b^{\lambda\mu}_{i\beta}+n^{(k)}_3 b^{\lambda\mu}_{i3}\right]\!\!\right]=0 \quad (b^{\lambda\mu}_{ij}\leftrightarrow b^{*\lambda\mu}_{ij}\leftrightarrow q^{\alpha\lambda\mu}_{ij}\leftrightarrow r^{\alpha\lambda\mu}_{ij}), \end{aligned} \tag{18.32}$$

where $n^{(k)}_i$ denotes the unit normal at the surface of discontinuity, related to the coordinate system y_1, y_2, z. This is in contrast to the $z=z^{\pm}$ conditions in the local problems (18.14)–(18.16) and (18.29), (18.30), where $n^{\pm}_i$, the unit normals to the surfaces $S^{\pm}$, are related to the coordinate system x_1, x_2, x_3. In actually solving these problems, however, it proves more convenient to rewrite them in a form similar to (15.58) and (15.59) and to employ unit normals $n^{\pm(y)}_i$ related to y_1, y_2, z.

The local problems (18.9)–(18.32) are linear in the unknown functions, and their solutions are unique up to constant terms. This ambiguity is removed by imposing conditions in a form analogous to (15.57). That is,

$$\langle U^{\lambda\mu}_k\rangle_y=0 \quad \text{when } z=0 \quad (U^{\lambda\mu}_k\leftrightarrow V^{\lambda\mu}_k\leftrightarrow Q^{\alpha\lambda\mu}_k\leftrightarrow R^{\alpha\lambda\mu}_k), \tag{18.33}$$

where $\langle\cdots\rangle_y$ indicates average with respect to y_1 and y_2 only.

Note that the problems for the functions $U^{\lambda\mu}_k$ and $V^{\lambda\mu}_k$ are identical in form to the corresponding local problems of the linear theory of elasticity, equations (15.41), (15.53) and (15.55)–(15.61); in the remaining two problems these functions are considered as known.

Substituting (18.28) into (18.25) and using the notation of (18.31) we arrive at

$$\sigma^{(1)}_{ij}=b^{\lambda\mu}_{ij}\varepsilon^{(1)}_{\lambda\mu}+b^{*\lambda\mu}_{ij}\tau_{\lambda\mu}+q^{\alpha\lambda\mu}_{ij}w_{,\alpha}\varepsilon^{(1)}_{\lambda\mu}+r^{\alpha\lambda\mu}_{ij}w_{,\alpha}\tau_{\lambda\mu}. \tag{18.34}$$

Returning now to equations (18.29) and (18.30), we average their left-hand sides after first multiplying them by z and z^2 and we take into account, in doing so, the $z=z^{\pm}$ boundary conditions and periodicity in y_1 and y_2. This gives, in addition to (15.62), the following relations for the effective elastic moduli of the homogenized shell:

$$\begin{aligned} &\langle b^{\lambda\mu}_{i3}\rangle=\langle zb^{\lambda\mu}_{i3}\rangle=0, \qquad (b^{\lambda\mu}_{i3}\leftrightarrow b^{*\lambda\mu}_{i3}\leftrightarrow q^{\alpha\lambda\mu}_{i3}),\\ &\langle r^{\alpha\lambda\mu}_{i3}\rangle=\langle z\rangle\langle b^{\alpha\lambda}_{i\mu}\rangle-\langle zb^{\alpha\lambda}_{i\mu}\rangle. \end{aligned} \tag{18.35}$$

Clearly, the symmetry properties (15.75) also retain their truth.

Finally, from (18.11) and (18.26):

$$\begin{aligned}
&\langle t_{\alpha\beta}^{(0)}\rangle = \langle \sigma_{\alpha\beta}^{(0)}\rangle, \qquad \langle t_{\alpha 3}^{(0)}\rangle = \langle \sigma_{\alpha\beta}^{(0)}\rangle w_{,\beta},\\
&\langle t_{\alpha\beta}^{(1)}\rangle = \langle \sigma_{\alpha\beta}^{(1)}\rangle, \qquad \langle z t_{\alpha\beta}^{(1)}\rangle = \langle z\sigma_{\alpha\beta}^{(1)}\rangle,\\
&\langle t_{\alpha 3}^{(1)}\rangle = \langle \sigma_{\alpha\beta}^{(1)}\rangle w_{,\beta} + \langle r_{\alpha 3}^{\beta\lambda\mu}\rangle w_{,\beta}\tau_{\lambda\mu},
\end{aligned} \tag{18.36}$$

where equations (18.22)–(18.24), (18.34) and (18.35) have also been used.

18.2. Governing equations for homogenized shell behaviour

The problem we address next is to derive a system of equations for $v_1^{(1)}(\boldsymbol{x})$, $v_2^{(1)}(\boldsymbol{x})$ and $w(\boldsymbol{x})$, three functions of the 'slow' coordinates, which enter equations (18.24), (18.27) and (18.34) and in terms of which the components of the displacement vector are expressed through (18.5), (18.22), (18.23) and (18.28). We begin by writing the following terms in (18.8):

$$\begin{aligned}
&\langle H_\beta^{(0)}\rangle + \delta H_\beta^{(1)} = 0 \quad (\beta = 1,2),\\
&\langle H_3^{(0)}\rangle + \delta H_3^{(1)} + \delta^2 H_3^{(2)} = 0,
\end{aligned} \tag{18.37}$$

where (18.12) and (18.20) have been used. Applying the averaging operator, and using conditions (18.9), relations (18.8) and (18.21) and the periodicity in y_1 and y_2, we find that

$$\begin{aligned}
&\langle t_{\alpha\beta}^{(0)}\rangle_{,\alpha} + \delta(\langle t_{\alpha\beta}^{(1)}\rangle_{,\alpha} + g_\beta^* + \langle f_\beta^*\rangle) = 0,\\
&\langle t_{\alpha 3}^{(0)}\rangle_{,\alpha} + \delta\langle t_{\alpha 3}^{(1)}\rangle_{,\alpha} + \delta^2(\langle t_{\alpha 3}^{(2)}\rangle_{,\alpha} + g_3^* + \langle f_3^*\rangle) = 0,
\end{aligned} \tag{18.38}$$

where

$$g_j^* = \int_{-1/2}^{1/2}\int_{-1/2}^{1/2} (g_j^{*+}\omega^+ + g_j^{*-}\omega^-)\mathrm{d}y_1\mathrm{d}y_2,$$

and where

$$\omega^\pm = \left[1 + \frac{1}{h_1^2}\left(\frac{\partial F^\pm}{\partial y_1}\right)^2 + \frac{1}{h_2^2}\left(\frac{\partial F^\pm}{\partial y_2}\right)^2\right]^{1/2}$$

is the Cartesian coordinate representation of (15.26).

From the first of equations (18.37), multiplying by z and averaging, we obtain:

$$\langle z\rangle\langle t_{\alpha\beta}^{(0)}\rangle_{,\alpha} + \delta(\langle z t_{\alpha\beta}^{(1)}\rangle_{,\alpha} - \langle t_{3\beta}^{(2)}\rangle + m_\beta^* + \langle z f_\beta^*\rangle) = 0, \tag{18.39}$$

where

$$m_\beta^* = \int_{-1/2}^{1/2}\int_{-1/2}^{1/2} (z^+ g_\beta^{*+}\omega^+ + z^- g_\beta^{*-}\omega^-)\,\mathrm{d}y_1\,\mathrm{d}y_2.$$

Note that the functions $g_j^*(\boldsymbol{x})$ and $m_\beta^*(\boldsymbol{x})$, associated with external loads, coincide in a linear approximation with the functions $r_\lambda(\boldsymbol{\alpha})$, $q_3(\boldsymbol{\alpha})$ and $\rho_\mu(\boldsymbol{\alpha})$ defined by (15.67) and (15.69).

Eliminating $\langle t_{\alpha 3}^{(2)}\rangle_{,\alpha}$ from (18.39) and the second of (18.38) and noting that $\langle t_{3\alpha}^{(2)}\rangle = \langle t_{\alpha 3}^{(2)}\rangle$ within the accuracy of the calculation, we find

$$\begin{aligned}&\langle t_{\alpha 3}^{(0)}\rangle_{,\alpha} + \delta\langle t_{\alpha 3}^{(1)}\rangle_{,\alpha} + \delta[\langle z\rangle\langle t_{\alpha\beta}^{(0)}\rangle_{,\alpha}\\ &\quad + \delta(\langle zt_{\alpha\beta}^{(1)}\rangle_{,\alpha} + m_\beta^* + \langle zf_\beta^*\rangle)]_{,\beta} + \delta^2(g_3^* + \langle f^*\rangle) = 0.\end{aligned} \tag{18.40}$$

Now from (18.24), (18.27), (18.35) and (18.36) it follows that

$$\langle t_{\alpha 3}^{(1)}\rangle + \langle z\rangle\langle t_{\alpha\beta}^{(0)}\rangle_{,\beta} = \langle\sigma_{\alpha\beta}^{(1)}\rangle w_{,\beta} + \langle z\sigma_{\alpha\beta}^{(0)}\rangle_{,\beta}, \tag{18.41}$$

which, combined with (18.36), reduces (18.40) to the form

$$\begin{aligned}&[(\langle\sigma_{\alpha\beta}^{(0)}\rangle + \delta\langle\sigma_{\alpha\beta}^{(1)}\rangle)w_{,\beta}]_{,\alpha} + \delta[(\langle z\sigma_{\alpha\beta}^{(0)}\rangle\\ &\quad + \delta\langle z\sigma_{\alpha\beta}^{(1)}\rangle)_{,\alpha} + \delta(m_\beta^* + \langle zf_\beta^*\rangle)]_{,\beta} + \delta^2(g_3^* + \langle f_3^*\rangle) = 0.\end{aligned} \tag{18.42}$$

Using (18.36), the first of equations (18.38) can be put into the form

$$(\langle\sigma_{\alpha\beta}^{(0)}\rangle + \delta\langle\sigma_{\alpha\beta}^{(1)}\rangle)_{,\alpha} + \delta(g_\beta^* + \langle f_\beta^*\rangle) = 0 \quad (\beta = 1, 2). \tag{18.43}$$

The system (18.42), (18.43) must of course be complemented by the elastic relations of the homogenized shell.

Introducing

$$v_\lambda(\boldsymbol{x}) = \delta v_\lambda^{(1)}(\boldsymbol{x}), \qquad \varepsilon_{\lambda\mu} = \delta\varepsilon_{\lambda\mu}^{(1)} = v_{\lambda,\mu} \tag{18.44}$$

in accordance with (15.86) and (18.27), and averaging (18.24) and (18.34), we obtain the elastic relations of the form

$$\begin{aligned}\langle z^l\sigma_{\alpha\beta}^{(0)}\rangle + \delta\langle z^l\sigma_{\alpha\beta}^{(1)}\rangle &= \langle z^l b_{\alpha\beta}^{\lambda\mu}\rangle(\varepsilon_{\lambda\mu} + \tfrac{1}{2}w_{,\lambda}w_{,\mu})\\ &\quad + \delta\langle z^l b_{\alpha\beta}^{*\lambda\mu}\rangle\tau_{\lambda\mu} + \langle z^l q_{\alpha\beta}^{\theta\lambda\mu}\rangle w_{,\theta}\varepsilon_{\lambda\mu}\\ &\quad + \delta\langle z^l r_{\alpha\beta}^{\theta\lambda\mu}\rangle w_{,\theta}\tau_{\lambda\mu} \quad (l = 0, 1).\end{aligned} \tag{18.45}$$

This, when substituted into (18.42) and (18.43), now yields the desired system of three governing equations for the functions $v_1(\boldsymbol{x}), v_2(\boldsymbol{x})$ and $w(\boldsymbol{x})$. Together with the solutions of the local problems (18.29)–(18.33), these functions enable us to calculate very accurately the displacement vector components which, using (18.5), (18.22), (18.23) and (18.28) along with (18.44), are found to be given by

$$\begin{aligned}u_\alpha &= v_\alpha(\boldsymbol{x}) - \gamma w_{,\alpha} + \delta U_\alpha^{\lambda\mu}(\varepsilon_{\lambda\mu} + \tfrac{1}{2}w_{,\lambda}w_{,\mu})\\ &\quad + \delta Q_\alpha^{\beta\lambda\mu}w_{,\beta}\varepsilon_{\lambda\mu} + \delta^2 V_\alpha^{\lambda\mu}\tau_{\lambda\mu} + \delta^2 R_\alpha^{\beta\lambda\mu}w_{,\beta}\tau_{\lambda\mu} + \cdots \quad (\alpha = 1, 2),\\ u_3 &= w(\boldsymbol{x}) - \frac{\gamma}{2}w_{,\beta}w_{,\beta} + \delta U_3^{\lambda\mu}(\varepsilon_{\lambda\mu} + \tfrac{1}{2}w_{,\lambda}w_{,\mu})\\ &\quad + \delta Q_3^{\beta\lambda\mu}w_{,\beta}\varepsilon_{\lambda\mu} + \delta^2 V_3^{\lambda\mu}\tau_{\lambda\mu} + \delta^2 R_3^{\beta\lambda\mu}w_{,\beta}\tau_{\lambda\mu} + \cdots.\end{aligned} \tag{18.46}$$

It will be recognized that equations (18.45) and (18.46), derived in the framework of the geometrically non-linear theory of elasticity, generalize the corresponding linear results, namely equations (15.71), (15.72) and (15.87).

18.3. Homogenization versus geometrically non-linear shell theory

At this point it appears useful to discuss how the above homogenization model relates to the geometrically non-linear theory of shells.

We begin by writing down, with reference to (18.7), the leading order expressions for the stress and moment resultants:

$$\begin{aligned} N_{\alpha\beta} &= \delta \langle \sigma_{\alpha\beta}^{(0)} \rangle + \delta^2 \langle \sigma_{\alpha\beta}^{(1)} \rangle, \\ M_{\alpha\beta} &= \delta^2 \langle z\sigma_{\alpha\beta}^{(0)} \rangle + \delta^3 \langle z\sigma_{\alpha\beta}^{(1)} \rangle. \end{aligned} \tag{18.47}$$

Using these and (18.6) in equations (18.42) and (18.43), it is seen that these latter coincide with the equations of motion written in terms of projections onto non-deformed axes in the framework of the non-linear mean–flexure shell theory as discussed, for example, in Grigolyuk and Kabanov (1978).

We thus see that both the material inhomogeneity and varying thickness have their effect on the values of the effective elastic moduli that appear as coefficients in the elastic relations (18.45).

In the limiting case of a homogeneous isotropic shell of constant thickness, $c_{ijkl} = \text{const.}$, $F^{\pm} \equiv 0$, and the elastic coefficients of interest are given by (15.91). The coordinates y_1 and y_2 do not in fact play a role, and the local problems (18.29)–(18.33) are solvable exactly. As far as the functions $U_k^{\lambda\mu}$ and $V_k^{\lambda\mu}$ are concerned, the non-zero solutions of the local problems and the corresponding effective moduli have already be found to be given by (15.92) and (15.93). As for the functions $Q_k^{\alpha\lambda\mu}$ and $R_k^{\alpha\lambda\mu}$, the set of non-zero solutions of the pertinent local problems are as follows:

$$\begin{aligned} &Q_1^{111} = Q_2^{222} = \frac{z}{1-\nu}, \qquad Q_1^{122} = Q_2^{211} = \frac{\nu z}{1-\nu}, \\ &Q_1^{221} = Q_2^{112} = z, \qquad R_1^{122} = R_2^{211} = \frac{\nu z^2}{2(1-\nu)}, \\ &R_1^{221} = R_2^{112} = -\frac{\nu z^2}{2(1-\nu)}. \end{aligned} \tag{18.48}$$

Substituting this into (18.31) we obtain:

$$q_{\alpha\beta}^{\theta\lambda\mu} = 0, \qquad r_{\alpha\beta}^{\theta\lambda\mu} = 0. \tag{18.49}$$

If we now substitute (15.93) and (18.49) into (18.45) and (18.47), we arrive at elastic relations of the form usually adopted in geometrically non-linear mean–flexure shell theory. For the mid-surface strains, we have, by (18.47) and (18.44),

$$\begin{aligned} &\varepsilon_1 = v_{1,1} + \tfrac{1}{2}w_{,1}^2, \qquad \varepsilon_2 = v_{2,2} + \tfrac{1}{2}w_{,2}^2, \\ &\omega = v_{1,2} + v_{2,1} + w_{,1}w_{,2}, \\ &\kappa_1 = -w_{,11}, \qquad \kappa_2 = -w_{,22}, \qquad \tau = -w_{,12}, \end{aligned} \tag{18.50}$$

which, using (18.46), are readily shown to coincide with von Karman's formulae (Novozhilov 1948, Ciarlet and Rabier 1980).

In conclusion, it should be noted that the non-linear homogenized shell model that we have developed may be useful in the study of elastic stability under conditions of small precritical deformations.

Although the mathematical analysis of this kind usually involves stability equations (expressed in terms of forces and moments) and the usual mid-surface strain expressions, one must employ the elastic relations (18.45) and (18.47) to describe the relationships between the forces, moments and strains. The problem can be considerably simplified, however, by noting that the products $w_{,\theta}\varepsilon_{\lambda\mu}$ and $w_{,\theta}\tau_{\lambda\mu}$ are smaller than other terms and may therefore be dropped from (18.45) and (18.47). Note also that, as was the case in Section 11, the above approach to the stability problem is based on the assumption that, at the moment of buckling, the stress and strain characteristics of the body vary on a length scale longer than the tangential dimensions of the unit cell (which are, of course, the same order of magnitude as the thickness of the layer).

6 Structurally Non-homogeneous Periodic Shells and Plates

There are many applications in a number of technical areas which may and do benefit from the advantages offered by homogeneous thin-walled structural elements with a regular structure. The unit cell of the material system consists in this case of several thin-walled parts made of homogeneous materials (cf. Fig. 15.2), and the non-homogeneity in question is associated with the structural features of the object and is also determined by the specific form of the reinforcement used (the reader is reminded that voids may also be considered as reinforcement). The structurally non-homogeneous elements of these type include the rib and wafer types of reinforced plates and shells (see Figs. 20.1, 20.4 and 23.1), three-layered shells with a honeycomb filler (Figs. 21.1 and 21.3) and corrugated surface shells (Figs 24.2, 24.6, and 24.11). All these structures are considered in this chapter on the basis of the general theory developed above for regularly non-homogeneous shells with rapidly oscillating thickness.

19. LOCAL PROBLEM FORMULATION FOR STRUCTURALLY NON-HOMOGENEOUS PLATES AND SHELLS OF ORTHOTROPIC MATERIAL

We consider a structurally non-homogeneous shell (or plate), each element of which is made of a homogeneously orthotropic material. By the definition of orthotropy, such a material possesses elastic symmetry with respect to three mutually orthogonal planes, so that its tensor of elastic moduli, or stiffness tensor, has nine independent components and may be represented as a symmetric 6×6 matrix as follows (cf. (1.82)):

$$[c] = \begin{bmatrix} c_{11} & c_{12} & c_{13} & 0 & 0 & 0 \\ c_{12} & c_{22} & c_{23} & 0 & 0 & 0 \\ c_{13} & c_{23} & c_{33} & 0 & 0 & 0 \\ 0 & 0 & 0 & c_{44} & 0 & 0 \\ 0 & 0 & 0 & 0 & c_{55} & 0 \\ 0 & 0 & 0 & 0 & 0 & c_{66} \end{bmatrix}. \tag{19.1}$$

In terms of the engineering properties of the material, its compliance tensor may also be represented as a 6×6 matrix, namely

$$[J]=[c]^{-1}=\begin{bmatrix} \frac{1}{E_1} & -\frac{\nu_{12}}{E_2} & -\frac{\nu_{13}}{E_3} & 0 & 0 & 0 \\ -\frac{\nu_{21}}{E_1} & \frac{1}{E_2} & -\frac{\nu_{23}}{E_3} & 0 & 0 & 0 \\ -\frac{\nu_{31}}{E_1} & -\frac{\nu_{32}}{E_2} & \frac{1}{E_3} & 0 & 0 & 0 \\ 0 & 0 & 0 & \frac{1}{G_{12}} & 0 & 0 \\ 0 & 0 & 0 & 0 & \frac{1}{G_{13}} & 0 \\ 0 & 0 & 0 & 0 & 0 & \frac{1}{G_{23}} \end{bmatrix}, \tag{19.2}$$

where E_1, E_2 and E_3 are the Young's moduli corresponding to the three principal elastic directions; G_{12}, G_{13} and G_{23} are the shear moduli; and ν_{12}, ν_{21}, ν_{31}, ν_{13}, ν_{23} and ν_{32} are Poisson's ratios.

Note that

$$\frac{\nu_{12}}{E_2}=\frac{\nu_{21}}{E_1}, \qquad \frac{\nu_{13}}{E_3}=\frac{\nu_{31}}{E_1}, \qquad \frac{\nu_{23}}{E_3}=\frac{\nu_{32}}{E_2}, \tag{19.3}$$

in view of the symmetry of the matrix (19.2).

The relationships between the elastic moduli and engineering properties follow from (19.1)–(19.3):

$$\begin{aligned}
E_1 &= c_{11}+\frac{2c_{12}c_{13}c_{23}-c_{12}^2c_{33}-c_{13}^2c_{22}}{c_{22}c_{33}-c_{23}^2},\\
E_2 &= c_{22}+\frac{2c_{12}c_{13}c_{23}-c_{12}^2c_{33}-c_{23}^2c_{11}}{c_{11}c_{33}-c_{13}^2},\\
E_3 &= c_{33}+\frac{2c_{12}c_{13}c_{23}-c_{13}^2c_{22}-c_{23}^2c_{11}}{c_{11}c_{22}-c_{12}^2},\\
\nu_{12} &= \frac{c_{12}c_{33}-c_{13}c_{23}}{c_{11}c_{33}-c_{13}^2}, \qquad \nu_{21}=\frac{c_{12}c_{33}-c_{13}c_{23}}{c_{22}c_{33}-c_{23}^2},\\
\nu_{13} &= \frac{c_{22}c_{13}-c_{12}c_{23}}{c_{11}c_{22}-c_{12}^2}, \qquad \nu_{31}=\frac{c_{22}c_{13}-c_{12}c_{23}}{c_{22}c_{33}-c_{23}^2},\\
\nu_{23} &= \frac{c_{11}c_{23}-c_{13}c_{12}}{c_{11}c_{22}-c_{12}^2}, \qquad \nu_{32}=\frac{c_{11}c_{23}-c_{13}c_{12}}{c_{11}c_{33}-c_{13}^2},\\
G_{12} &= c_{66}=c_{1212}, \qquad G_{13}=c_{55}=c_{1313},\\
G_{23} &= c_{44}=c_{2323}.
\end{aligned} \tag{19.4}$$

If the material is a transversely isotropic one, that is, if it is elastically symmetric with respect to all directions perpendicular to the third coordinate axis, then equations (19.1)–(19.4) are also true, but in this case

$$
\begin{gathered}
c_{11}=c_{22}, \qquad c_{13}=c_{23}, \qquad c_{55}=c_{44}, \\
c_{66}=\tfrac{1}{2}(c_{11}-c_{12}), \qquad E_1=E_2, \qquad \nu_{13}=\nu_{23}, \\
G_{13}=G_{23}, \qquad G_{12}=\frac{E_1}{2(1+\nu_{12})},
\end{gathered} \tag{19.5}
$$

and we are left with only five independent components.

Finally, for a totally isotropic material,

$$
\begin{gathered}
E_1=E_2=E_3=E, \qquad \nu_{13}=\nu_{23}=\nu_{12}=\nu, \\
G_{12}=G_{13}=G_{23}=\frac{E}{2(1+\nu)},
\end{gathered}
$$

and the relationship between the elastic moduli and the constants E and ν is given by (15.91).

19.1. Three-dimensional local problems

It was shown in Section 15 that in order to evaluate the local stress distributions as given by equations (15.74), and to determine the effective stiffness moduli involved in the homogenized relations (15.88), the corresponding local problems must be solved for the functions $b_{ij}^{\lambda\mu}(\xi, z)$ and $b_{ij}^{*\lambda\mu}(\xi, z)$ assumed to be periodic in ξ_1 and ξ_2 with respective periods A_1 and A_2. There are two groups of elastic problems that arise from the above local problems and are set in the orthotropic region Ω. One of these (to be referred to as the (b) group) is expressed by equations (15.55), (15.57), (15.58) and (15.60), and the other (the (b^*) group) by equations (15.56), (15.57), (15.59) and (15.61).

If the material is homogeneous, simplification of these groups of problems will result.

(b)-type local problems. In each of three ($b\lambda\mu$) problems ($\lambda\mu=11, 22, 12$) we determine the functions $U_1^{\lambda\mu}(\xi, z)$, $U_2^{\lambda\mu}(\xi, z)$ and $u_3^{\lambda\mu}(\xi, z)$ which are periodic in ξ_1 and ξ_2 (with respective periods A_1 and A_2) and satisfy the following system of equations:

$$
\frac{1}{h_\beta}\frac{\partial \tau_{i\beta}^{\lambda\mu}}{\partial \xi_\beta}+\frac{\partial \tau_{i3}^{\lambda\mu}}{\partial z}=0, \tag{19.6}
$$

$$
\begin{aligned}
\tau_{11}^{\lambda\mu} &= \frac{1}{h_1}c_{11}\frac{\partial U_1^{\lambda\mu}}{\partial \xi_1}+\frac{1}{h_2}c_{12}\frac{\partial U_2^{\lambda\mu}}{\partial \xi_2}+c_{13}\frac{\partial U_3^{\lambda\mu}}{\partial z}, \\
\tau_{22}^{\lambda\mu} &= \frac{1}{h_1}c_{21}\frac{\partial U_1^{\lambda\mu}}{\partial \xi_1}+\frac{1}{h_2}c_{22}\frac{\partial U_2^{\lambda\mu}}{\partial \xi_2}+c_{23}\frac{\partial U_3^{\lambda\mu}}{\partial z}, \\
\tau_{33}^{\lambda\mu} &= \frac{1}{h_1}c_{31}\frac{\partial U_1^{\lambda\mu}}{\partial \xi_1}+\frac{1}{h_2}c_{32}\frac{\partial U_2^{\lambda\mu}}{\partial \xi_2}+c_{33}\frac{\partial U_3^{\lambda\mu}}{\partial z},
\end{aligned} \tag{19.7}
$$

$$\tau_{23}^{\lambda\mu} = c_{44}\left(\frac{1}{h_2}\frac{\partial U_3^{\lambda\mu}}{\partial \xi_2} + \frac{\partial U_2^{\lambda\mu}}{\partial z}\right),$$

$$\tau_{13}^{\lambda\mu} = c_{55}\left(\frac{1}{h_1}\frac{\partial U_3^{\lambda\mu}}{\partial \xi_1} + \frac{\partial U_1^{\lambda\mu}}{\partial z}\right),$$

$$\tau_{12}^{\lambda\mu} = c_{66}\left(\frac{1}{h_2}\frac{\partial U_1^{\lambda\mu}}{\partial \xi_2} + \frac{1}{h_1}\frac{\partial U_2^{\lambda\mu}}{\partial \xi_1}\right),$$

together with the outer surface boundary conditions:

$$\begin{aligned}
t_1^{11} &= -c_{11}\frac{n_1}{h_1}, & t_2^{11} &= -c_{21}\frac{n_2}{h_2}, & t_3^{11} &= -c_{31}n_3, \\
t_1^{22} &= -c_{12}\frac{n_1}{h_1}, & t_2^{22} &= -c_{22}\frac{n_2}{h_2}, & t_3^{22} &= -c_{32}n_3, \\
t_1^{12} &= -c_{66}\frac{n_2}{h_2}, & t_2^{12} &= c_{66}\frac{n_1}{h_1}, & t_3^{12} &= 0,
\end{aligned} \tag{19.8}$$

where

$$\begin{aligned}
t_1^{\lambda\mu} &= \tau_{11}^{\lambda\mu}\frac{n_1}{h_1} + \tau_{12}^{\lambda\mu}\frac{n_2}{h_2} + \tau_{13}^{\lambda\mu}n_3, \\
t_2^{\lambda\mu} &= \tau_{12}^{\lambda\mu}\frac{n_1}{h_1} + \tau_{22}^{\lambda\mu}\frac{n_2}{h_2} + \tau_{23}^{\lambda\mu}n_3, \\
t_3^{\lambda\mu} &= \tau_{13}^{\lambda\mu}\frac{n_1}{h_1} + \tau_{23}^{\lambda\mu}\frac{n_2}{h_2} + \tau_{33}^{\lambda\mu}n_3,
\end{aligned} \tag{19.9}$$

where n_1, n_2 and n_3 are the components of the unit vector normal to the bounding surface of the region Ω.

The functions that are determined from the group (*b*) problems are

$$\begin{aligned}
b_{11}^{11} &= \tau_{11}^{11} + c_{11}, & b_{12}^{11} &= \tau_{12}^{11}, & b_{22}^{11} &= \tau_{22}^{11} + c_{21}, \\
b_{11}^{22} &= \tau_{11}^{22} + c_{21}, & b_{12}^{22} &= \tau_{12}^{22}, & b_{22}^{22} &= \tau_{22}^{22} + c_{22}, \\
b_{11}^{12} &= \tau_{11}^{12}, & b_{12}^{12} &= \tau_{12}^{12} + c_{66}, & b_{22}^{12} &= \tau_{22}^{12}, \\
b_{33}^{11} &= \tau_{33}^{11} + c_{31}, & b_{33}^{22} &= \tau_{33}^{22} + c_{32}, & b_{3\beta}^{\lambda\mu} &= \tau_{\beta 3}^{\lambda\mu}.
\end{aligned} \tag{19.10}$$

(*b**)-*type local problems*. From the problems (*b**11), (*b**22) and (*b**12) we determine the functions $V_1^{\lambda\mu}(\xi, z)$, $V_2^{\lambda\mu}(\xi, z)$ and $V_3^{\lambda\mu}(\xi, z)$, also periodic in ξ_1 and ξ_2 and satisfying the following equations:

$$\frac{1}{h_\beta}\frac{\partial \bar{\tau}_{i\beta}^{\lambda\mu}}{\partial \xi_\beta} + \frac{\partial \bar{\tau}_{i3}^{\lambda\mu}}{\partial z} = -c_{33\lambda\mu}\delta_{i3}, \tag{19.11}$$

$$\begin{aligned}
\bar{\tau}_{11}^{\lambda\mu} &= \frac{1}{h_1} c_{11} \frac{\partial V_1^{\lambda\mu}}{\partial \xi_1} + \frac{1}{h_2} c_{12} \frac{\partial V_2^{\lambda\mu}}{\partial \xi_2} + c_{13} \frac{\partial V_3^{\lambda\mu}}{\partial z}, \\
\bar{\tau}_{22}^{\lambda\mu} &= \frac{1}{h_1} c_{21} \frac{\partial V_1^{\lambda\mu}}{\partial \xi_1} + \frac{1}{h_2} c_{22} \frac{\partial V_2^{\lambda\mu}}{\partial \xi_2} + c_{23} \frac{\partial V_3^{\lambda\mu}}{\partial z}, \\
\bar{\tau}_{33}^{\lambda\mu} &= \frac{1}{h_1} c_{31} \frac{\partial V_1^{\lambda\mu}}{\partial \xi_1} + \frac{1}{h_2} c_{32} \frac{\partial V_2^{\lambda\mu}}{\partial \xi_2} + c_{33} \frac{\partial V_3^{\lambda\mu}}{\partial z}, \\
\bar{\tau}_{23}^{\lambda\mu} &= c_{44} \left(\frac{1}{h_2} \frac{\partial V_3^{\lambda\mu}}{\partial \xi_2} + \frac{\partial V_2^{\lambda\mu}}{\partial z} \right), \\
\bar{\tau}_{13}^{\lambda\mu} &= c_{55} \left(\frac{1}{h_1} \frac{\partial V_3^{\lambda\mu}}{\partial \xi_1} + \frac{\partial V_1^{\lambda\mu}}{\partial z} \right), \\
\bar{\tau}_{12}^{\lambda\mu} &= c_{66} \left(\frac{1}{h_2} \frac{\partial V_1^{\lambda\mu}}{\partial \xi_2} + \frac{1}{h_1} \frac{\partial V_2^{\lambda\mu}}{\partial \xi_1} \right),
\end{aligned} \tag{19.12}$$

and the boundary conditions on the other surfaces of Ω:

$$\begin{aligned}
&\bar{t}_1^{11} = -zc_{11} \frac{n_1}{h_1}, \qquad \bar{t}_2^{11} = -zc_{21} \frac{n_2}{h_2}, \qquad \bar{t}_3^{11} = -zc_{31} n_3, \\
&\bar{t}_1^{22} = -zc_{12} \frac{n_1}{h_1}, \qquad \bar{t}_2^{22} = -zc_{22} \frac{n_2}{h_2}, \qquad \bar{t}_3^{22} = -zc_{32} n_3, \\
&\bar{t}_1^{12} = -zc_{66} \frac{n_2}{h_2}, \qquad \bar{t}_2^{12} = -zc_{66} \frac{n_1}{h_1}, \qquad \bar{t}_3^{12} = 0,
\end{aligned} \tag{19.13}$$

where

$$\begin{aligned}
\bar{t}_1^{\lambda\mu} &= \bar{\tau}_{11}^{\lambda\mu} \frac{n_1}{h_1} + \bar{\tau}_{12}^{\lambda\mu} \frac{n_2}{h_2} + \bar{\tau}_{13}^{\lambda\mu} n_3, \\
\bar{t}_2^{\lambda\mu} &= \bar{\tau}_{12}^{\lambda\mu} \frac{n_1}{h_1} + \bar{\tau}_{22}^{\lambda\mu} \frac{n_2}{h_2} + \bar{\tau}_{23}^{\lambda\mu} n_3, \\
\bar{t}_3^{\lambda\mu} &= \bar{\tau}_{13}^{\lambda\mu} \frac{n_1}{h_1} + \bar{\tau}_{23}^{\lambda\mu} \frac{n_2}{h_2} + \bar{\tau}_{33}^{\lambda\mu} n_3.
\end{aligned} \tag{19.14}$$

Having solved the (b^*)-type problems, we are in a position to obtain the following functions:

$$\begin{aligned}
&b_{11}^{*11} = \bar{\tau}_{11}^{11} + zc_{11}, \qquad b_{12}^{*11} = \bar{\tau}_{12}^{11}, \qquad b_{22}^{*11} = \bar{\tau}_{22}^{11} + zc_{21}, \\
&b_{11}^{*22} = \bar{\tau}_{11}^{22} + zc_{21}, \qquad b_{12}^{*22} = \bar{\tau}_{12}^{22}, \qquad b_{22}^{*22} = \bar{\tau}_{22}^{22} + zc_{22}, \\
&b_{11}^{*12} = \bar{\tau}_{11}^{12}, \qquad b_{12}^{*12} = \bar{\tau}_{12}^{12} + zc_{66}, \qquad b_{22}^{*12} = \bar{\tau}_{22}^{12}, \\
&b_{33}^{*11} = \bar{\tau}_{33}^{11} + zc_{31}, \qquad b_{33}^{*22} = \bar{\tau}_{33}^{22} + zc_{32}, \qquad b_{3\beta}^{*\lambda\mu} = \bar{\tau}_{\beta 3}^{\lambda\mu}.
\end{aligned} \tag{19.15}$$

19.2. Local problems in a two-dimensional formulation

Consider a shell strengthened by a set of ribs or cavities located along one of the coordinate axes, say α_1. The functions $U_i^{\lambda\mu}$ and $V_i^{\lambda\mu}$ will then depend on the variables ξ_2 and z, and the local problems of the group $(b\lambda\mu)$ $(\lambda\mu = 11, 22, 12)$ reduce to the determination of the functions $U_2^{\lambda\mu}(\xi_2, z)$, $U_3^{\lambda\mu}(\xi_2, z)$ $(\lambda\mu = 11, 22)$ and $U_1^{12}(\xi_2, z)$ from the system of equations

$$\frac{1}{h_2}\frac{\partial \tau_{i2}^{\lambda\mu}}{\partial \xi_2} + \frac{\partial \tau_{i3}^{\lambda\mu}}{\partial z} = 0 \quad (i = 2, 3;\ \lambda\mu = 11, 22), \tag{19.16}$$

$$\frac{1}{h_2}\frac{\partial \tau_{12}^{12}}{\partial \xi_2} + \frac{\partial \tau_{13}^{12}}{\partial z} = 0, \tag{19.17}$$

$$\begin{gathered}
\tau_{11}^{\lambda\mu} = \frac{1}{h_2} c_{12} \frac{\partial U_2^{\lambda\mu}}{\partial \xi_2} + c_{13} \frac{\partial U_3^{\lambda\mu}}{\partial z}, \\
\tau_{22}^{\lambda\mu} = \frac{1}{h_2} c_{22} \frac{\partial U_2^{\lambda\mu}}{\partial \xi_2} + c_{23} \frac{\partial U_3^{\lambda\mu}}{\partial z}, \\
\tau_{33}^{\lambda\mu} = \frac{1}{h_2} c_{32} \frac{\partial U_2^{\lambda\mu}}{\partial \xi_2} + c_{33} \frac{\partial U_3^{\lambda\mu}}{\partial z}, \\
\tau_{23}^{\lambda\mu} = c_{44}\left(\frac{1}{h_2}\frac{\partial U_3^{\lambda\mu}}{\partial \xi_2} + \frac{\partial U_2^{\lambda\mu}}{\partial z}\right) \quad (\lambda\mu = 11, 22), \\
\tau_{13}^{12} = c_{55}\frac{\partial U_1^{12}}{\partial z}, \qquad \tau_{12}^{12} = c_{66}\frac{1}{h_2}\frac{\partial U_1^{12}}{\partial \xi_2},
\end{gathered} \tag{19.18}$$

subject to the following conditions on the outer surfaces of Ω:

$$\begin{gathered}
t_2^{11} = -c_{21}\frac{n_2}{h_2}, \qquad t_3^{11} = -c_{31} n_3, \\
t_2^{22} = -c_{22}\frac{n_2}{h_2}, \qquad t_3^{22} = -c_{32} n_3, \\
t_1^{12} = -c_{66}\frac{n_2}{h_2},
\end{gathered} \tag{19.19}$$

where

$$\begin{gathered}
t_2^{\lambda\mu} = \tau_{22}^{\lambda\mu}\frac{n_2}{h_2} + \tau_{23}^{\lambda\mu} n_3, \\
t_3^{\lambda\mu} = \tau_{23}^{\lambda\mu}\frac{n_2}{h_2} + \tau_{33}^{\lambda\mu} n_3 \quad (\lambda\mu = 11, 22), \\
t_1^{12} = \tau_{12}^{12}\frac{n_2}{h_2} + \tau_{13}^{12} n_3.
\end{gathered} \tag{19.20}$$

The above problems yield

$$\begin{aligned}
&b_{11}^{11}=\tau_{11}^{11}+c_{11}, &&b_{22}^{11}=\tau_{22}^{11}+c_{21}, &&b_{12}^{12}=\tau_{12}^{12}+c_{66},\\
&b_{11}^{22}=\tau_{11}^{22}+c_{21}, &&b_{22}^{22}=\tau_{22}^{22}+c_{22}, && \\
&b_{33}^{22}=\tau_{33}^{22}+c_{32}, &&b_{3\beta}^{\lambda\mu}=\tau_{\beta 3}^{\lambda\mu}, &&b_{33}^{11}=\tau_{33}^{11}+c_{31}.
\end{aligned} \tag{19.21}$$

The important point to be made about the two-dimensional analysis is that each of the local problems divide itself into a plane strain problem, equation (19.16), and an antiplane strain problem, equation (19.17), either one being solved independently of the other. A similar situation exists with regard to the group (b^*) problems, the two-dimensional forms of which are

$$\frac{1}{h_2}\frac{\partial \bar{\tau}_{i2}^{\lambda\mu}}{\partial \xi_2}+\frac{\partial \bar{\tau}_{i3}^{\lambda\mu}}{\partial z}=-c_{33\lambda\mu}\delta_{i3} \quad (i=2,3;\ \lambda\mu=11,22), \tag{19.22}$$

$$\frac{1}{h_2}\frac{\partial \bar{\tau}_{12}^{12}}{\partial \xi_2}+\frac{\partial \bar{\tau}_{13}^{12}}{\partial z}=0, \tag{19.23}$$

$$\begin{aligned}
\bar{\tau}_{11}^{\lambda\mu}&=\frac{1}{h_2}c_{12}\frac{\partial V_2^{\lambda\mu}}{\partial \xi_2}+c_{13}\frac{\partial V_3^{\lambda\mu}}{\partial z},\\
\bar{\tau}_{22}^{\lambda\mu}&=\frac{1}{h_2}c_{22}\frac{\partial V_2^{\lambda\mu}}{\partial \xi_2}+c_{23}\frac{\partial V_3^{\lambda\mu}}{\partial z},\\
\bar{\tau}_{33}^{\lambda\mu}&=\frac{1}{h_2}c_{32}\frac{\partial V_2^{\lambda\mu}}{\partial \xi_2}+c_{33}\frac{\partial V_3^{\lambda\mu}}{\partial z},\\
\bar{\tau}_{23}^{\lambda\mu}&=c_{44}\left(\frac{1}{h_2}\frac{\partial V_3^{\lambda\mu}}{\partial \xi_2}+\frac{\partial V_2^{\lambda\mu}}{\partial z}\right)\\
&(\lambda\mu=11,22),\\
\bar{\tau}_{13}^{12}&=c_{55}\frac{\partial V_1^{12}}{\partial z}, \qquad \bar{\tau}_{12}^{12}=c_{66}\frac{1}{h_2}\frac{\partial V_1^{12}}{\partial \xi_2}.
\end{aligned} \tag{19.24}$$

To these we have to add the following boundary conditions:

$$\begin{aligned}
&\bar{t}_2^{11}=-zc_{21}\frac{n_2}{h_2}, &&\bar{t}_3^{11}=-zc_{31}n_3,\\
&\bar{t}_2^{22}=-zc_{22}\frac{n_2}{h_2}, &&\bar{t}_3^{22}=-zc_{32}n_3,\\
&\bar{t}_1^{12}=-zc_{66}\frac{n_2}{h_2}, &&
\end{aligned} \tag{19.25}$$

where

$$\begin{aligned}
&\bar{t}_2^{\lambda\mu}=\bar{\tau}_{22}^{\lambda\mu}\frac{n_2}{h_2}+\bar{\tau}_{23}^{\lambda\mu}n_3, \qquad \bar{t}_3^{\lambda\mu}=\bar{\tau}_{23}^{\lambda\mu}\frac{n_2}{h_2}+\bar{\tau}_{33}^{\lambda\mu}n_3\\
&(\lambda\mu=11,22),\\
&\bar{t}_1^{12}=\bar{\tau}_{12}^{12}\frac{n_2}{h_2}+\bar{\tau}_{13}^{12}n_3.
\end{aligned} \tag{19.26}$$

The solutions of the group (b^*) problems having been found, we then calculate the following functions:

$$\begin{aligned}
&b^{*11}_{11} = \bar{\tau}^{11}_{11} + zc_{11}, &&b^{*11}_{22} = \bar{\tau}^{11}_{22} + zc_{21},\\
&b^{*22}_{11} = \bar{\tau}^{22}_{11} + zc_{21}, &&b^{*22}_{22} = \bar{\tau}^{22}_{22} + zc_{22},\\
&b^{*12}_{12} = \bar{\tau}^{12}_{12} + zc_{66}, &&b^{*11}_{33} = \bar{\tau}^{11}_{31} + zc_{31},\\
&b^{*22}_{33} = \bar{\tau}^{22}_{33} + zc_{32}, &&b^{*\lambda\mu}_{3\beta} = \bar{\tau}^{\lambda\mu}_{\beta 3},\\
&b^{*12}_{11} = b^{*12}_{22} = b^{*11}_{21} = b^{*22}_{12} = 0.
\end{aligned} \tag{19.27}$$

19.3. General features of the effective elastic moduli of ribbed shell and plate structures

On the basis of the analysis of the two-dimensional local problems (19.16)–(19.20) and (19.22)–(19.26), combined with relations (19.21) and (19.27), some general predictions concerning the effective elastic properties of ribbed plates or shells can be made that are independent of the specific shape of the unit cell. In particular, it turns out that in the case of a homogeneous orthotropic material we have

$$\begin{aligned}
&\langle b^{11}_{11}\rangle = \nu^2_{21}\langle b^{22}_{22}\rangle + E_1\langle 1\rangle, &&\langle b^{22}_{11}\rangle = \langle b^{11}_{22}\rangle = \nu_{21}\langle b^{22}_{22}\rangle,\\
&\langle b^{*11}_{11}\rangle = \nu^2_{21}\langle b^{*22}_{22}\rangle + E_1\langle z\rangle, &&\langle b^{*22}_{11}\rangle = \langle b^{*11}_{22}\rangle = \nu_{21}\langle b^{*22}_{22}\rangle,\\
&\langle zb^{*11}_{11}\rangle = \nu^2_{21}\langle zb^{*22}_{22}\rangle + E_1\langle z^2\rangle, &&\langle zb^{*22}_{11}\rangle = \langle zb^{*11}_{22}\rangle = \nu_{21}\langle zb^{*22}_{22}\rangle.
\end{aligned} \tag{19.28}$$

The proof of these relations is as follows. Using expressions for $\tau^{\lambda\mu}_{22}$ and $\tau^{\lambda\mu}_{33}$ from (19.18), it is easily verified that

$$\frac{1}{h_2}\frac{\partial U^{\lambda\mu}_{22}}{\partial \xi_2} = \frac{c_{33}\tau^{\lambda\mu}_{22} - c_{23}\tau^{\lambda\mu}_{33}}{c_{22}c_{33} - c^2_{23}}, \qquad \frac{\partial U^{\lambda\mu}_{3}}{\partial z} = \frac{c_{22}\tau^{\lambda\mu}_{33} - c_{23}\tau^{\lambda\mu}_{22}}{c_{22}c_{33} - c^2_{23}}, \tag{19.29}$$

which when substituted into (19.18) for $\tau^{\lambda\mu}_{11}$ gives

$$\tau^{\lambda\mu}_{11} = \nu_{21}\tau^{\lambda\mu}_{22} + \nu_{31}\tau^{\lambda\mu}_{33}, \tag{19.30}$$

with the help of expressions for ν_{21} and ν_{31} in (19.4).

From (19.24), proceeding in a precisely similar manner, we find

$$\bar{\tau}^{\lambda\mu}_{11} = \nu_{21}\bar{\tau}^{\lambda\mu}_{22} + \nu_{31}\tau^{\lambda\mu}_{33}. \tag{19.31}$$

Now using (19.21) and (19.30) along with (19.4) we find that

$$\begin{aligned}
b^{11}_{11} &= c_{11} + \nu_{21}\tau^{11}_{22} + \nu_{31}\tau^{11}_{33} = c_{11} - \nu_{21}c_{21}\\
&\quad - \nu_{31}c_{31} + \nu_{21}b^{11}_{22} + \nu_{31}b^{11}_{33} = E_1 + \nu_{21}b^{11}_{22} + \nu_{31}b^{11}_{33},\\
b^{22}_{11} &= c_{12} + \nu_{21}\tau^{22}_{22} + \nu_{31}\tau^{22}_{33} = c_{12} - \nu_{21}c_{22}\\
&\quad - \nu_{31}c_{32} + \nu_{21}b^{22}_{22} + \nu_{31}b^{22}_{33} = \nu_{21}b^{22}_{22} + \nu_{31}b^{22}_{33}.
\end{aligned}$$

Using (19.27) and (19.31) we find that, for the general case of a two-dimensional region Ω.

$$b_{11}^{11} = E_1 + \nu_{21} b_{22}^{11} + \nu_{31} b_{33}^{11}, \tag{19.32}$$

$$b_{11}^{22} = \nu_{21} b_{22}^{22} + \nu_{31} b_{33}^{22}, \tag{19.33}$$

$$b_{11}^{*11} = zE_1 + \nu_{21} b_{22}^{*11} + \nu_{31} b_{33}^{*11}, \tag{19.34}$$

$$b_{11}^{*22} = \nu_{21} b_{22}^{*22} + \nu_{31} b_{33}^{*22}. \tag{19.35}$$

Letting (15.27) act on (19.32) and using the general formula (15.62), now gives, for a homogeneous material,

$$\langle b_{11}^{11} \rangle = \nu_{21} \langle b_{22}^{11} \rangle + E_1 \langle 1 \rangle, \qquad \langle b_{11}^{22} \rangle = \nu_{21} \langle b_{22}^{22} \rangle, \tag{19.36}$$

which when combined with the general properties of the effective elastic moduli, yields the first line in equations (19.28).

Averaging (19.33) after first multiplying by z, and using (15.62), we find

$$\langle z b_{11}^{22} \rangle = \nu_{21} \langle z b_{22}^{22} \rangle,$$

which, by (15.76), may be written as

$$\langle b_{22}^{*11} \rangle = \nu_{21} \langle b_{22}^{*22} \rangle. \tag{19.37}$$

Next we average (19.34) and (19.35), again using (15.76). This gives

$$\langle b_{11}^{*11} \rangle = \nu_{21} \langle b_{22}^{*11} \rangle + E_1 \langle z \rangle, \qquad \langle b_{11}^{*22} \rangle = \nu_{21} \langle b_{22}^{*22} \rangle, \tag{19.38}$$

and by combining (19.37) and (19.38) we obtain the second line in (19.28).

The third and the last line in (19.28) follows by averaging (19.24) and (19.35) after first multiplying them by z and with the aid of (15.62) and (15.76).

We note that formulae (19.28) are perfectly general as far as two-dimensional problems are concerned, and that they lead to considerable simplification in the homogenized shell elastic relations (15.88).

19.4. Formulation of local heat conduction and thermoelastic problems

The effective thermal conductivities occurring in the homogenized heat conduction relations (16.35) are found by solving the local heat conduction problems for the functions $l_{i\mu}(\xi, z)$ and $l_{i\mu}^{*}(\xi, z)$, periodic in ξ_1 and ξ_2 with respective periods A_1 and A_2. These local problems divide themselves into two groups of problems for the orthotropic region Ω, (l) and (l^*), posed by equations (16.24), (16.26) and (16.28)–(16.30), and (16.24), (16.27)–(16.29) and (16.31), respectively.

The heat conductivity tensor of an orthotropic material is of the form

$$\lambda_{ij} = \begin{bmatrix} \lambda_1 & 0 & 0 \\ 0 & \lambda_2 & 0 \\ 0 & 0 & \lambda_3 \end{bmatrix}, \tag{19.39}$$

and from (16.24) it therefore follows that

$$l_{11}=\lambda_1+\frac{\lambda_1}{h_1}\frac{\partial W_1}{\partial \xi_1},\qquad l_{21}=\frac{\lambda_2}{h_2}\frac{\partial W_1}{\partial \xi_2},\qquad l_{31}=\lambda_3\frac{\partial W_1}{\partial z},$$
$$l_{12}=\frac{\lambda_1}{h_1}\frac{\partial W_2}{\partial \xi_1},\qquad l_{22}=\lambda_2+\frac{\lambda_2}{h_2}\frac{\partial W_2}{\partial \xi_2},\qquad l_{32}=\lambda_3\frac{\partial W_2}{\partial z},\tag{19.40}$$

$$l^*_{11}=z\lambda_1+\frac{\lambda_1}{h_1}\frac{\partial W^*_1}{\partial \xi_1},\qquad l^*_{21}=\frac{\lambda_2}{h_2}\frac{\partial W^*_1}{\partial \xi_2},\qquad l^*_{31}=\lambda_3\frac{\partial W^*_1}{\partial z},$$
$$l^*_{12}=\frac{\lambda_1}{h_1}\frac{\partial W^*_2}{\partial \xi_1},\qquad l^*_{22}=z\lambda_2+\frac{\lambda_2}{h_2}\frac{\partial W^*_2}{\partial \xi_2},\qquad l^*_{32}=\lambda_3\frac{\partial W^*_2}{\partial z}.\tag{19.41}$$

Substitution of (19.40) into (16.26) yields

$$\frac{\lambda_\beta}{h^2_\beta}\frac{\partial^2 W_1}{\partial \xi^2_\beta}+\lambda_3\frac{\partial^2 W_1}{\partial z^2}=0,\tag{19.42}$$

$$\frac{\lambda_\beta}{h^2_\beta}\frac{\partial^2 W_2}{\partial \xi^2_\beta}+\lambda_3\frac{\partial^2 W_2}{\partial z^2}=0,\tag{19.43}$$

and from (16.29), (16.30) and (19.40) the conditions to be satisfied on the boundary surfaces are

$$\frac{\lambda_\beta}{h_\beta}\frac{\partial W_1}{\partial \xi_\beta}n_\beta+\lambda_3\frac{\partial W_1}{\partial z}n_3=-\frac{\lambda_1}{h_1}n_1,\tag{19.44}$$

$$\frac{\lambda_\beta}{h^2_\beta}\frac{\partial W_2}{\partial \xi_\beta}n_\beta+\lambda_3\frac{\partial W_1}{\partial z}n_3=-\frac{\lambda_2}{h_2}n_2,\tag{19.45}$$

The local problem (l) thus divides itself into two uncoupled problems, ($l1$) (equations (19.42) and (19.44)), and ($l2$) (equations (19.43) and (19.45)), the solutions of which determine the functions $W_1(\boldsymbol{\xi},z)$ and $W_2(\boldsymbol{\xi},z)$, periodic in ξ_1 and ξ_2 with respective periods A_1 and A_2.

Similarly, from (19.41), (16.27), (16.29) and (16.31) we obtain the formulation of the problem (l^*) which also decouples into two independent problems, (l^*1) for the function $W^*_1(\boldsymbol{\xi},z)$, and ($l^*2$) for $W^*_2(\boldsymbol{\xi},z)$. That is,

$$\frac{\lambda_\beta}{h^2_\beta}\frac{\partial^2 W^*_1}{\partial \xi^2_\beta}+\lambda_3\frac{\partial^2 W^*_1}{\partial z^2}=0,$$
$$\frac{\lambda_\beta}{h^2_\beta}\frac{\partial W^*_1}{\partial \xi_\beta}n_\beta+\lambda_3\frac{\partial W^*_1}{\partial z}n_3=-z\frac{\lambda_1}{h_1}n_1,\tag{19.46}$$

$$\frac{\lambda_\beta}{h^2_\beta}\frac{\partial^2 W^*_2}{\partial \xi^2_\beta}+\lambda_3\frac{\partial^2 W^*_2}{\partial z^2}=0,$$
$$\frac{\lambda_\beta}{h^2_\beta}\frac{\partial W^*_2}{\partial \xi_\beta}n_\beta+\lambda_3\frac{\partial W^*_2}{\partial z}=-z\frac{\lambda_2}{h_2}n_2,\tag{19.47}$$

where $W_1^*(\xi, z)$ and $W_2^*(\xi, z)$ are also periodic in ξ_1 and ξ_2 with respective periods A_1 and A_2.

The above local problem formulations should be complemented by relations (16.28) in order to secure the uniqueness of their solutions.

The local thermoelastic problems (17.9), (17.11), (17.15) and (17.16) give us the functions $s_{ij}(\xi, z)$ and $s_{ij}^*(\xi, z)$, from which the effective thermoelastic coefficients involved in the constitutive relations (17.23) are calculated. The local problems for these functions again form two groups of problems, (s) and (s^*), for the functions $S_m(\xi, z)$ and $S_m^*(\xi, z)$, respectively. These local problems may be represented in a form entirely analogous to the local elastic problems (19.6)–(19.10) and (19.11)–(19.15), with appropriate modifications in relations (15.41) and (15.53) for elastic problems, and in (17.9) for thermoelastic problems.

The reader will be reminded, with reference to Section 17, that evaluation of the effective thermoelastic properties does not actually presuppose the solution of the above problems and can be carried out instead by formulas (17.17), in which only the local elastic solutions are needed.

20. AN APPROXIMATE METHOD FOR DETERMINING THE EFFECTIVE PROPERTIES OF RIBBED AND WAFER SHELLS

20.1. Effective elastic moduli of ribbed and wafer shells of orthotropic materials

The structure we are concerned with in this section is an eccentrically reinforced wafer type shell (or plate), the unit cell of which is shown schematically in Fig. 20.1. Ignoring unavoidable inter-elemental fillets, as we do in the figure, it is seen that the unit cell consists of three mutually perpendicular plane elements, or plates.

An approximate solution of the local problems relevant to this kind of geometry may

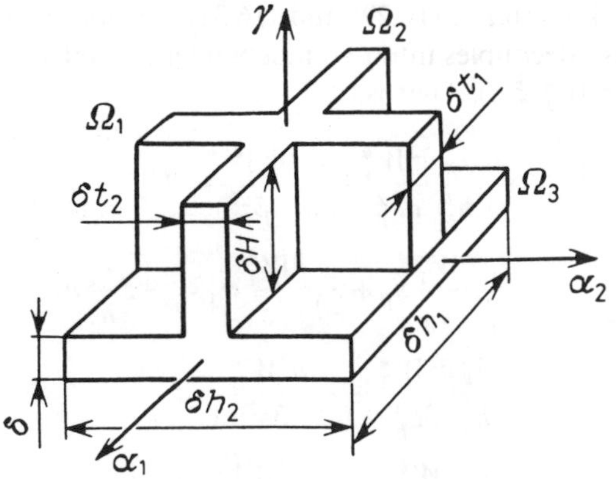

Fig. 20.1 *Unit cell of a wafer shell.*

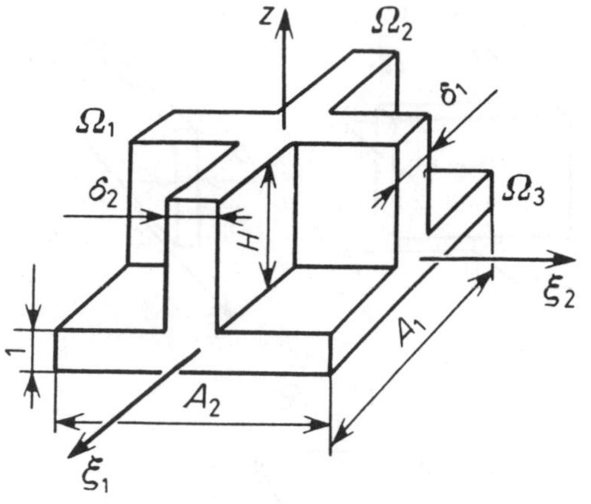

Fig. 20.2 *Unit cell of a wafer shell in the coordinate system ξ_1, ξ_2, z.*

be found under the assumption that the thickness of each of the three elements is small compared with its other two dimensions, i.e.

$$t_1 \ll h_2, \qquad t_2 \ll h_1, \qquad H \sim h_1, h_2. \tag{20.1}$$

The local problems can then be approximately solved for each of the cell elements separately and, although the near-joint stress concentrations clearly remain unaccounted for in this approach, one feels confident in believing that the overall properties of the shell will be predicted accurately enough owing to the fact that the troublesome regions are highly localized at the joints and contribute little to the integrals over the unit cell. The error introduced in this way will be estimated in a more refined analysis in Section 22, with proper attention to all the interactions involved.

If we change to the coordinate system ξ_1, ξ_2, z, the unit cell of the problem will be as shown in Fig. 20.2, where

$$\delta_1 = \frac{t_1 A_1}{h_1}, \qquad \delta_2 = \frac{t_2 A_2}{h_2}. \tag{20.2}$$

and if conditions (20.1) are fulfilled, the local problems of interest can be solved independently for the elements Ω_1, Ω_2 and Ω_3 shown in Fig. 20.3. We begin our discussion with the problem ($b11$) posed by equations (19.6)–(19.9) and (15.57) in which $\lambda\mu = 11$.

(a) In region Ω_3, defined by $\{|\xi_1| < A_1/2, |\xi_2| < A_2/2, |z| < \frac{1}{2}\}$, the functions τ_{13}^{11}, τ_{23}^{11} and τ_{33}^{11} are assumed to satisfy the following conditions on the surfaces $z = \pm\frac{1}{2}$:

$$\tau_{13}^{11} = \tau_{23}^{11} = 0, \qquad \tau_{33}^{11} = -c_{31}.$$

Because of the periodicity in ξ_1 and ξ_2, this implies that

$$U_1^{11} = U_2^{11} = 0, \qquad U_3^{11} = -\frac{c_{31}}{c_{33}} z, \tag{20.3}$$

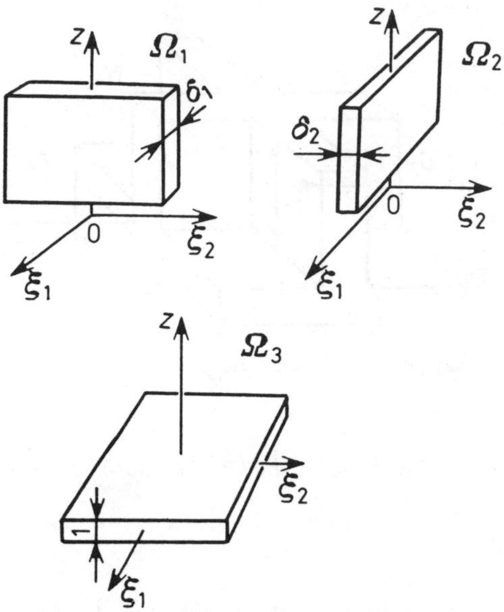

Fig. 20.3 *Individual elements of the unit cell of a wafer shell.*

and hence

$$\tau_{11}^{11} = -\frac{c_{13}^2}{c_{33}}, \qquad \tau_{22}^{11} = -\frac{c_{23}c_{13}}{c_{33}},$$
$$\tau_{33}^{11} = -c_{13}, \qquad \tau_{23}^{11} = \tau_{13}^{11} = \tau_{12}^{11} = 0 \tag{20.4}$$

everywhere in Ω_3.

(b) In region Ω_1, defined by $\{|\xi_1| < \delta_1/2, |\xi_2| < A_2/2, \frac{1}{2} < z < \frac{1}{2} + H\}$, the boundary conditions (19.8) and (19.9) take the form

$$\tau_{11}^{11} = -c_{11}, \qquad \tau_{12}^{11} = \tau_{13}^{11} = 0 \quad (\xi_1 = \pm\delta_1/2)$$
$$\tau_{13}^{11} = \tau_{23}^{11} = 0, \qquad \tau_{33}^{11} = -c_{13} \quad (z = \tfrac{1}{2}, z = \tfrac{1}{2} + H),$$

giving

$$U_1^{11} = -h_1\xi_1, \qquad U_2^{11} = U_3^{11} = 0 \tag{20.5}$$

in view of the periodicity in ξ_2, so that everywhere in Ω_1 we have

$$\tau_{11}^{11} = -c_{11}, \qquad \tau_{22}^{11} = -c_{12}, \qquad \tau_{33}^{11} = -c_{13},$$
$$\tau_{23}^{11} = \tau_{13}^{11} = \tau_{12}^{11} = 0. \tag{20.6}$$

(c) In region Ω_2, defined by $\{|\xi_1| < A_1/2, |\xi_2| < \delta_2/2, \frac{1}{2} < z < \frac{1}{2} + H\}$, the face conditions are

$$\tau_{12}^{11} = \tau_{23}^{11} = 0, \qquad \tau_{22}^{11} = -c_{12} \quad (\xi_2 = \pm\delta_2/2),$$
$$\tau_{13}^{11} = \tau_{23}^{11} = 0, \qquad \tau_{33}^{11} = -c_{13} \quad (z = \tfrac{1}{2}, z = \tfrac{1}{2} + H),$$

and because of the periodicity in ξ_1,

$$U_1^{11}=0, \qquad U_2^{11}=h_2\frac{c_{13}c_{23}-c_{12}c_{33}}{c_{22}c_{33}-c_{23}^2}\xi_2,$$
$$U_3^{11}=\frac{c_{12}c_{23}-c_{13}c_{22}}{c_{22}c_{33}-c_{23}^2}z, \tag{20.7}$$

which means that

$$\tau_{11}^{11}=\frac{2c_{12}c_{13}c_{23}-c_{12}^2c_{33}-c_{13}^2c_{22}}{c_{22}c_{33}-c_{23}^2},$$
$$\tau_{22}^{11}=-c_{12}, \qquad \tau_{33}^{11}=-c_{13}, \qquad \tau_{23}^{11}=\tau_{13}^{11}=\tau_{12}^{11}=0 \tag{20.8}$$

in region Ω_2.

Combining the solutions (20.4), (20.6) and (20.8), and using (19.4) and (19.10) we find

$$b_{11}^{11}=\begin{cases} c_{11}-\dfrac{c_{13}^2}{c_{33}}=\dfrac{E_1}{1-\nu_{12}\nu_{21}}, & \text{in } \Omega_3,\\ 0, & \text{in } \Omega_1\\ c_{11}+\dfrac{2c_{12}c_{13}c_{23}-c_{12}^2c_{33}-c_{13}^2c_{22}}{c_{22}c_{33}-c_{23}^2}=E_1, & \text{in } \Omega_2,\end{cases} \tag{20.9}$$

$$b_{22}^{11}=\begin{cases} c_{12}-\dfrac{c_{23}c_{13}}{c_{33}}=\dfrac{\nu_{12}E_1}{1-\nu_{12}\nu_{21}}=\dfrac{\nu_{21}E_2}{1-\nu_{12}\nu_{21}}, & \text{in } \Omega_3,\\ 0, & \text{in } \Omega_1,\Omega_2,\end{cases}$$

$$b_{12}^{11}=b_{33}^{11}=b_{32}^{11}=b_{31}^{11}=0.$$

Solutions of the problems ($b22$) and ($b12$) are obtained in a precisely similar manner. We give the solution of the problem ($b22$) first.

(a) In region Ω_3,

$$U_1^{22}=U_2^{22}=0, \qquad U_3^{22}=-\frac{c_{32}}{c_{33}}z,$$
$$\tau_{11}^{22}=-\frac{c_{13}c_{23}}{c_{33}}, \qquad \tau_{22}^{22}=-\frac{c_{23}^2}{c_{33}}, \tag{20.10}$$
$$\tau_{33}^{22}=-c_{32}, \qquad \tau_{23}^{22}=\tau_{13}^{22}=\tau_{12}^{22}=0.$$

(b) In region Ω_1,

$$U_1^{22}=h_1\frac{c_{13}c_{23}-c_{12}c_{33}}{c_{11}c_{33}-c_{13}^2}\xi_1, \qquad U_2^{22}=0,$$
$$U_3^{22}=\frac{c_{13}c_{12}-c_{11}c_{23}}{c_{11}c_{33}-c_{13}^2}z,$$
$$\tau_{11}^{22}=-c_{12}, \qquad \tau_{22}^{22}=\frac{2c_{12}c_{13}c_{23}-c_{12}^2c_{33}-c_{11}c_{23}^2}{c_{11}c_{33}-c_{13}^2}, \tag{20.11}$$
$$\tau_{33}^{22}=-c_{23}, \qquad \tau_{23}^{22}=\tau_{13}^{22}=\tau_{12}^{22}=0.$$

(c) In region Ω_2,

$$\begin{aligned} &U_1^{22} = U_3^{22} = 0, \qquad U_2^{22} = -h_2\xi_2, \\ &\tau_{11}^{22} = -c_{12}, \qquad \tau_{22}^{22} = -c_{22}, \qquad \tau_{33}^{22} = -c_{23}, \\ &\tau_{23}^{22} = \tau_{13}^{22} = \tau_{12}^{22} = 0. \end{aligned} \tag{20.12}$$

Using (20.10)–(20.12) together with (19.3) and (19.10), it is found that

$$\begin{aligned} &b_{22}^{22} = \begin{cases} c_{22} - \dfrac{c_{23}^2}{c_{33}} = \dfrac{E_2}{1 - \nu_{12}\nu_{21}}, & \text{in } \Omega_3, \\ c_{22} + \dfrac{2c_{12}c_{13}c_{23} - c_{12}^2 c_{33} - c_{11}c_{23}^2}{c_{11}c_{33} - c_{13}^2} = E_2, & \text{in } \Omega_1, \\ 0, & \text{in } \Omega_2, \end{cases} \\ &b_{11}^{22} = \begin{cases} c_{12} - \dfrac{c_{13}c_{23}}{c_{33}} = \dfrac{\nu_{12}E_1}{1 - \nu_{12}\nu_{21}} = \dfrac{\nu_{21}E_2}{1 - \nu_{12}\nu_{21}}, & \text{in } \Omega_3, \\ 0, & \text{in } \Omega_1, \Omega_2, \end{cases} \\ &b_{12}^{22} = b_{33}^{22} = b_{32}^{22} = b_{31}^{22} = 0. \end{aligned} \tag{20.13}$$

The solution of the problem ($b12$) is as follows:

(a) In region Ω_3,

$$\begin{aligned} &U_1^{12} = U_2^{12} = U_3^{12} = 0, \\ &\tau_{11}^{12} = \tau_{22}^{12} = \tau_{33}^{12} = \tau_{23}^{12} = \tau_{13}^{12} = \tau_{12}^{12} = 0. \end{aligned} \tag{20.14}$$

(b) In region Ω_1,

$$\begin{aligned} &U_1^{12} = U_3^{12} = 0, \qquad U_2^{12} = -h_1\xi_1, \\ &\tau_{12}^{12} = -c_{66}, \qquad \tau_{11}^{12} = \tau_{22}^{12} = \tau_{33}^{12} = \tau_{23}^{12} = \tau_{13}^{12} = 0. \end{aligned} \tag{20.15}$$

(c) In region Ω_2,

$$\begin{aligned} &U_1^{12} = -h_2\xi_2, \qquad U_2^{12} = U_3^{12} = 0, \\ &\tau_{12}^{12} = -c_{66}, \qquad \tau_{11}^{12} = \tau_{22}^{12} = \tau_{33}^{12} = \tau_{23}^{12} = \tau_{13}^{12} = 0, \end{aligned} \tag{20.16}$$

Hence by (19.4) and (19.10):

$$\begin{aligned} &b_{12}^{12} = \begin{cases} c_{66} = G_{12}, & \text{in } \Omega_3, \\ 0, & \text{in } \Omega_1, \Omega_2, \end{cases} \\ &b_{11}^{12} = b_{22}^{12} = b_{33}^{12} = b_{23}^{12} = b_{13}^{12} = 0. \end{aligned} \tag{20.17}$$

To proceed further it is necessary to construct approximate solutions for the local problems of group (b^*). Here again they are obtained by superposing the solutions for regions Ω_1, Ω_2 and Ω_3.

The approximate solution for the problem (b^*11) is as follows.

(a) In region Ω_3,

$$V_1^{11} = V_2^{11} = 0, \qquad V_3^{11} = -\frac{c_{31}}{c_{33}}\frac{z^2}{2},$$
$$\bar{\tau}_{11}^{11} = -z\frac{c_{13}^2}{c_{33}}, \qquad \bar{\tau}_{22}^{11} = -\frac{c_{23}c_{13}}{c_{33}}z, \tag{20.18}$$
$$\bar{\tau}_{33}^{11} = -zc_{13}, \qquad \bar{\tau}_{23}^{11} = \bar{\tau}_{13}^{11} = \bar{\tau}_{12}^{11} = 0.$$

(b) In region Ω_1,

$$V_1^{11} = -h_1\xi_1 z, \qquad V_2^{11} = 0, \qquad V_3^{11} = \frac{h_1^2}{2}\left(\xi^2 - \frac{\delta_1^2}{12}\right),$$
$$\bar{\tau}_{11}^{11} = -zc_{11}, \qquad \bar{\tau}_{22}^{11} = -zc_{21}, \qquad \bar{\tau}_{33}^{11} = -zc_{31}, \tag{20.19}$$
$$\bar{\tau}_{23}^{11} = \bar{\tau}_{13}^{11} = \bar{\tau}_{12}^{11} = 0.$$

(c) In region Ω_2,

$$V_1^{11} = 0, \qquad V_2^{11} = h_2\frac{c_{13}c_{23} - c_{12}c_{33}}{c_{22}c_{33} - c_{23}^2}z\xi_2,$$
$$V_3^{11} = \frac{c_{12}c_{23} - c_{13}c_{22}}{c_{22}c_{33} - c_{23}^2}\frac{z^2}{2} - \frac{h_2^2}{2}\frac{c_{13}c_{23} - c_{12}c_{33}}{c_{22}c_{33} - c_{23}^2}\left(\xi_2^2 - \frac{\delta_2^2}{12}\right),$$
$$\bar{\tau}_{11}^{11} = \frac{2c_{12}c_{13}c_{23} - c_{12}^2c_{33} - c_{13}^2c_{22}}{c_{22}c_{33} - c_{23}^2}z, \tag{20.20}$$
$$\bar{\tau}_{22}^{11} = -zc_{12}, \qquad \bar{\tau}_{33}^{11} = -zc_{13}, \qquad \bar{\tau}_{23}^{11} = \bar{\tau}_{13}^{11} = \bar{\tau}_{12}^{11} = 0.$$

Using (20.18)–(20.20) along with (19.4) and (19.15) now gives expressions for the functions b_{ij}^{*11},which when compared with (20.9) lead to the result

$$b_{ij}^{*11} = zb_{ij}^{11}. \tag{20.21}$$

Next follows the solution of the problem (b^*22).

(a) In region Ω_3,

$$V_1^{22} = V_2^{22} = 0, \qquad V_3^{22} = -\frac{c_{23}}{c_{33}}\frac{z^2}{2},$$
$$\bar{\tau}_{11}^{22} = -\frac{c_{13}c_{23}}{c_{33}}z, \qquad \bar{\tau}_{22}^{22} = -\frac{c_{23}^2}{c_{23}}z, \tag{20.22}$$
$$\bar{\tau}_{33}^{22} = -zc_{23}, \qquad \bar{\tau}_{23}^{22} = \bar{\tau}_{13}^{22} = \bar{\tau}_{12}^{22} = 0.$$

(b) In region Ω_1,

$$V_1^{22} = h_1 \frac{c_{13}c_{23} - c_{12}c_{33}}{c_{11}c_{33} - c_{13}^2} z\xi_1, \qquad V_2^{22} = 0,$$

$$V_3^{22} = \frac{c_{13}c_{12} - c_{11}c_{23}}{c_{11}c_{33} - c_{13}^2}\frac{z^2}{2} - \frac{h_1^2}{2}\frac{c_{13}c_{23} - c_{12}c_{33}}{c_{11}c_{33} - c_{13}^2}\left(\xi_1^2 - \frac{\delta_1^2}{12}\right), \tag{20.23}$$

$$\bar{\tau}_{11}^{22} = -zc_{12}, \qquad \bar{\tau}_{22}^{22} = \frac{2c_{12}c_{13}c_{23} - c_{12}^2c_{33} - c_{11}c_{23}^2}{c_{11}c_{33} - c_{13}^2} z,$$

$$\bar{\tau}_{33}^{22} = -zc_{23}, \qquad \bar{\tau}_{23}^{22} = \bar{\tau}_{13}^{22} = \bar{\tau}_{12}^{22} = 0.$$

(c) In region Ω_2,

$$V_1^{22} = 0, \qquad V_2^{22} = -h_2\xi_2 z, \qquad V_3^{22} = \frac{h_2^2}{2}\left(\xi_2^2 - \frac{\delta_2^2}{12}\right),$$

$$\bar{\tau}_{11}^{22} = -zc_{12}, \qquad \bar{\tau}_{22}^{22} = -zc_{22}, \qquad \bar{\tau}_{33}^{22} = -zc_{23}, \tag{20.24}$$

$$\bar{\tau}_{23}^{22} = \bar{\tau}_{13}^{22} = \bar{\tau}_{12}^{22} = 0.$$

From (20.22)–(20.24), (19.4) and (19.15) we may now obtain expressions for the functions b_{ij}^{*22} and, by the use of (20.13), we find that

$$b_{ij}^{*22} = z b_{ij}^{22}. \tag{20.25}$$

Turning now to the problem (b*12) we must admit that its solution is not so elementary as we have seen to be the case with problems (b*11) and (b*22).

(a) In region Ω_3 we have

$$V_1^{12} = V_2^{12} = V_3^{12} = 0,$$

$$\bar{\tau}_{11}^{12} = \bar{\tau}_{22}^{12} = \bar{\tau}_{33}^{12} = \bar{\tau}_{23}^{12} = \bar{\tau}_{13}^{12} = \bar{\tau}_{12}^{12} = 0. \tag{20.26}$$

(b) In region Ω_1, taking into account the boundary conditions

$$\bar{\tau}_{11}^{12} = 0, \qquad \bar{\tau}_{12}^{12} = -zc_{66}, \qquad \bar{\tau}_{13}^{12} = 0 \quad (\xi_1 = \pm\delta_1/2),$$

$$\bar{\tau}_{13}^{12} = 0, \qquad \bar{\tau}_{23}^{12} = 0, \qquad \bar{\tau}_{33}^{12} = 0 \quad (z = \tfrac{1}{2},\ z = \tfrac{1}{2} + H),$$

and the periodicity in ξ_2, we may set

$$\bar{\tau}_{11}^{12} = \bar{\tau}_{33}^{12} = \bar{\tau}_{13}^{12} = 0.$$

But then $V_1^{12} = 0$, $V_2^{12} = V_2^{12}(\xi_1, z)$ and $V_3^{12} = 0$, and the only non-zero functions are

$$\bar{\tau}_{12}^{12} = \frac{c_{66}}{h_1}\frac{\partial V_2^{12}(\xi_1, z)}{\partial \xi_1}, \qquad \bar{\tau}_{23}^{12} = c_{44}\frac{\partial V_2^{12}}{\partial z}, \tag{20.27}$$

which must be related by

$$\frac{1}{h_1}\frac{\partial \bar{\tau}_{12}^{12}}{\partial \xi_1} + \frac{\partial \bar{\tau}_{23}^{12}}{\partial z} = 0$$

in the region $|\xi_1| < \delta_1/2$, $\frac{1}{2} < z < \frac{1}{2} + H$, and must satisfy the boundary conditions

$$\bar{\tau}_{12}^{12} = -zc_{66} \quad (\xi_1 = \pm\delta_1/2),$$
$$\bar{\tau}_{23}^{12} = 0 \qquad (z = \tfrac{1}{2},\ z = \tfrac{1}{2} + H).$$

A boundary value problem for V_2^{12} (ξ_1, z), can now be obtained from (20.27) and (19.4) as

$$\begin{aligned} &\frac{G_{12}}{h_1^2}\frac{\partial^2 V_2^{12}}{\partial \xi_1^2} + G_{23}\frac{\partial^2 V_2^{12}}{\partial z^2} = 0, \\ &\frac{1}{h_1}\frac{\partial V_2^{12}}{\partial \xi_1} = -z \quad (\xi_1 = \pm\delta_1/2), \\ &\frac{\partial V_2^{12}}{\partial z} = 0 \quad (z = \tfrac{1}{2},\ z = \tfrac{1}{2} + H), \end{aligned} \tag{20.28}$$

and the solution satisfying (15.57) is found to be given by

$$V_2^{12} = -\frac{h_1}{2}(H+1)\xi_1 + \sqrt{\frac{G_{12}}{G_{23}}}\frac{2H^2}{\pi^3}\sum_{n=1}^{\infty}\frac{[1-(-1)^n]\sinh\left(\dfrac{\pi n h_1 \xi_1}{H}\sqrt{\dfrac{G_{23}}{G_{12}}}\right)}{n^3\cosh\left(\dfrac{\pi n h_1 \delta_1}{2H}\sqrt{\dfrac{G_{23}}{G_{12}}}\right)}\cos\frac{\pi n(z-\frac{1}{2})}{H}. \tag{20.29}$$

Finally, in region Ω_1 we have

$$\bar{\tau}_{11}^{12} = \bar{\tau}_{22}^{12} = \bar{\tau}_{33}^{12} = \bar{\tau}_{13}^{12} = 0,$$

$$\bar{\tau}_{12}^{12} = -G_{12}\frac{H+1}{2} + G_{12}\frac{2H}{\pi^2}\sum_{n=1}^{\infty}\frac{[1-(-1)^n]\cosh\left(\dfrac{\pi n h_1 \xi_1}{H}\sqrt{\dfrac{G_{23}}{G_{12}}}\right)}{n^2\cosh\left(\dfrac{\pi n h_1 \delta_1}{2H}\sqrt{\dfrac{G_{23}}{G_{12}}}\right)}\cos\frac{\pi n(z-\frac{1}{2})}{H}, \tag{20.30}$$

$$\bar{\tau}_{23}^{12} = -\sqrt{G_{23}G_{12}}\frac{2H}{\pi^2}\sum_{n=1}^{\infty}\frac{[1-(-1)^n]\sinh\left(\dfrac{\pi n h_1 \xi_1}{H}\sqrt{\dfrac{G_{23}}{G_{12}}}\right)}{n^2\cosh\left(\dfrac{\pi n h_1 \delta_1}{2H}\sqrt{\dfrac{G_{23}}{G_{12}}}\right)}\sin\frac{\pi n(z-\frac{1}{2})}{H}.$$

(c) The solution for region Ω_2 is carried out in much the same way as for Ω_1. The result is

$$V_1^{12} = -\frac{h_2}{2}(H+1)\xi_2 + \sqrt{\frac{G_{12}}{G_{13}}}\frac{2H^2}{\pi^3}\sum_{n=1}^{\infty}\frac{[1-(-1)^n]\sinh\left(\dfrac{\pi n h_2 \xi_2}{H}\sqrt{\dfrac{G_{13}}{G_{12}}}\right)}{n^3\cosh\left(\dfrac{\pi n h_2 \delta_2}{2H}\sqrt{\dfrac{G_{13}}{G_{12}}}\right)}\cos\frac{\pi n(z-\frac{1}{2})}{H},$$

$$V_2^{12} = 0, \qquad V_3^{12} = 0,$$

$$\bar{\tau}_{11}^{12} = \bar{\tau}_{22}^{12} = \bar{\tau}_{33}^{12} = \bar{\tau}_{23}^{12} = 0, \tag{20.31}$$

$$\bar{\tau}_{12}^{12} = -G_{12}\frac{H+1}{2} + G_{12}\frac{2H}{\pi^2}\sum_{n=1}^{\infty}\frac{[1-(-1)^n]\cosh\left(\dfrac{\pi n h_2 \xi_2}{H}\sqrt{\dfrac{G_{13}}{G_{12}}}\right)}{n^2\cosh\left(\dfrac{\pi n h_2 \delta_2}{2H}\sqrt{\dfrac{G_{13}}{G_{12}}}\right)}\cos\frac{\pi n(z-\frac{1}{2})}{H},$$

$$\bar{\tau}_{13}^{12} = \sqrt{G_{13}G_{12}}\frac{2H}{\pi^2}\sum_{n=1}^{\infty}\frac{[1-(-1)^n]\sinh\left(\dfrac{\pi n h_2 \xi_2}{H}\sqrt{\dfrac{G_{13}}{G_{12}}}\right)}{n^2\cosh\left(\dfrac{\pi n h_2 \delta_2}{2H}\sqrt{\dfrac{G_{13}}{G_{12}}}\right)}\sin\frac{\pi n(z-\frac{1}{2})}{H},$$

From (19.15), (20.26), (20.30) and (20.31) we find

$$b_{11}^{*12} = b_{22}^{*12} = b_{33}^{*12} = 0,$$

$$b_{12}^{*12} = \begin{cases} zG_{12}, & \text{in } \Omega_3, \\ G_{12}\left(z-\dfrac{H+1}{2}\right) + G_{12}\dfrac{2H}{\pi^2}\displaystyle\sum_{n=1}^{\infty}\dfrac{[1-(-1)^n]}{n^2}\dfrac{\cosh\left(\dfrac{\pi n h_1 \xi_1}{H}\sqrt{\dfrac{G_{23}}{G_{12}}}\right)}{\cosh\left(\dfrac{\pi n h_1 \delta_1}{2H}\sqrt{\dfrac{G_{23}}{G_{12}}}\right)}\cos\dfrac{\pi n(z-\frac{1}{2})}{H}, & \text{in } \Omega_1, \\ G_{12}\left(z-\dfrac{H+1}{2}\right) + G_{12}\dfrac{2H}{\pi^2}\displaystyle\sum_{n=1}^{\infty}\dfrac{[1-(-1)^n]}{n^2}\dfrac{\cosh\left(\dfrac{\pi n h_2 \xi_2}{H}\sqrt{\dfrac{G_{13}}{G_{12}}}\right)}{\cosh\left(\dfrac{\pi n h_2 \delta_2}{2H}\sqrt{\dfrac{G_{13}}{G_{12}}}\right)}\cos\dfrac{\pi n(z-\frac{1}{2})}{H}, & \text{in } \Omega_2, \end{cases} \tag{20.32}$$

$$b_{13}^{*12} = \begin{cases} \bar{\tau}_{13}^{12}, & \text{in } \Omega_2, \\ 0, & \text{in } \Omega_1, \Omega_3, \end{cases}$$

$$b_{23}^{*12} = \begin{cases} \bar{\tau}_{23}^{12}, & \text{in } \Omega_1, \\ 0, & \text{in } \Omega_2, \Omega_3. \end{cases}$$

Note that since $\langle \bar{\tau}_{13}^{12} \rangle_{\Omega_2} = \langle \bar{\tau}_{23}^{12} \rangle_{\Omega_1} = 0$ in the obvious notation, it follows that we have not violated the truth of the relation (15.62) earlier proved in a general fashion.

Relations (20.9), (20.13), (20.17), (20.21), (20.25) and (20.32) enable us to obtain explicit expressions for the entire set of effective elastic moduli of the wafer-type shell with the unit cell shown in Fig. 20.1.

Noting that averaging in the sense of (15.27) is actually integration over the coordinates y_1, y_2 and z, the following auxiliary formulas can be proved (cf. Fig. 20.1):

$$\begin{aligned}
&\langle 1\rangle_{\Omega_1} = t_1 H/h_1 = F_1^{(w)}, \qquad \langle z\rangle_{\Omega_1} = (H^2+H)t_1/(2h_1) = S_1^{(w)},\\
&\langle z^2\rangle_{\Omega_1} = t_1(4H^3+6H^2+3H)/(12h_1) = J_1^{(w)}(1\leftrightarrow 2),\\
&\langle 1\rangle_{\Omega_3} = 1, \qquad \langle z\rangle_{\Omega_3} = 0, \qquad \langle z^2\rangle_{\Omega_3} = \tfrac{1}{12},
\end{aligned} \tag{20.33}$$

where $F_1^{(w)}$ and $F_2^{(w)}$ are the cross-section areas; $S_1^{(w)}$ and $S_2^{(w)}$ the static moments; and $J_1^{(w)}$ and $J_2^{(w)}$, the inertia moments of the cross-sections of the reinforcing elements Ω_1 and Ω_2 relative to the middle surface of the shell, are calculated in the coordinate system y_1, y_2, z.

If each of the elements Ω_k $(k = 1, 2, 3)$ is made of an orthotropic material with the engineering properties $E_1^{(k)}$, $E_2^{(k)}$, $\nu_{21}^{(k)}$, $G_{12}^{(k)}$, $G_{13}^{(k)}$ and $G_{23}^{(k)}$, the non-zero effective moduli of the wafer-type shell of Fig. 20.1 are given by the following expressions:

$$\begin{aligned}
&\langle b_{11}^{11}\rangle = \frac{E_1^{(3)}}{1-\nu_{12}^{(3)}\nu_{21}^{(3)}} + E_1^{(2)}F_2^{(w)}, \qquad \langle b_{22}^{22}\rangle = \frac{E_2^{(3)}}{1-\nu_{12}^{(3)}\nu_{21}^{(3)}} + E_2^{(1)}F_1^{(w)},\\
&\langle b_{22}^{11}\rangle = \langle b_{11}^{22}\rangle = \frac{\nu_{12}^{(3)}E_1^{(3)}}{1-\nu_{12}^{(3)}\nu_{21}^{(3)}}, \qquad \langle b_{12}^{12}\rangle = G_{12}^{(3)},\\
&\langle zb_{11}^{11}\rangle = \langle b_{11}^{*11}\rangle = E_1^{(2)}S_2^{(w)}, \qquad \langle zb_{22}^{22}\rangle = \langle b_{22}^{*22}\rangle = E_2^{(1)}S_1^{(w)},\\
&\langle zb_{11}^{*11}\rangle = \frac{E_1^{(3)}}{12(1-\nu_{12}^{(3)}\nu_{21}^{(3)})} + E_1^{(2)}J_2^{(w)},\\
&\langle zb_{22}^{*22}\rangle = \frac{E_2^{(3)}}{12(1-\nu_{12}^{(3)}\nu_{21}^{(3)})} + E_2^{(1)}J_1^{(w)},\\
&\langle zb_{22}^{*11}\rangle = \langle zb_{11}^{*22}\rangle = \frac{\nu_{12}^{(3)}E_1^{(3)}}{12(1-\nu_{12}^{(3)}\nu_{21}^{(3)})},\\
&\langle zb_{12}^{*12}\rangle = \frac{G_{12}^{(3)}}{12} + \frac{G_{12}^{(1)}}{12}\left(\frac{H^3 t_1}{h_1} - K_1\right) + \frac{G_{12}^{(2)}}{12}\left(\frac{H^3 t_2}{h_2} - K_2\right),
\end{aligned} \tag{20.34}$$

where K_1 and K_2 are defined by

$$\begin{aligned}
K_1 &= \frac{96H^4}{\pi^5 A_1 h_1}\sqrt{\frac{G_{12}^{(1)}}{G_{23}^{(1)}}}\sum_{n=1}^{\infty}\frac{[1-(-1)^n]}{n^5}\tanh\left(\frac{\pi n A_1 t_1}{2H}\sqrt{\frac{G_{23}^{(1)}}{G_{12}^{(1)}}}\right),\\
K_2 &= \frac{96H^4}{\pi^5 A_2 h_2}\sqrt{\frac{G_{12}^{(2)}}{G_{13}^{(2)}}}\sum_{n=1}^{\infty}\frac{[1-(-1)^n]}{n^5}\tanh\left(\frac{\pi n A_2 t_2}{2H}\sqrt{\frac{G_{13}^{(2)}}{G_{12}^{(2)}}}\right).
\end{aligned} \tag{20.35}$$

Equations (20.34) may be employed to calculate the effective stiffness moduli of ribbed shells. In particular, if the strengthening ribs are directed along the coordinate axis α_1, the strengthening element, Ω_1, will be absent in Fig. 20.1, and in (20.33) we set $t_1 = 0$ thereby also turning the quantities $F_1^{(w)}$, $S_1^{(w)}$, $J_1^{(w)}$ and K_1 to zero.

If the material is isotropic, equations (20.34) take a simpler form and agree closely with results known from the structurally anisotropic theory of strengthened plates and shells (see Kalamkarov *et al.* 1987a). A notable exception is the torsional stiffness $\langle zb_{12}^{*12} \rangle$, which in terms of structural theory is given by

$$\langle zb_{12}^{*12} \rangle = \frac{E}{24(1+\nu)}\left(1 + H\frac{t_1^3}{h_1} + H\frac{t_2^3}{h_2}\right). \tag{20.36}$$

The difference between (20.34) and (20.35) on the one hand and (20.36) on the other turns out to be considerable for high ribs. If, for example, we take $A_1 = A_2 = 1$, $\nu = 0.3$, $H = 20$, $h_1 = h_2 = 60$ and $t_1 = t_2 = 2$, then from (20.34) and (20.35) we obtain $\langle zb_{12}^{*12} \rangle / E = 0.192$, whereas (20.36) gives a result of 0.203, i.e. 5.4% more. If $H = 10$, $h_1 = h_2 = 10$ and $t_1 = t_2 = 1$, then formulas (20.34) and (20.35) yield $\langle zb_{12}^{*12} \rangle / E = 0.0921$, compared with 0.0962 from (20.35). The percentage change in this case is 4.3%.

We now apply the above approximating scheme to the case of T-shaped ribs shown in Fig. 20.4. Since the solutions for regions Ω_2 and Ω_3 are already known, clearly only the local problems associated with the element Ω_4 remain to be considered.

By the same procedure used in the Ω_2 and Ω_3 problems, the non-zero moduli for an isotropic material are found to be

$$\langle b_{11}^{11} \rangle = \frac{E}{1-\nu^2} + E(F_2^{(w)} + F_4), \qquad \langle b_{22}^{22} \rangle = \frac{E}{1-\nu^2},$$

$$\langle b_{22}^{11} \rangle = \langle b_{11}^{22} \rangle = \frac{\nu E}{1-\nu^2}, \qquad \langle b_{12}^{12} \rangle = G,$$

$$\langle zb_{11}^{11} \rangle = \langle b_{11}^{*11} \rangle = E(S_2^{(w)} + S_4), \qquad \langle zb_{11}^{*11} \rangle = \frac{E}{12(1-\nu^2)} + E(J_2^{(w)} + J_4), \tag{20.37}$$

$$\langle zb_{22}^{*22} \rangle = \frac{E}{12(1-\nu^2)}, \qquad \langle zb_{22}^{*11} \rangle = \langle zb_{11}^{*22} \rangle = \frac{\nu E}{12(1-\nu^2)},$$

$$\langle zb_{12}^{*12} \rangle = \frac{G}{12}\left\{1 + \left(\frac{H^3 t_2}{h_2} - K_2\right) + \left[\frac{p^3 t_4}{h_2} - \frac{96 p^4}{\pi^5 A_2 h_2}\sum_{n=1}^{\infty}\frac{[1-(-1)^n]}{n^5}\tanh\left(\frac{\pi n A_2 t_4}{2p}\right)\right]\right\},$$

where the quantities F_4, S_4 snd J_4 are given by formulae analogous to (20.33).

Changing to the coordinate system y_1, y_2, z (cf. Fig. 20.4) we find

$$F_4 = pt_4/h_2, \qquad S_4 = pt_4(1 + 2H + p)/(2h_2),$$
$$J_4 = pt_4[3(1 + 2H)(1 + 2H + 2p) + 4p^3]/(12h_2).$$

Again with the exception of the torsional stiffness expression, equations (20.37) are identical to the corresponding results of the above-mentioned structurally anisotropic theory (see, for example, Korolev 1971). The difference in the $\langle zb_{12}^{*12} \rangle$ expressions is assoc-

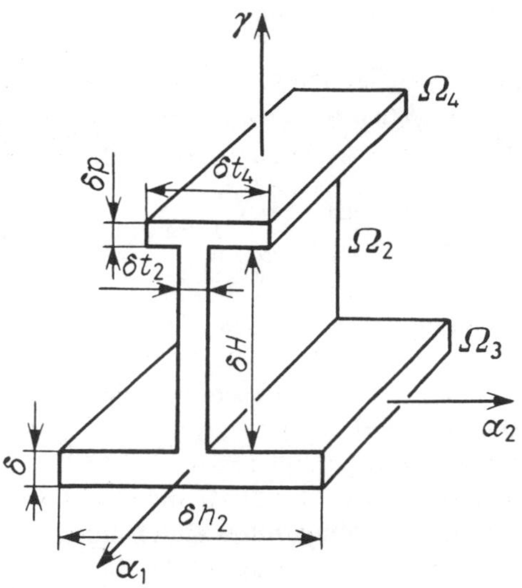

Fig. 20.4 *Unit cell of a ribbed shell for the case of T-shaped ribs.*

iated with the term in square brackets, which describes the contribution from the 'shelf' (element Ω_4) and in which the second term (the sum) is absent in the structurally anisotropic theory. Calculations show, however, that this term changes the result substantially at large values of p and t_4. If, for example, we take $H = 20$, $h_2 = 60$, $t_2 = 2$, $t_4 = 15$, $p = 1.5$ and $A_2 = 1$, then for the value of $\langle zb_{12}^{*12} \rangle (G/12)^{-1}$ we have 4.510 from the structurally anisotropic theory and 4.289 (i.e. 4.9% less) from (20.37). For $H = 30$, $h_2 = 80$, $t_2 = 3$, $t_4 = 20$, and $p = 2$, a 5.8% reduction is obtained, i.e. from 13.125 to 12.362.

20.2. *Effective thermoelastic and heat conduction properties of wafer-type and ribbed shells of orthotropic materials*

In this section we apply equations (17.17) and the approximate solutions of the local problems to calculate the effective thermoelastic properties of the wafer shell of Fig. 20.1.

For an orthotropic material we have

$$\alpha_{ij}^{T} = \begin{pmatrix} \alpha_1^T & 0 & 0 \\ 0 & \alpha_2^T & 0 \\ 0 & 0 & \alpha_3^T \end{pmatrix}. \tag{20.38}$$

With this, using (17.17), (20.9), (20.13), (20.17), (20.21), (20.25) and (20.32), the non-zero properties of the structure are found to be

$$\delta \langle s_{11} \rangle = \frac{\alpha_1^{T(3)} E_1^{(3)} + \alpha_2^{T(3)} \nu_{12}^{(3)} E_1^{(3)}}{1 - \nu_{12}^{(3)} \nu_{21}^{(3)}} + \alpha_1^{T(2)} E_1^{(2)} F_2^{(w)},$$

$$\delta \langle s_{22} \rangle = \frac{\alpha_2^{T(3)} E_2^{(3)} + \alpha_1^{T(3)} \nu_{21}^{(3)} E_2^{(3)}}{1 - \nu_{12}^{(3)} \nu_{21}^{(3)}} + \alpha_2^{T(1)} E_2^{(1)} F_1^{(w)},$$

$$\delta\langle zs_{11}\rangle = \delta\langle s_{11}^{*}\rangle = \alpha_1^{\mathrm{T}(2)} E_1^{(2)} S_2^{(\mathrm{w})}, \tag{20.39}$$

$$\delta\langle zs_{22}\rangle = \delta\langle s_{22}^{*}\rangle = \alpha_2^{\mathrm{T}(1)} E_2^{(1)} S_1^{(\mathrm{w})},$$

$$\delta\langle zs_{11}^{*}\rangle = \frac{\alpha_1^{\mathrm{T}(3)} E_1^{(3)} + \alpha_2^{\mathrm{T}(3)} \nu_{12}^{(3)} E_1^{(3)}}{12(1-\nu_{12}^{(3)}\nu_{21}^{(3)})} + \alpha_1^{\mathrm{T}(2)} E_1^{(2)} J_2^{(\mathrm{w})},$$

$$\delta\langle zs_{22}^{*}\rangle = \frac{\alpha_2^{\mathrm{T}(3)} E_2^{(3)} + \alpha_1^{\mathrm{T}(3)} \nu_{21}^{(3)} E_2^{(3)}}{12(1-\nu_{12}^{(3)}\nu_{21}^{(3)})} + \alpha_2^{\mathrm{T}(1)} E_2^{(1)} J_1^{(\mathrm{w})}.$$

Turning now to heat conduction properties of the same shell, the local problems to be considered are

($l1$): equations (19.42) and (19.44);
($l2$): equations (19.43) and (19.45);
(l^*1): equations (19.46); and
(l^*2): equations (19.47), all in combination with (16.28).

Approximate solutions for these problems may be found for each of the elements Ω_1, Ω_2 and Ω_3 independently (see Fig. 20.3) by proceeding in much the same way as was done with the local problems of elasticity. We omit the details of the solution and present here only the final expressions for the functions $l_{i\mu}$ and $l_{i\mu}^{*}$ as obtained from (19.40) and (19.41). We have

$$l_{11} = \begin{cases} \lambda_1, & \text{in } \Omega_3, \Omega_2, \\ 0, & \text{in } \Omega_1, \end{cases} \qquad l_{22} = \begin{cases} \lambda_2, & \text{in } \Omega_3, \Omega_1, \\ 0, & \text{in } \Omega_2, \end{cases}$$

$$l_{12} = l_{21} = l_{31} = l_{32} = 0,$$

$$l_{11}^{*} = \begin{cases} \lambda_1 z, & \text{in } \Omega_3, \Omega_2, \\ \lambda_1 z - \lambda_1 (H+1)/2 + \lambda_1 \dfrac{2H}{\pi^2} \displaystyle\sum_{n=1}^{\infty} \dfrac{[1-(-1)^n]\cosh\left(\dfrac{\pi n h_1 \xi_1}{H}\sqrt{\dfrac{\lambda_3}{\lambda_1}}\right)}{n^2 \cosh\left(\dfrac{\pi n h_1 \delta_1}{2H}\sqrt{\dfrac{\lambda_3}{\lambda_1}}\right)} \cos\dfrac{\pi n (z-\frac{1}{2})}{H}, & \text{in } \Omega_1, \end{cases} \tag{20.40}$$

$$l_{22}^{*} = \begin{cases} \lambda_1 z, & \text{in } \Omega_3, \Omega_1, \\ \lambda_2 z - \lambda_2 (H+1)/2 + \lambda_2 \dfrac{2H}{\pi^2} \displaystyle\sum_{n=1}^{\infty} \dfrac{[1-(-1)^n]\cosh\left(\dfrac{\pi n h_2 \xi_2}{H}\sqrt{\dfrac{\lambda_3}{\lambda_2}}\right)}{n^2 \cosh\left(\dfrac{\pi n h_2 \delta_2}{2H}\sqrt{\dfrac{\lambda_3}{\lambda_2}}\right)} \cos\dfrac{\pi n (z-\frac{1}{2})}{H}, & \text{in } \Omega_2, \end{cases}$$

$$l_{12}^{*} = l_{21}^{*} = 0$$

Note that although the functions l_{31}^{*} and l_{32}^{*} are also different from zero, they do tend to zero after the averaging operation, in full agreement with equations (16.36) which we proved earlier for the general case.

From (20.40), the non-zero effective heat conduction properties of the wafer shell of Fig. 20.1 are found to be given by the following equations:

$$\begin{aligned}
&\langle l_{11}\rangle = \lambda_1^{(3)} + \lambda_1^{(2)} F_2^{(w)}, \qquad \langle l_{22}\rangle = \lambda_2^{(3)} + \lambda_2^{(1)} F_1^{(w)},\\
&\langle z l_{11}\rangle = \langle l^*_{11}\rangle = \lambda_1^{(2)} S_2^{(w)}, \qquad \langle z l_{22}\rangle = \langle l^*_{22}\rangle = \lambda_2^{(1)} S_1^{(w)},\\
&\langle z l^*_{11}\rangle = \frac{\lambda_1^{(3)}}{12} + \lambda_1^{(2)} J_2^{(w)} + \frac{\lambda_1^{(1)}}{12}\left(\frac{H^3 t_1}{h_1} - L_1\right),\\
&\langle z l^*_{22}\rangle = \frac{\lambda_2^{(3)}}{12} + \lambda_2^{(1)} J_1^{(w)} + \frac{\lambda_2^{(2)}}{12}\left(\frac{H^3 t_2}{h_2} - L_2\right),
\end{aligned} \tag{20.41}$$

which together with the definitions

$$L_1 = \sqrt{\frac{\lambda_1}{\lambda_3}}\frac{96H^4}{\pi^5 A_1 h_1}\sum_{n=1}^{\infty}\frac{[1-(-1)^n]}{n^5}\tanh\left(\frac{\pi n A_1 t_1}{2H}\sqrt{\frac{\lambda_3}{\lambda_1}}\right)$$

and

$$L_2 = \sqrt{\frac{\lambda_2}{\lambda_3}}\frac{96H^4}{\pi^5 A_2 h_2}\sum_{n=1}^{\infty}\frac{[1-(-1)^n]}{n^5}\tanh\left(\frac{\pi n A_2 t_2}{2H}\sqrt{\frac{\lambda_3}{\lambda_2}}\right) \tag{20.42}$$

determine all the coefficients occurring in the heat conduction constitutive relations of the homogenized shell, i.e. equations (16.35). The system of heat conduction governing equations of the homogenized shell, equations (16.32) and (16.34), also involves the coefficients $\mathfrak{J}_m (m = 0, 1, 2)$, R_m, and $Z_m (m = 0, 1)$, the values of which are given by equations (16.33) and (15.21). For the unit cell of Fig. 20.1 we find

$$\begin{aligned}
\mathfrak{J}_0 &= \alpha_S^+\left[1 + \frac{2H(h_1 - t_1)}{h_1 h_2 A_2} + \frac{2H(h_2 - t_2)}{h_1 h_2 A_1}\right] + \alpha_S^-,\\
\mathfrak{J}_1 &= \alpha_S^+\left[\frac{1}{2} + \frac{H t_2}{h_2} + \frac{H^2 + H}{h_2 A_2}\right] - \frac{\alpha_S^-}{2},\\
\mathfrak{J}_2 &= \alpha_S^+\left[\tfrac{1}{4} + (H^2 + H)\left(\frac{t_2}{h_2} + \frac{t_1}{h_1} - \frac{t_1 t_2}{h_1 h_2}\right)\right.\\
&\qquad \left. + \frac{(4H^3 + 6H^2 + 3H)}{6 h_1 h_2}\left(\frac{h_1 - t_1}{A_2} + \frac{h_2 - t_2}{A_1}\right)\right] + \frac{\alpha_S^-}{4}.
\end{aligned} \tag{20.43}$$

Expressions for R_0 and R_1 are obtained from $\mathfrak{J}_0$ and $\mathfrak{J}_1$ by replacing $\alpha_S^\pm$ by $g_S^\pm$. Finally, $Z_0 = Z_1 = 0$ because in the case we consider the quantity $\lambda_{3i} N_i^\pm = \lambda_3 N_3^\pm$ differs from zero only on the surfaces parallel to the middle surface of the shell, but there we have $z^\pm = \text{const.}$, and hence $\zeta^\pm = 0$.

Equations (20.41) may be useful in determining the effective heat conduction properties of ribbed shells. If the ribs are directed along the coordinate line α_1, for example, we set $t_1 = 0$ in (20.41) and, accordingly, $F_1 = S_1 = \mathfrak{J}_1 = L_1 = 0$. The parameters $\mathfrak{J}_m$ of the ribbed shell are given by

$$\mathfrak{I}_0 = \alpha_S^+ \left(1 + \frac{2H}{h_2 A_2}\right) + \alpha_S^-,$$

$$\mathfrak{I}_1 = \alpha_S^+ \left(\frac{1}{2} + \frac{Ht_2}{h_2} + \frac{H^2 + H}{h_2 A_2}\right) - \frac{\alpha_S^-}{2},$$

$$\mathfrak{I}_2 = \alpha_S^+ \left[\frac{1}{4} + \frac{(H^2 + H)t_2}{h_2} + \frac{4H^3 + 6H^2 + 3H}{6h_2 A_2}\right] + \frac{\alpha_S^-}{4}. \tag{20.44}$$

To determine R_0 and R_1, we again have to replace $\alpha_S^{\pm}$ by $g_S^{\pm}$ in $\mathfrak{I}_0$ and $\mathfrak{I}_1$, respectively. Note that equations (20.44) cannot be deduced from (20.43) by merely setting $t_1 = 0$.

21. EFFECTIVE PROPERTIES OF A THREE-LAYERED SHELL WITH A HONEYCOMB FILLER

21.1. Effective properties of a honeycomb shell of tetrahedral structure

The problem we consider here is that of a three-layered shell (or plate) composed of a honeycomb filler of tetrahedral structure sandwiched between two carrying layers, as shown in Fig. 21.1.

The effective properties of this system may be evaluated by employing the approximate method based on the independent solution of the local problems for each of the elements of the unit cell of the system (Ω_1, Ω_2, Ω_3, Ω_4; see Fig. 21.2). We find that the solutions

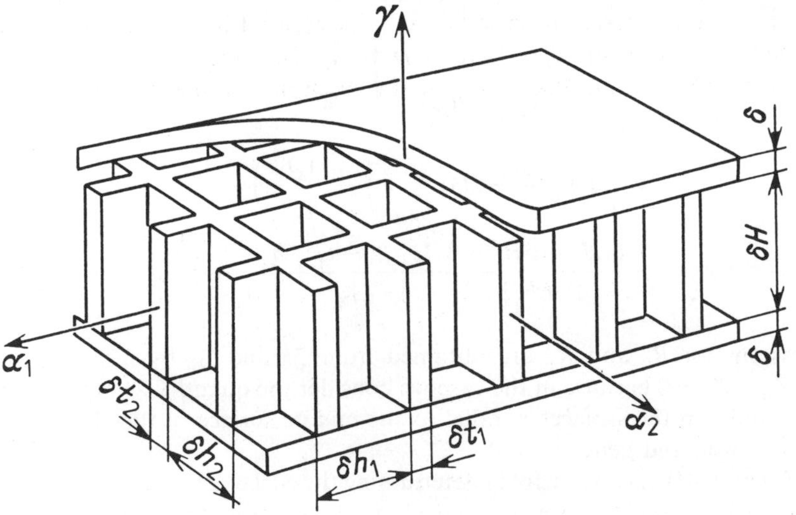

Fig. 21.1 *Three-layered shell with a tetrahedral honeycomb filler.*

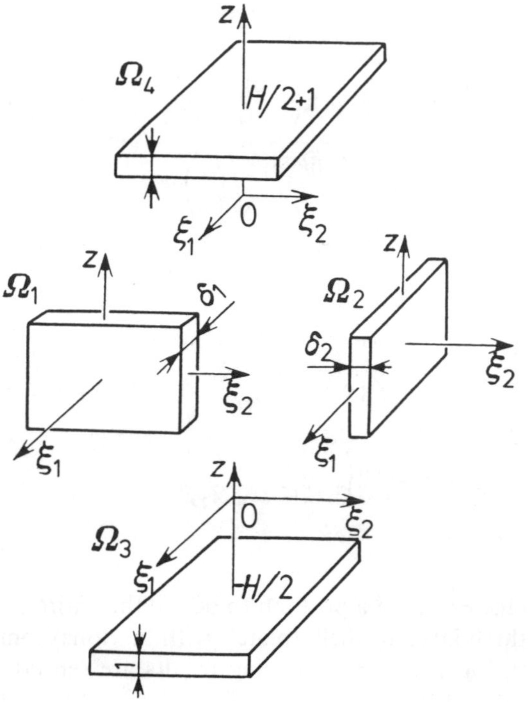

Fig. 21.2 *Unit cell elements of a three-layered shell with a tetrahedral honeycomb filler.*

of the local problems (*b*), (*b**11), (*b**22) and (*l*), obtained above for the elements Ω_1, Ω_2 and Ω_3 of the wafer-type shell of Fig. 20.3, may be utilized without any modifications in the analysis of the honeycomb structure, while the solution for Ω_4 simply coincides with those of the corresponding problems for Ω_3. The same is true for the local problems (*b**12), (*l**1) and (*l**2) for the elements Ω_3 and Ω_4. With regard to the elements Ω_1 and Ω_2, however, some modification is needed, associated with the change in the location of the elements with respect to the *z* axis (cf. Figs 20.3 and 21.2).

Omitting all the intermediate steps of the solution, the final results are

$$b_{12}^{*12} = \begin{cases} zG_{12}, & \text{in } \Omega_3, \Omega_4, \\[2ex] zG_{12} + G_{12}\dfrac{2H}{\pi^2}\displaystyle\sum_{n=1}^{\infty}\frac{[1-(-1)^n]}{n^2}\,\frac{\cosh\left(\dfrac{\pi n h_1 \xi_1}{H}\sqrt{\dfrac{G_{23}}{G_{12}}}\right)}{\cosh\left(\dfrac{\pi n h_1 \delta_1}{2H}\sqrt{\dfrac{G_{23}}{G_{12}}}\right)}\cos\dfrac{\pi n\left(z+\dfrac{H}{2}\right)}{H}, & \text{in } \Omega_1, \\[2ex] zG_{12} + G_{12}\dfrac{2H}{\pi^2}\displaystyle\sum_{n=1}^{\infty}\frac{[1-(-1)^n]}{n^2}\,\frac{\cosh\left(\dfrac{\pi n h_2 \xi_2}{H}\sqrt{\dfrac{G_{13}}{G_{12}}}\right)}{\cosh\left(\dfrac{\pi n h_2 \delta_2}{2H}\sqrt{\dfrac{G_{13}}{G_{12}}}\right)}\cos\dfrac{\pi n\left(z+\dfrac{H}{2}\right)}{H}, & \text{in } \Omega_2, \end{cases}$$

$$b_{11}^{*12} = b_{22}^{*12} = b_{33}^{*12} = 0, \tag{21.1}$$

$$l_{11}^{*} = \begin{cases} \lambda_1 z, & \text{in } \Omega_2, \Omega_3, \Omega_4, \\ \lambda_1 z + \lambda_1 \dfrac{2H}{\pi^2} \displaystyle\sum_{n=1}^{\infty} \dfrac{[1-(-1)^n] \cosh\left(\dfrac{\pi n h_1 \xi_1}{H} \sqrt{\dfrac{\lambda_3}{\lambda_1}}\right)}{n^2 \cosh\left(\dfrac{\pi n h_1 \delta_1}{2H} \sqrt{\dfrac{\lambda_3}{\lambda_1}}\right)} \cos \dfrac{\pi n \left(z + \dfrac{H}{2}\right)}{H}, & \text{in } \Omega_1, \end{cases}$$

$$l_{22}^{*} = \begin{cases} \lambda_2 z, & \text{in } \Omega_1, \Omega_3, \Omega_4, \\ \lambda_2 z + \lambda_2 \dfrac{2H}{\pi^2} \displaystyle\sum_{n=1}^{\infty} \dfrac{[1-(-1)^n] \cosh\left(\dfrac{\pi n h_2 \xi_2}{H} \sqrt{\dfrac{\lambda_3}{\lambda_2}}\right)}{n^2 \cosh\left(\dfrac{\pi n h_2 \delta_2}{2H} \sqrt{\dfrac{\lambda_3}{\lambda_2}}\right)} \cos \dfrac{\pi n \left(z + \dfrac{H}{2}\right)}{H}, & \text{in } \Omega_2, \end{cases}$$

$$l_{12}^{*} = l_{21}^{*} = 0.$$

With these solutions, we are in a position to derive the entire set of non-zero effective properties of the three-layered shell (plate) with a honeycomb filling element of tetrahedral structure, Fig. 21.1. The following results are reproduced from a paper by Kalamkarov (1989).

Effective moduli:

$$\begin{aligned}
&\langle b_{11}^{11} \rangle = \frac{E_1^{(3)}}{1 - \nu_{12}^{(3)}\nu_{21}^{(3)}} + \frac{E_1^{(4)}}{1 - \nu_{12}^{(4)}\nu_{21}^{(4)}} + E_1^{(2)} F_2^{(\mathrm{h})} \quad (1 \leftrightarrow 2), \\
&\langle b_{11}^{22} \rangle = \langle b_{22}^{11} \rangle = \frac{\nu_{12}^{(3)} E_1^{(3)}}{1 - \nu_{12}^{(3)}\nu_{21}^{(3)}} + \frac{\nu_{12}^{(4)} E_1^{(4)}}{1 - \nu_{12}^{(4)}\nu_{21}^{(4)}}, \\
&\langle b_{12}^{12} \rangle = G_{12}^{(3)} + G_{12}^{(4)}, \qquad \langle b_{12}^{*12} \rangle = (G_{12}^{(3)} - G_{12}^{(4)}) S_3^{(\mathrm{h})}, \\
&\langle z b_{11}^{11} \rangle = \langle b_{11}^{*11} \rangle = \left(\frac{E_1^{(3)}}{1 - \nu_{12}^{(3)}\nu_{21}^{(3)}} - \frac{E_1^{(4)}}{1 - \nu_{12}^{(4)}\nu_{21}^{(4)}} \right) S_3^{(\mathrm{h})} \quad (1 \leftrightarrow 2), \\
&\langle z b_{11}^{22} \rangle = \langle z b_{22}^{11} \rangle = \langle b_{11}^{*22} \rangle = \langle b_{22}^{*11} \rangle = \left(\frac{\nu_{12}^{(3)} E_1^{(3)}}{1 - \nu_{12}^{(3)}\nu_{21}^{(3)}} - \frac{\nu_{12}^{(4)} E_1^{(4)}}{1 - \nu_{12}^{(4)}\nu_{21}^{(4)}} \right) S_3^{(\mathrm{h})}, \\
&\langle z b_{11}^{*11} \rangle = \left(\frac{E_1^{(3)}}{1 - \nu_{12}^{(3)}\nu_{21}^{(3)}} + \frac{E_1^{(4)}}{1 - \nu_{12}^{(4)}\nu_{21}^{(4)}} \right) J_3^{(\mathrm{h})} + E_1^{(2)} J_2^{(\mathrm{h})} \quad (1 \leftrightarrow 2), \\
&\langle z b_{22}^{*11} \rangle = \langle z b_{11}^{*22} \rangle = \left(\frac{\nu_{12}^{(3)} E_1^{(3)}}{1 - \nu_{12}^{(3)}\nu_{21}^{(3)}} + \frac{\nu_{12}^{(4)} E_1^{(4)}}{1 - \nu_{12}^{(4)}\nu_{21}^{(4)}} \right) J_3^{(\mathrm{h})}, \\
&\langle z b_{12}^{*12} \rangle = (G_{12}^{(3)} + G_{12}^{(4)}) J_3^{(\mathrm{h})} + \frac{G_{12}^{(1)}}{12} \left(\frac{H^3 t_1}{h_1} - K_1 \right) + \frac{G_{12}^{(2)}}{12} \left(\frac{H^3 t_2}{h_2} - K_2 \right).
\end{aligned} \tag{21.2}$$

Effective thermoelastic properties:

$$\langle s_{11}\rangle = \frac{\alpha_1^{T(3)}E_1^{(3)} + \alpha_2^{T(3)}\nu_{12}^{(3)}E_1^{(3)}}{1-\nu_{12}^{(3)}\nu_{21}^{(3)}} + \frac{\alpha_1^{T(4)}E_1^{(4)} + \alpha_2^{T(4)}\nu_{12}^{(4)}E_1^{(4)}}{1-\nu_{12}^{(4)}\nu_{21}^{(4)}} + \alpha_1^{T(2)}E_1^{(2)}F_2^{(h)} \quad (1\leftrightarrow 2),$$

$$\langle zs_{11}\rangle = \langle s_{11}^*\rangle = \left(\frac{\alpha_1^{T(3)}E_1^{(3)} + \alpha_2^{T(3)}\nu_{12}^{(3)}E_1^{(3)}}{1-\nu_{12}^{(3)}\nu_{21}^{(3)}} - \frac{\alpha_1^{T(4)}E_1^{(4)} + \alpha_2^{T(4)}\nu_{12}^{(4)}E_1^{(4)}}{1-\nu_{12}^{(4)}\nu_{21}^{(4)}}\right)S_3^{(h)} \quad (1\leftrightarrow 2),$$

$$\langle zs_{11}^*\rangle = \frac{\alpha_1^{T(3)}E_1^{(3)} + \alpha_2^{T(3)}\nu_{12}^{(3)}E_1^{(3)}}{1-\nu_{12}^{(3)}\nu_{21}^{(3)}}J_3^{(h)} + \frac{\alpha_1^{T(4)}E_1^{(4)} + \alpha_2^{T(4)}\nu_{12}^{(4)}E_1^{(4)}}{1-\nu_{12}^{(4)}\nu_{21}^{(4)}}J_4^{(h)} + \alpha_1^{T(2)}E_1^{(2)}J_2^{(h)} \quad (1\leftrightarrow 2). \qquad (21.3)$$

Effective heat conduction coefficients:

$$\langle l_{11}\rangle = \lambda_1^{(3)} + \lambda_1^{(4)} + \lambda_1^{(2)}F_2^{(h)}, \qquad \langle l_{22}\rangle = \lambda_2^{(3)} + \lambda_2^{(4)} + \lambda_2^{(1)}F_1^{(h)},$$

$$\langle l_{11}^*\rangle = (\lambda_1^{(3)} - \lambda_1^{(4)})S_3^{(h)}, \qquad \langle l_{22}^*\rangle = (\lambda_2^{(3)} - \lambda_2^{(4)})S_3^{(h)}, \qquad (21.4)$$

$$\langle zl_{11}^*\rangle = \lambda_1^{(2)}J_2^{(h)} + \lambda_1^{(3)}J_3^{(h)} + \lambda_1^{(4)}J_4^{(h)} + \frac{\lambda_1^{(1)}}{12}\left(\frac{H^3t_1}{h_1} - L_1\right),$$

$$\langle zl_{22}^*\rangle = \lambda_2^{(1)}J_1^{(h)} + \lambda_1^{(3)}J_3^{(h)} + \lambda_2^{(4)}J_4^{(h)} + \frac{\lambda_2^{(2)}}{12}\left(\frac{H^3t_2}{h_2} - L_2\right).$$

In the above, the elements of the unit cell are referenced by the appropriate superscript on material properties, and the quantities K_1, K_2, L_1 and L_2 are defined in (20.35) and (20.42). We have also introduced the following notations:

$$\langle 1\rangle_{\Omega_1} = t_1H/h_1 = F_1^{(h)}, \qquad \langle z^2\rangle_{\Omega_1} = t_1H^3/(12h_1) = J_1^{(h)} \quad (1\leftrightarrow 2),$$

$$\langle z\rangle_{\Omega_3} = -(H+1)/2 = S_3^{(h)}, \qquad \langle z\rangle_{\Omega_4} = (H+1)/2 = -S_3^{(h)}, \qquad (21.5)$$

$$\langle z^2\rangle_{\Omega_3} = \langle z^2\rangle_{\Omega_4} = (3H^2 + 6H + 4)/12 = J_3^{(h)} = J_4^{(h)},$$

and recalled that, for the honeycomb shell of Fig. 21.1,

$$\langle 1\rangle_{\Omega_3} = \langle 1\rangle_{\Omega_4} = 1, \qquad \langle z\rangle_{\Omega_1} = \langle z\rangle_{\Omega_2} = 0.$$

It should be noted that since the honeycomb shell we are considering is geometrically symmetric with respect to its middle surface, the presence of non-zero skew-symmetric properties in equations (21.2)–(21.4), that is $\langle b_{11}^{*11}\rangle$, $\langle b_{22}^{*22}\rangle$, $\langle b_{11}^{*22}\rangle$, $\langle s_{11}^*\rangle$, $\langle s_{22}^*\rangle$, $\langle l_{11}^*\rangle$ and $\langle l_{22}^*\rangle$, is due entirely to the difference in the material properties of the upper and lower carrying layers. Clearly, all these coefficients turn to zero if the corresponding properties are the same in the elements Ω_3 and Ω_4.

Note also that if $t_1 = 0$ (and hence $F_1^{(h)} = S_1^{(h)} = J_1^{(h)} = K_1 = L_1 = 0$), what equations (21.2)–(21.4) yield is the effective properties of a shell containing within itself a system of grooves of square cross-section directed along the coordinate axis α_1 (see Fig. 21.1 for $t_1 = 0$).

Finally, the coefficients in the heat conduction equations (16.32) and (16.34) should be written down for the case of a honeycomb shell. These are

$$\begin{aligned} \mathfrak{J}_0 &= \alpha_S^+ + \alpha_S^-, \qquad \mathfrak{J}_1 = (H/2+1)(\alpha_S^+ - \alpha_S^-), \\ \mathfrak{J}_2 &= (H/2+1)^2(\alpha_S^+ + \alpha_S^-), \qquad R_0 = g_S^+ + g_S^-, \\ R_1 &= (H/2+1)(g_S^+ - g_S^-), \qquad Z_0 = Z_1 = 0. \end{aligned} \tag{21.6}$$

Expressions (21.2) for the effective elastic moduli of a honeycomb shell agree quite well with the corresponding results of Aleksandrov (1965) obtained for an isotropic material by the application of the energy balance approach. Note, however, that the expression for the torsional stiffness, $\langle zb_{12}^{*12} \rangle$, as given by (21.2), is more accurate than Aleksandrov's counterpart derived by neglecting the contribution from the honeycomb filler. This neglect is indeed a justifiable assumption in many practical situations when the elastic properties of the carrying layers and the filler are of the same order of magnitude; the contribution from the filler, as estimated by (21.2), is then about only one-tenth percent of the contributions from the carrying layers. The error may be significant, however, if the filler material is much stiffer than that of the layers.

21.2. Effective elastic moduli of a three-layered shell with a honeycomb filler of hexagonal structure

Referring to Fig. 21.3, which shows the geometry of the problem, we assume that

$$t_0 \ll H, \qquad t \ll H, \tag{21.7}$$

that is, the vertical dimension of the honeycomb filler is much larger than the thickness of the carrying layers and of the honeycomb cells themselves. These conditions, as well as those in (20.1), are typically fulfilled in practical honeycomb plate and shell structures (see, for example, Endogur *et al.* 1986) and provide sufficient justification for the approximating scheme of the previous sections, in which the pertinent local problems are solved independently for each of the unit cell elements. We confine our attention to the case in which $A_1 = A_2 = A$.

If we change to the coordinates ξ_1, ξ_2, z in terms of which the local problems are formulated, the unit cell of our structure deforms into the shape shown in Fig. 21.4,

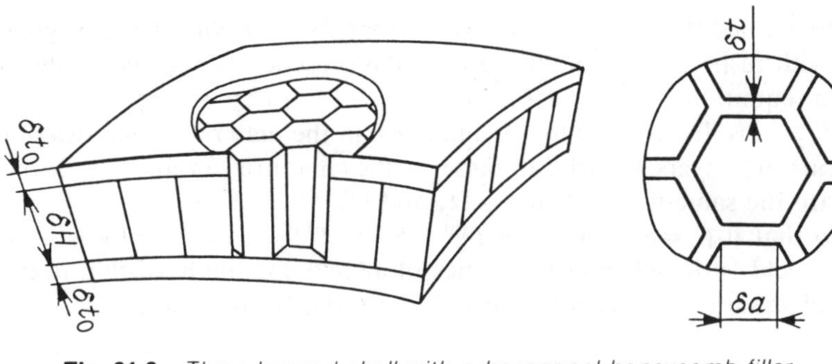

Fig. 21.3 *Three-layered shell with a hexagonal honeycomb filler.*

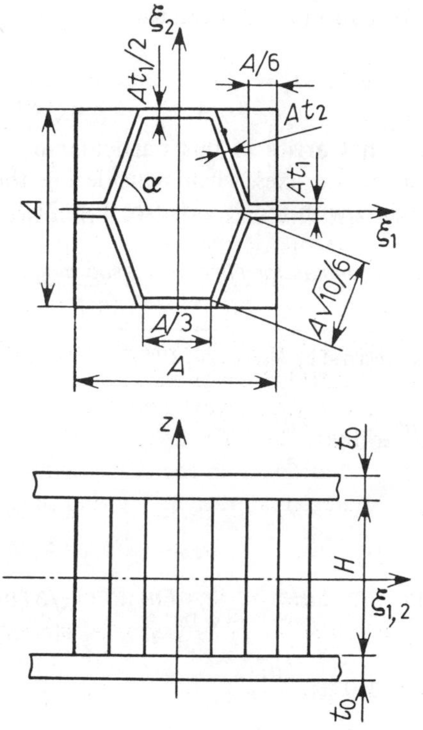

Fig. 21.4 *Unit cell of a three-layered shell with a hexagonal honeycomb filler; coordinate system* ξ_1, ξ_2, z.

Table 21.1 *Effective elastic moduli of the three-layered shell with a honeycomb filler of hexagonal structure (Fig. 21.3):* $H = 25,\ a = 2.5,\ t_0 = 1,\ \nu_0 = \nu = 0.3,\ A = 1.$

Effective elastic moduli	The insertion of both carrying layers	The insertion of the filler foil	
		$t = 0.1$	$t = 0.5$
$\langle b_{11}^{11} \rangle$	$2.1978\ E_0$	$0.4330\ E$	$2.1651\ E$
$\langle b_{22}^{11} \rangle$	$0.6593\ E_0$	$0.1443\ E$	$0.7217\ E$
$\langle b_{12}^{12} \rangle$	$0.7692\ E_0$	$0.1443\ E$	$0.7217\ E$
$\langle zb_{11}^{*11} \rangle$	$371.6117\ E_0$	$22.5527\ E$	$112.7637\ E$
$\langle zb_{22}^{*11} \rangle$	$111.4835\ E_0$	$7.5176\ E$	$37.5879\ E$
$\langle zb_{12}^{*12} \rangle$	$130.0641\ E_0$	$3.6643\ E$	$18.3540\ E$

and, by the rules of transformation of coordinates,

$$h_1 = 3a, \qquad h_2 = \sqrt{3}a, \qquad t_1 = \frac{t}{a\sqrt{3}}, \qquad t_2 = \frac{2t}{a\sqrt{30}}, \qquad \sin\alpha = \frac{3}{\sqrt{10}}.$$

The additional difficulty that arises in this particular example is that we have to perform a rotation of coordinate axes when considering the local problems of cell elements that make an angle α with the ξ_1 axis (see Fig. 21.4).

Table 21.2 *Comparision of the formulas for effective elastic moduli of the honeycomb filler of hexagonal structure.*

	Formulas obtained by Aleksandrov (1965)	**Formulas (21.8)**
$\langle b_{11}^{11}\rangle$	$\frac{2.5}{3\sqrt{3}}\frac{EtH}{a} \approx 0.481\frac{EtH}{a}$	$\frac{\sqrt{3}}{4}\frac{EtH}{a} \approx 0.433\frac{EtH}{a}$
$\langle b_{22}^{22}\rangle$	$0.5\frac{EtH}{a}$	$\frac{\sqrt{3}}{4}\frac{EtH}{a} \approx 0.433\frac{EtH}{a}$
$\langle b_{11}^{22}\rangle = \langle b_{22}^{11}\rangle$	$\frac{2.5}{(15\sqrt{3})^{1/2}}\sqrt{\nu_0}\frac{EtH}{a} \approx 0.49\sqrt{\nu_0}\frac{EtH}{a}$ for $\nu_0 = 0.3$: $0.268\frac{EtH}{a}$ for $\nu_0 = 0.25$: $0.245\frac{EtH}{a}$	$\frac{\sqrt{3}}{12}\frac{EtH}{a} \approx 0.144\frac{EtH}{a}$
$\langle b_{12}^{12}\rangle$	$\frac{\sqrt{3}}{12}\frac{EtH}{a}$	$\frac{\sqrt{3}}{12}\frac{EtH}{a}$
$\langle zb_{11}^{*11}\rangle$	$\frac{2.5}{36\sqrt{3}}\frac{EtH^3}{a} \approx 0.04\frac{EtH^3}{a}$	$\frac{\sqrt{3}}{48}\frac{EtH^3}{a} \approx 0.036\frac{EtH^3}{a}$
$\langle zb_{22}^{*22}\rangle$	$\frac{1}{24}\frac{EtH^3}{a} \approx 0.042\frac{EtH^3}{a}$	$\frac{\sqrt{3}}{48}\frac{EtH^3}{a} \approx 0.036\frac{EtH^3}{a}$
$\langle zb_{22}^{*11}\rangle = \langle zb_{11}^{*22}\rangle$	$\frac{2.5\sqrt{\nu_0}}{12(15\sqrt{3})^{1/2}}\frac{EtH^3}{a} \approx \sqrt{\nu_0}\,0.041\frac{EtH^3}{a}$ for $\nu_0 = 0.3$: $0.022\frac{EtH^3}{a}$ for $\nu_0 = 0.25$: $0.021\frac{EtH^3}{a}$	$\frac{\sqrt{3}}{144}\frac{EtH^3}{a} \approx 0.012\frac{EtH^3}{a}$
$\langle zb_{12}^{*12}\rangle$	$\frac{\sqrt{3}}{288}\frac{EtH^3}{a} \approx 0.006\frac{EtH^3}{a}$	when $H = 25$, $\nu = 0.3$, $A = 1$ for $t = 0.1$: $0.00586\frac{EtH^3}{a}$ for $t = 0.5$: $0.00587\frac{EtH^3}{a}$

The calculation of the non-vanishing effective elastic moduli of the shell of Fig. 21.3 includes the solution of the local problems on each of the cell elements and is somewhat lengthy to be reproduced here, so we only quote the final results of the calculation. For an isotropic material, both in the carrying layers (E_0, ν_0) and in the filler foil (E, ν), we have

$$
\begin{aligned}
&\langle b_{11}^{11}\rangle = \langle b_{22}^{22}\rangle = \frac{2E_0t_0}{1-\nu_0^2} + \frac{\sqrt{3}}{4}\frac{EHt}{a}, \qquad \langle b_{12}^{12}\rangle = \frac{E_0t_0}{1+\nu_0} + \frac{\sqrt{3}}{12}\frac{EHt}{a},\\
&\langle b_{11}^{22}\rangle = \langle b_{22}^{11}\rangle = \frac{2\nu_0E_0t_0}{1-\nu_0^2} + \frac{\sqrt{3}}{12}\frac{EHt}{a},\\
&\langle zb_{11}^{*11}\rangle = \langle zb_{22}^{*22}\rangle = \frac{E_0}{1-\nu_0^2}\left(\frac{H^2t_0}{2} + Ht_0^2 + \tfrac{2}{3}t_0^3\right) + \frac{\sqrt{3}}{48}\frac{EH^3t}{a},\\
&\langle zb_{11}^{*22}\rangle = \langle zb_{22}^{*11}\rangle = \frac{\nu_0E_0}{1-\nu_0^2}\left(\frac{H^2t_0}{2} + Ht_0^2 + \tfrac{2}{3}t_0^3\right) + \frac{\sqrt{3}}{144}\frac{EH^3t}{a},\\
&\langle zb_{12}^{*12}\rangle = \frac{E_0}{2(1+\nu_0)}\left(\frac{H^2t_0}{2} + Ht_0^2 + \frac{2t_0^3}{3}\right)\\
&\qquad + \frac{EH^3t}{12(1+\nu)a}\left\{\frac{3+\nu}{4\sqrt{3}} - \frac{128H}{\sqrt{3}\pi^5At}\sum_{n=1}^{\infty}\frac{\tanh[\pi(2n-1)At/(2H)]}{(2n-1)^5}\right\},
\end{aligned}
\tag{21.8}
$$

where the first terms everywhere describe the contribution from the carrying layers, and the second terms describe the contribution from the filler. It is seen that the latter contribution may be made comparable with, or even greater than, the former by appropriately varying the parameters E, H, t and a. This is clearly shown in Table 21.1, where the results based on (21.8) are listed.

It is of interest to compare (21.8) with the elastic moduli results obtained by different methods in earlier work on the subject. Comparison with the work of Aleksandrov (1965), see Table 21.2, shows that the greatest corrections occur in the elastic moduli $\langle b_{11}^{22}\rangle$ and $\langle zb_{11}^{*22}\rangle$. Although the formulas of Aleksandrov *et al.* (1960) are more accurate when compared with Aleksandrov (1965) (very much at the expense of simplicity), the calculations show that even in this case the corrections given by (21.8) are quite appreciable, as illustrated by Table 21.3 for $\langle b_{11}^{22}\rangle$.

Table 21.3 *Comparison of the different formulas for effective elastic modulus $\langle b_{11}^{22}\rangle$ of the honeycomb filler of hexagonal structure.*

		$\langle b_{11}^{22}\rangle\left(\frac{EtH}{a}\right)^{-1}$	
$H=25$, $a=2.5$, $A=1$		**Formulas by Aleksandrov *et al.* (1960)**	**Formulas (21.8)**
$t=0.1$	$\nu=0.3$	0.179	0.144
	$\nu=0.25$	0.173	0.144
$t=0.5$	$\nu=0.3$	0.225	0.144
	$\nu=0.25$	0.218	0.144

22. ELASTIC MODULI AND LOCAL STRESSES IN WAFER-TYPE PLATES AND SHELLS, INCLUDING THE INTERACTION BETWEEN CELL ELEMENTS

We remind the reader that the elastic analysis of wafer-type plate and shell structures (Fig. 20.1)—and hence the derivation of the elastic properties of such structures, equations (20.34)—were performed in Section 20 under the assumption of there being no interaction between the individual elements of the unit cell of the problem, Ω_1, Ω_2 and Ω_3 (see Figs 20.1 and 20.3). The removal of this assumption in the present section will enable us to obtain more accurate solutions as well as to estimate the accuracy of the solutions given by (20.34). We shall again assume that conditions (20.1) are satisfied and we limit ourselves to shells made of isotropic materials.

Based on (20.1), the pertinent local problems will be solved for each cell element separately. The functions sought will be averaged over the thickness of an element, and the interaction between the elements will be modelled by introducing certain stress singularities which have to be found in the course of the solution. In particular, referring to Fig. 22.1 and in accordance with the boundary local problems posed by (19.8), (19.9)

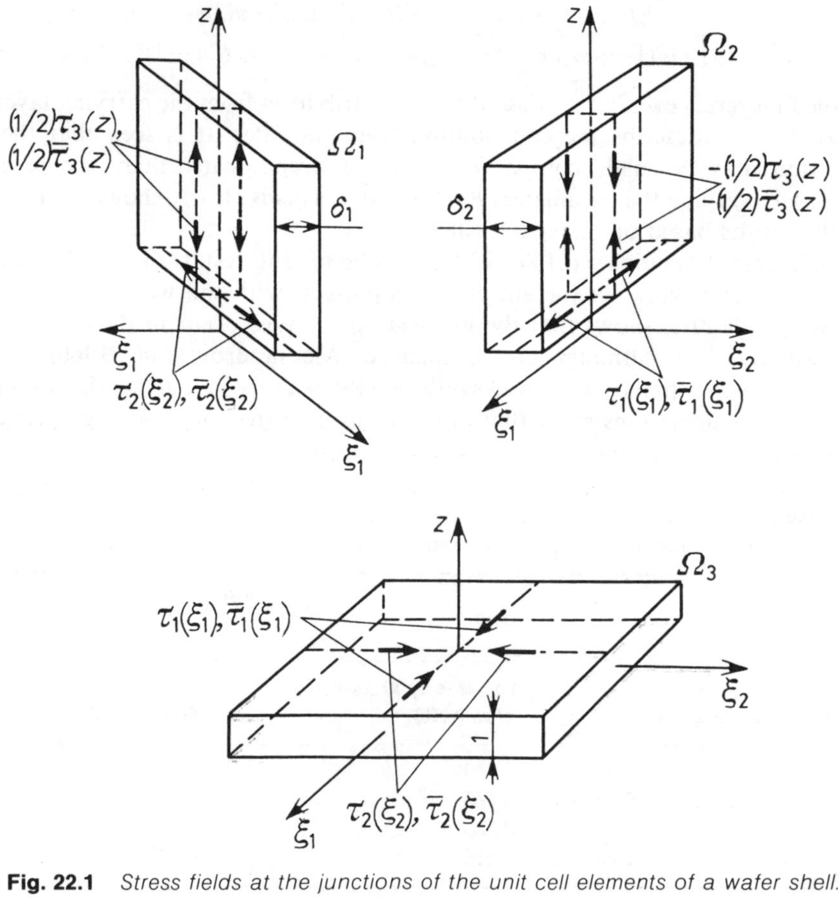

Fig. 22.1 *Stress fields at the junctions of the unit cell elements of a wafer shell.*

and (19.13), (19.14), the boundary conditions to be satisfied along Ω_1, Ω_2 junctions are

$$
\begin{cases} \dfrac{1}{h_1}\tau_{13}^{\lambda\mu} = \pm\tfrac{1}{2}\delta(\xi_2)\tau_3^{\lambda\mu}(z) \\ \dfrac{1}{h_1}\bar{\tau}_{13}^{\lambda\mu} = \pm\tfrac{1}{2}\delta(\xi_2)\bar{\tau}_3^{\lambda\mu}(z) \end{cases} \quad \left(\xi_1 = \pm\frac{\delta_1}{2}\right),
$$
$$
\begin{cases} \dfrac{1}{h_2}\tau_{23}^{\lambda\mu} = \mp\tfrac{1}{2}\delta(\xi_1)\tau_3^{\lambda\mu}(z) \\ \dfrac{1}{h_2}\bar{\tau}_{23}^{\lambda\mu} = \mp\tfrac{1}{2}\delta(\xi_1)\bar{\tau}_3^{\lambda\mu}(z) \end{cases} \quad \left(\xi_2 = \pm\frac{\delta_2}{2}\right). \tag{22.1}
$$

Along junctions between the strengthening elements Ω_1 and Ω_2 and the carrying element Ω_3, i.e. at $z = \frac{1}{2}$, we have the following conditions (cf. Fig. 22.1):

$$
\begin{aligned} \tau_{23}^{\lambda\mu} &= \delta(\xi_1)\tau_2^{\lambda\mu}(\xi_2), \qquad & \bar{\tau}_{23}^{\lambda\mu} &= \delta(\xi_1)\bar{\tau}_2^{\lambda\mu}(\xi_2), \\ \tau_{13}^{\lambda\mu} &= \delta(\xi_2)\tau_1^{\lambda\mu}(\xi_1), \qquad & \bar{\tau}_{13}^{\lambda\mu} &= \delta(\xi_2)\bar{\tau}_1^{\lambda\mu}(\xi_1). \end{aligned} \tag{22.2}
$$

The stresses occurring at the junctions of the elements (see Fig. 22.1) are found by requiring that the material displacements remain continuous as we pass from one element to another, that is,

$$
\begin{aligned} \tilde{U}_1^{\lambda\mu}\big|_{z=\frac{1}{2}}^{\Omega_2} &= \tilde{U}_1^{\lambda\mu}\big|_{\xi_2=0}^{\Omega_3}, \qquad & \tilde{U}_2^{\lambda\mu}\big|_{z=\frac{1}{2}}^{\Omega_1} &= \tilde{U}_2^{\lambda\mu}\big|_{\xi_1=0}^{\Omega_3}, \\ \tilde{U}_3^{\lambda\mu}\big|_{\xi_2=0}^{\Omega_1} &= \tilde{U}_3^{\lambda\mu}\big|_{\xi_1=0}^{\Omega_3} \qquad & (\tilde{U}_i^{\lambda\mu} &\leftrightarrow \tilde{V}_i^{\lambda\mu}), \end{aligned} \tag{22.3}
$$

where a tilde denotes thickness average over the corresponding element.

It can be shown that all six types of the local problem $(b\lambda\mu)$ and $(b^*\lambda\mu)$ may be solved analytically by applying the above approximating scheme and expanding the solutions in trigonometric series with respect to ξ_1 and ξ_2 (which secures in a very natural fashion the periodicity conditions in these coordinates). With regard to the coordinate z, which is 'non-periodic', the local problems associated with the elements Ω_1 and Ω_2 were solved by expanding in terms of the variable $z' = z - \frac{1}{2}$ on the segment $[0, H]$.

For the stresses along the junctions of the elements, taking into account the symmetry of each particular problem for different cell elements, the necessary trigonometric expansions may be written as

$$
\begin{aligned} \tau_1^{\lambda\mu}(\xi_1) &= \sum_{m=1}^{\infty} \tau_{1m}^{\lambda\mu} \sin\left(\frac{2\pi m}{A_1}\xi_1\right), \qquad & \tau_2^{\lambda\mu}(\xi_2) &= \sum_{m=1}^{\infty} \tau_{2m}^{\lambda\mu} \sin\left(\frac{2\pi m}{A_2}\xi_2\right), \\ \bar{\tau}_1^{\lambda\mu}(\xi_1) &= \sum_{m=1}^{\infty} \bar{\tau}_{1m}^{\lambda\mu} \sin\left(\frac{2\pi m}{A_1}\xi_1\right), \qquad & \bar{\tau}_2^{\lambda\mu}(\xi_2) &= \sum_{m=1}^{\infty} \bar{\tau}_{2m}^{\lambda\mu} \sin\left(\frac{2\pi m}{A_2}\xi_2\right), \\ \tau_3^{\lambda\mu}(z) &= \sum_{k=1}^{\infty} \tau_{3k}^{\lambda\mu} \cos\left(\frac{\pi k z'}{H}\right), \qquad & \bar{\tau}_3^{\lambda\mu}(z) &= \sum_{k=1}^{\infty} \bar{\tau}_{3k}^{\lambda\mu} \cos\left(\frac{\pi k z'}{H}\right). \end{aligned} \tag{22.4}
$$

The solution of the entire set of local problems is tedious, and we give only the results here. The non-zero effective elastic moduli are as follows:

$$\langle b_{11}^{11}\rangle = \frac{E}{1-\nu^2} + E(F_2^{(w)} + K\Sigma_1), \qquad \langle b_{22}^{22}\rangle = \frac{E}{1-\nu^2} + E(F_1^{(w)} + K\Sigma_1),$$

$$\langle b_{11}^{22}\rangle = \langle b_{22}^{11}\rangle = \frac{\nu E}{1-\nu^2} - EK\Sigma_1, \qquad \langle b_{12}^{12}\rangle = G,$$

$$\langle zb_{11}^{11}\rangle = \langle b_{11}^{*11}\rangle = E(S_2^{(w)} + K\Sigma_2), \quad \langle zb_{22}^{22}\rangle = \langle b_{22}^{*22}\rangle = E(S_1^{(w)} + K\Sigma_2),$$

$$\langle zb_{11}^{22}\rangle = \langle zb_{22}^{11}\rangle = \langle b_{11}^{*22}\rangle = \langle b_{22}^{*11}\rangle = -EK\Sigma_2, \tag{22.5}$$

$$\langle zb_{11}^{*11}\rangle = \frac{E}{12(1-\nu^2)} + E(J_2^{(w)} + K\Sigma_3),$$

$$\langle zb_{22}^{*22}\rangle = \frac{E}{12(1-\nu^2)} + E(J_1^{(w)} + K\Sigma_3),$$

$$\langle zb_{22}^{*11}\rangle = \langle zb_{11}^{*22}\rangle = \frac{\nu E}{12(1-\nu^2)} - EK\Sigma_3.$$

Formula (20.34) for the torsional stiffness $\langle zb_{12}^{*12}\rangle$ remains unchanged and is not repeated here.

In the above equations we have defined, for $A_1 = A_2 = A$ and $h_1 = h_2 = h$,

$$K = \frac{8\nu^2 H^2 t_1 t_2}{\pi^3(1+\nu)Ah^2(t_1+t_2)}, \qquad \Sigma_1 = \sum_{k=1}^{\infty} \frac{1-(-1)^k}{k^2} x_k,$$

$$\Sigma_2 = \sum_{k=1}^{\infty} \frac{1-(-1)^k}{k^2} \bar{x}_k, \qquad \Sigma_3 = \sum_{k=1}^{\infty} \frac{[1-(-1)^k]/2 - H(-1)^k}{k^2} \bar{x}_k. \tag{22.6}$$

Here the coefficients x_k are determined from the following infinite system of simultaneous algebraic equations:

$$q_{0n}x_n + \sum_{k=1}^{\infty} [1-(-1)^{k+n}]\, q_{kn} x_k + \frac{16At_1t_2n}{\pi(1+\nu)h^2(t_1+t_2)} \sum_{m=1}^{\infty} r_{nm} y_m = \frac{1-(-1)^n}{n},$$

$$p_{0n} y_n + \sum_{m=1}^{\infty} p_{mn} y_m + \frac{8H^2(t_1+t_2)n}{\pi(1+\nu)^2 A^3 h t_1 t_2} \sum_{k=1}^{\infty} r_{kn} x_k = 0 \quad (n = 1, 2, 3, \ldots), \tag{22.7}$$

where

$$q_{0n} = (3-\nu)\coth\frac{\pi n}{a} - \frac{(1+\nu)\pi n}{a\sinh^2(\pi n/a)}, \qquad a = \frac{2H}{Ah},$$

$$q_{kn} = \frac{16a^2}{\pi^2(1+\nu)} \sum_{m=1}^{\infty} \frac{mn(\nu k^2 - a^2m^2)(\nu n^2 - a^2m^2)[\coth(\pi am) - (-1)^n]}{(k^2+a^2m^2)^2(n^2+a^2m^2)^2[\sinh(\pi am) - (-1)^n \pi am]},$$

$$r_{nm} = \frac{(\nu n^2 - a^2m^2)[\cosh(\pi am) - (-1)^n]}{(n^2+a^2m^2)^2[\sinh(\pi am) - (-1)^n \pi am]},$$

$$p_{0n} = \frac{3-\nu}{4(1+\nu)}\coth(\pi n) - \frac{\pi n}{4\sinh^2(\pi n)}$$
$$+ \frac{2[\sinh(\pi na)\cosh(\pi na) - \pi na]}{(1+\nu)^2 h[\sinh^2(\pi na) - (\pi na)^2]},$$
$$p_{mn} = \frac{mn^2}{\pi(n^2+m^2)^2}.$$

The infinite system of equations for the coefficients $\bar{x}_k$ in (22.6) differs from (22.7) only in having

$$[(1-(-1)^n)/2 - H(-1)^n]n^{-1}$$

instead of $[(1-(-1)^n)/n]$ in the right-hand side of the first equation and, clearly, we must replace x_k by $\bar{x}_k$ and y_k by $\bar{y}_k$ everywhere in (22.7).

The solutions of the above systems of equations and the coefficients in the expansions (22.4) for the stresses along the junctions (see eqs. (22.1) and (22.2)) are related by

$$\begin{aligned}
&\tau_{3k}^{11} = EK'x_k, \qquad \tau_{1m}^{11} - \tau_{2m}^{11} = EK'y_m,\\
&\bar{\tau}_{3k}^{11} = EK'\bar{x}_k, \qquad \bar{\tau}_{1m}^{11} - \bar{\tau}_{2m}^{11} = EK'\bar{y}_m,\\
&\tau_{3k}^{22} = -\tau_{3k}^{11}, \qquad \tau_{3k}^{12} = 0, \qquad \tau_{1m}^{22} = -\tau_{1m}^{11} \quad (\tau \leftrightarrow \bar{\tau}),\\
&\tau_{2m}^{22} = -\tau_{2m}^{11}, \qquad \tau_{1m}^{12} = \tau_{2m}^{12} = 0 \quad (\tau \leftrightarrow \bar{\tau})\\
&K' = \frac{8\nu A t_1 t_2}{\pi(1+\nu)h^2(t_1+t_2)}.
\end{aligned} \tag{22.8}$$

It should also be noted that

$$\tau_{2m}^{\lambda\mu} = -\tau_{1m}^{\lambda\mu}, \qquad \bar{\tau}_{2m}^{\lambda\mu} = -\bar{\tau}_{1m}^{\lambda\mu}$$

when $t_1 = t_2$.

For the more general case in which $A_1 \neq A_2$ and $h_1 \neq h_2$, the solution follows exactly the same pattern except for the fact that the system (22.7) will be augmented by one more group of equations and the corresponding coefficients will be more complex.

Expressions (22.5) for the effective elastic moduli of the wafer shell are a refined version of equations (20.34) and can be reduced to these latter by dropping the terms of the type $K\Sigma_1$ $(i = 1, 2, 3)$. The magnitude of the corrections was estimated by a series of calculations for different values of the parameters determining the dimensions of the unit cell of Fig. 20.1. The author showed numerically that, unless an accuracy of better than 10^{-4} is required, there is no need to consider more than 40 equations in the system (22.7).

In Table 22.1 we present both the numerical results obtained from (22.5) (rows I) and from (20.34) (rows II) for $A_1 = A_2 = 1$ and $\nu = 0.3$. The magnitudes of the percentage changes are shown in rows III. On the basis of the analysis of the table, the following conclusions are made:

(1) As found from formulae (22.5), the elastic moduli $\langle b_{11}^{11}\rangle$, $\langle b_{22}^{22}\rangle$, $\langle b_{11}^{*11}\rangle$, $\langle b_{22}^{*22}\rangle$,

Table 22.1 *Effective elastic moduli of the wafer shells and plates. Rows I: results obtained from (22.5); rows II: results obtained from (20.34); rows III: the magnitudes of the percentage changes.* $A_1 = A_2 = 1$, $\nu = 0.3$.

Wafer shell		$\langle b_{11}^{11}\rangle/E$	$\langle b_{22}^{22}\rangle/E$	$\langle b_{22}^{11}\rangle/E$	$\langle b_{11}^{*11}\rangle/E$	$\langle b_{22}^{*22}\rangle/E$	$\langle b_{22}^{*11}\rangle/E$	$\langle zb_{11}^{*11}\rangle/E$	$\langle zb_{22}^{*22}\rangle/E$	$\langle zb_{22}^{*11}\rangle/E$	$\langle zb_{12}^{*12}\rangle/E$
1 $H = 10$	I	2.1118	2.1118	0.3168	5.5648	5.5648	−0.0648	39.0728	39.0728	−0.3705	0.0921
$h = 10$	II	2.0989	2.0989	0.3297	5.5	5.5	0	38.6749	38.6749	0.0275	0.0921
$t_1 = 1$	III	0.6%	0.6%	−3.9%	1.2%	1.2%	—	1.03%	1.03%	—	—
$t_2 = 1$											
2 $H = 10$	I	1.6075	2.1075	0.3211	2.7932	5.5432	−0.0432	19.6485	38.9402	−0.2378	0.0660
$h = 10$	II	1.5989	2.0989	0.3297	2.75	5.5	0	19.3833	38.6749	0.0275	0.0660
$t_1 = 1$	III	0.54%	0.41%	−2.62%	1.57%	0.78%	—	1.37%	0.69%	—	—
$t_2 = 0.5$											
3 $H = 10$	I	1.3505	1.3505	0.3281	1.3828	1.3828	−0.0078	9.7856	9.7856	−0.0207	0.0359
$h = 20$	II	1.3489	1.3489	0.3297	1.375	1.375	0	9.7374	9.7374	0.0275	0.0359
$t_1 = 0.5$	III	0.12%	0.12%	−0.48%	0.57%	0.57%	—	0.49%	0.49%	—	—
$t_2 = 0.5$											
4 $H = 20$	I	1.6053	1.6053	0.3232	5.3148	5.3148	−0.0648	72.6514	72.6514	−0.7406	0.0400
$h = 20$	II	1.5989	1.5989	0.3297	5.25	5.25	0	71.8832	71.8832	0.0275	0.0400
$t_1 = 0.5$	III	0.4%	0.4%	−1.96%	1.23%	1.23%	—	1.07%	1.07%	—	—
$t_2 = 0.5$											

5	$H=2$	I	1.1550	1.1550	0.3296	0.0841	0.0841	−0.0001	0.2363	0.2363	0.0274	0.0334
	$h=25$	II	1.1549	1.1549	0.3297	0.084	0.084	0	0.2362	0.2362	0.0275	0.0334
	$t_1=0.7$	III	0.005%	0.005%	−0.03%	0.07%	0.07%	—	0.055%	0.055%	−0.33%	—
	$t_2=0.7$											
6	$H=5$	I	1.300	1.300	0.3291	0.6016	0.6016	−0.1561	2.3134	2.3134	0.0223	0.0394
	$h=20$	II	1.299	1.299	0.3297	0.6	0.6	0	2.3082	2.3082	0.0275	0.0394
	$t_1=0.8$	III	0.05%	0.05%	−0.2%	0.26%	0.26%	—	0.22%	0.22%	−18.75%	—
	$t_2=0.8$											
7	$H=8$	I	1.3129	1.3129	0.3290	0.9628	0.9628	−0.0028	5.5636	5.5636	0.0132	0.0403
	$h=30$	II	1.3122	1.3122	0.3297	0.96	0.96	0	5.5494	5.5494	0.0275	0.0403
	$t_1=0.8$	III	0.06%	0.06%	−0.22%	0.3%	0.3%	—	0.26%	0.26%	−51.96%	—
	$t_2=0.8$											
8	$H=8$	I	1.2060	1.3127	0.3292	0.4819	0.9619	−0.0019	2.8300	5.5589	0.0180	0.0367
	$h=30$	II	1.2056	1.3122	0.3297	0.48	0.96	0	2.8205	5.5494	0.0275	0.0367
	$t_1=0.8$	III	0.04%	0.04%	−0.15%	0.39%	0.2%	—	0.34%	0.17%	−34.62%	—
	$t_2=0.4$											
9	$H=20$	I	1.7683	1.7683	0.3269	7.0278	7.0278	−0.0277	96.1443	96.1443	−0.3030	0.1922
	$h=60$	II	1.7656	1.7656	0.3297	7.0	7.0	0	95.8138	95.8138	0.0275	0.1922
	$t_1=2$	III	0.15%	0.15%	−0.85%	0.4%	0.4%	—	0.34%	0.34%	—	—
	$t_2=2$											

$\langle zb_{11}^{*11}\rangle$ and $\langle zb_{22}^{*22}\rangle$ can be calculated from the simpler formulae (20.34). The percentage changes are less than 1%.

(2) For the modulus $\langle b_{22}^{11}\rangle$, somewhat greater, but also reasonable percentage changes are obtained (for example, 3.9% in the first example in Table 22.1).

(3) For the moduli $\langle b_{22}^{*11}\rangle$ and $\langle zb_{22}^{*11}\rangle$, more significant percentage changes occur. Note that while the change for $\langle zb_{22}^{*11}\rangle$ is only of a quantitative nature, that for $\langle b_{22}^{*11}\rangle$ is also interesting at a qualitative level because this modulus vanishes in view of (20.34) and is assumed to be zero in the framework of the structurally anisotropic theory of strengthened plates and shells.

One more point to be made is that the percentage changes invariably increase with the height of the ribs (parameter H) and decrease with the distance between the ribs (parameter h).

From (15.74), (19.10), (19.15), (22.1) (22.2), (22.4) and (22.8), using the solutions of the system (22.7), we are in a position to obtain the local stress distributions along the junctions of the cell elements (Figs 22.2 and 22.3).

In particular, if we take $h_1 = h_2 = h$, then in the case of simple tension ($\varepsilon_1 \neq 0$), the junctions of reinforcing elements Ω_1 and Ω_2 will be subjected to the stresses

$$\frac{\sigma_{13}}{\varepsilon_1}\bigg|_{\substack{\alpha_1 = \pm\delta t/2\\ \alpha_2 = 0}} = \pm\frac{h}{2}\tau_3(z'),$$

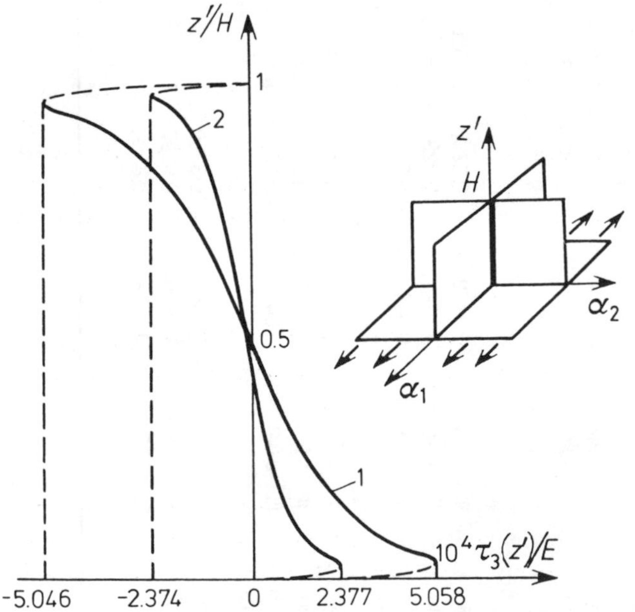

Fig. 22.2 *Stress fields at the junction of the strengthening elements of a wafer plate or shell under tension.*

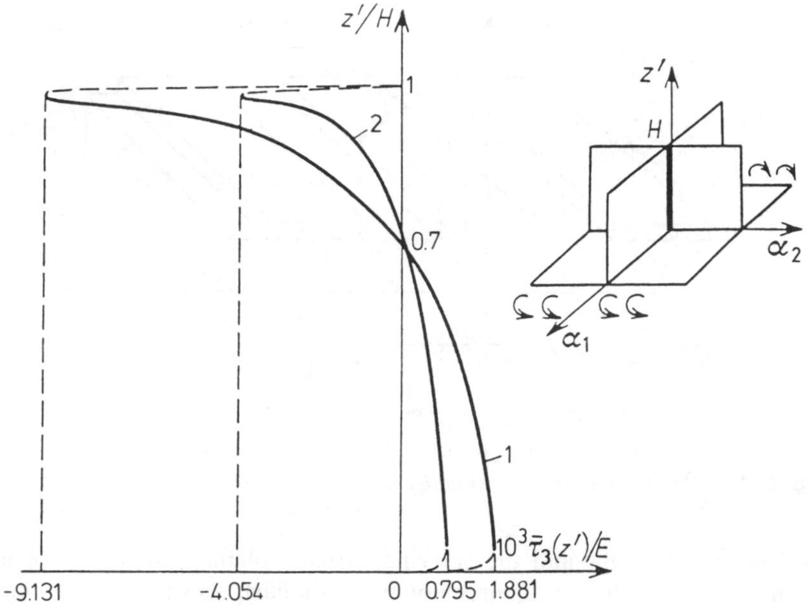

Fig. 22.3 *Stress fields at the junction of the strengthening elements of a wafer plate or shell under bending.*

and in the case of simple bending ($\kappa_1 \neq 0$)

$$\left.\frac{\sigma_{13}}{\delta\kappa_1}\right|_{\substack{\alpha_1 = \pm\delta t/2 \\ \alpha_2 = 0}} = \pm\frac{h}{2}\bar{\tau}_3(z').$$

The functions $\tau_3(z')$ and $\bar{\tau}_3(z')$ are shown graphically in Fig. 22.2 (for tension) and Fig. 22.3 (for bending). The curves marked 1 and 2 in these figures correspond, respectively, to cases 4 and 9 in Table 22.1. It will be noted that the above stresses turn to zero at $z' = H$ and $z' = 0$, but the Fourier cosine representation which we assumed for $\tau_3(z')$ and $\bar{\tau}_3(z')$ in (22.4) makes it difficult to approximate with reasonable accuracy the stress field in the immediate vicinity of these points.

23. STRETCHING OF PLATES OR SHELLS REINFORCED BY A REGULAR SYSTEM OF THIN SURFACE STRIPS

In the previous sections we obtained approximate solutions of the cell local problems for the case(s) when the thickness of the reinforcing element is small compared with its height. In the present section the opposite extreme will be considered. That is to say we will be concerned with shells (or plates) strengthened with a regular system of thin ribs arranged symmetrically with respect to the middle surface of the shell (see Fig. 23.1), and we will assume that the height of a rib is much smaller than its length. It will be seen that with this assumption the local problem admits of being treated analytically even in the case of a non-vanishing interaction between the ribs and the carrying surfaces.

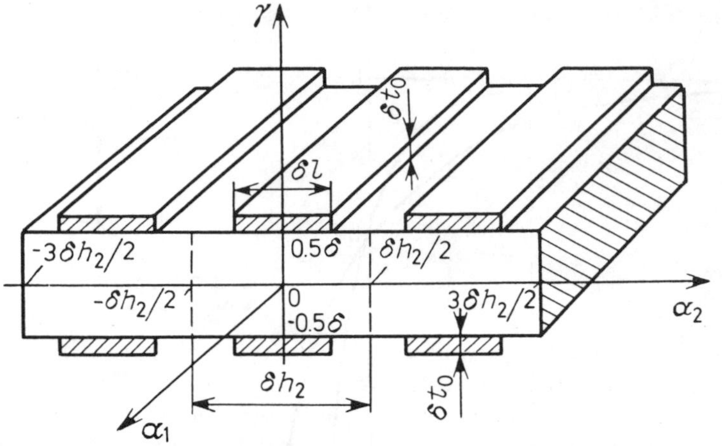

Fig. 23.1 *Plate or shell strengthened by a regular system of thin surface strips.*

We assume that the ribs and the carrying surfaces of the composite are made of isotropic materials (which may or may not be dissimilar); also that the strengthening elements within the unit cell of the structure are thin plates having no flexural stiffness and working in tension–compression only.

23.1. The solution of the local problems

The shape of the unit cell of our problem in the coordinate system ξ_2, z is shown in Fig. 23.2, in which $2\alpha = lA_2/h_2$ by the rules of coordinate transformation.

We begin with a discussion of the local problem ($b11$), posed by equations (19.16) and (19.18)–(19.20) for $\lambda\mu = 11$, and first consider the region $\Omega_3 = \{|z| \leqslant \frac{1}{2}, |\xi_2| \leqslant A_2/2\}$ on the carrying surface. In view of the symmetry with respect to z and taking into account the periodicity of the functions U_2^{11} and U_3^{11} in ξ_2 with period A_2, the solution of

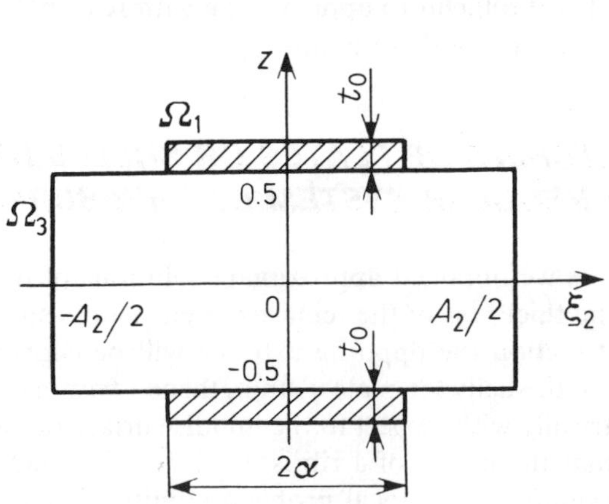

Fig. 23.2 *Unit cell for a plate or shell strengthened by a regular system of thin surface strips.*

the plane elastic problem that arises may be represented in the form of series expansions:

$$U_2^{11} = 2\sum_{n=1}^{\infty} [B_n \cosh(\lambda_n z) + zC_n \sinh(\lambda_n z)] \sin(\lambda_n h_2 \xi_2),$$

$$U_3^{11} = \Phi_0 z - 2\sum_{n=1}^{\infty} [B_n \sinh(\lambda_n z) + zC_n \cosh(\lambda_n z)] \cos(\lambda_n h_2 \xi_2), \tag{23.1}$$

where $\lambda_n = 2\pi n/(h_2 A_2)$.

The symmetry with respect to z now permits us to consider the half-plane $z > 0$, in which we mentally remove the ribs occupying the region $\Omega_1 = \{\frac{1}{2} \leqslant z \leqslant \frac{1}{2} + t_0, |\xi_2| \leqslant \alpha\}$ and replace their effect by a certain unknown displacement field. The assumptions we have introduced about the surface strips then make it possible to write the boundary conditions in the form

$$\begin{aligned} \tau_{33}^{11} &= -c_{12} \quad (z = \tfrac{1}{2}), \\ \tau_{23}^{11} &= 0 \quad (z = \tfrac{1}{2}, -A_2/2 \leqslant \xi_2 < -\alpha, \alpha < \xi_2 \leqslant A_2/2), \\ U_2^{11} &= \Psi(\xi_2) \quad (z = \tfrac{1}{2}, |\xi_2| < \alpha), \end{aligned} \tag{23.2}$$

where τ_{33}^{11} and τ_{23}^{11} are given in (19.18) in terms of the displacements U_2^{11} and U_3^{11}, and the function $\Psi(\xi_2)$ is as yet unknown and must be found in the course of the solution.

From the first of the conditions (23.2), we find that

$$\Phi_0 = -c_{12}/c_{11}, \qquad B_n = -C_n \lambda_n^{-1} \left[\frac{1-\nu}{1-2\nu} + \frac{\lambda_n}{2} \tanh(\lambda_n/2)\right],$$

whereas the remaining conditions lead to dual series relations which reduce, after some calculation, to

$$\begin{aligned} \sum_{n=1}^{\infty} F_n \sin(\lambda_n h_2 \xi_2) &= \frac{1-2\nu}{1-\nu} \Psi(\xi_2) \\ &\quad + \sum_{n=1}^{\infty} F_n R_n \sin(\lambda_n h_2 \xi_2) \quad (|\xi_2| < \alpha), \\ \sum_{n=1}^{\infty} \lambda_n F_n \sin(\lambda_n h_2 \xi_2) &= 0 \quad (-A_2/2 \leqslant \xi_2 < -\alpha, \alpha < \xi_2 \leqslant A/2), \end{aligned} \tag{23.3}$$

provided

$$F_n = C_n \frac{2(1-2\nu)\lambda_n - \sinh(\lambda_n)}{\lambda_n \cosh(\lambda_2/2)},$$

$$R_n = \frac{2\cosh^2(\lambda_n/2) - \sinh(\lambda_n) + 2(1-2\nu)\lambda_n}{2(1-2\nu)\lambda_n - \sinh(\lambda_n)}, \qquad \lim_{n\to\infty} R_n = 0.$$

Now if we introduce an auxiliary function $p(\xi_2)$ for which

$$\sum_{n=1}^{\infty} \beta_n F_n \sin(\beta_n \xi_2) = \begin{cases} p(\xi_2) & (|\xi_2| < \alpha), \\ 0 & (-A_2/2 \leqslant \xi_2 < -\alpha, \alpha < \xi_2 \leqslant A_2/2), \end{cases} \tag{23.4}$$

where $\beta_n = \lambda_n h_2$, then

$$F_n = \frac{1}{\beta_n} \int_{-\alpha}^{\alpha} p(t) \sin(\beta_n t)\, \mathrm{d}t \tag{23.5}$$

and we note that the function $p(\xi_2)$ and the contact stress $\tau_{23}^{11}|_{z=\frac{1}{2}}$ are related by

$$\tau_{23}^{11}|_{z=\frac{1}{2}} = p^*(\xi_2) = \frac{c_{44}}{(1-2\nu)h_2} p(\xi_2). \tag{23.6}$$

Now if we substitute (23.5) into the first of equations (23.3) we obtain the integral equation

$$\sum_{n=1}^{\infty} \frac{1}{\beta_n} \left(\int_{-\alpha}^{\alpha} p(t) \sin(\beta_n t)\, \mathrm{d}t \right) \sin(\beta_n \xi_2) = \frac{1-2\nu}{1-\nu} \Psi(\xi_2)$$
$$+ \sum_{n=1}^{\infty} \frac{R_n}{\beta_n} \left(\int_{-\alpha}^{\alpha} p(t) \sin(\beta_n t)\, \mathrm{d}t \right) \sin(\beta_n \xi_2) \quad (|\xi_2| < \alpha) \tag{23.7}$$

by means of which to calculate the function $p(t)$. The solution, in terms of the Chebyshev polynomials of the first kind, $T_{2m+1}(t)$, may be represented in the form

$$p(t) = \frac{1}{\sqrt{\alpha^2 - t^2}} \sum_{m=0}^{\infty} A_{2m+1} T_{2m+1}(t/\alpha), \tag{23.8}$$

and using the fact that

$$\int_{-\alpha}^{\alpha} \frac{\sin(\beta_n t) T_{2m+1}(t/\alpha)\, \mathrm{d}t}{\sqrt{\alpha^2 - t^2}} = (-1)^m \pi J_{2m+1}(\beta_n \alpha),$$

equation (23.7) can be put into the form

$$\pi \sum_{n=1}^{\infty} \frac{1-R_n}{\beta_n} \left[\sum_{m=0}^{\infty} (-1)^m A_{2m+1} J_{2m+1}(\beta_n \alpha) \right] \sin(\beta_n \xi_2)$$
$$= \frac{1-2\nu}{1-\nu} \Psi(\xi_2) \quad (|\xi_2| < \alpha). \tag{23.9}$$

A system of equations to determine the coefficients A_{2m+1} occurring in (23.8) can now be obtained by expressing $\Psi(\xi_2)$ in the form of an infinite series,

$$\Psi(\xi_2) = \sum_{k=0}^{\infty} d_{2k+1} T_{2k+1}(\xi_2/\alpha). \tag{23.10}$$

and recalling that

$$\sin(\beta_n \xi_2) = 2 \sum_{k=0}^{\infty} (-1)^k J_{2k+1}(\beta_n \alpha) T_{2k+1}(\xi_2/\alpha).$$

Substituting into (23.9) then results in

$$\sum_{m=0}^{\infty} (-1)^{m+k} a_{2m+1} \alpha_{2m+1,2k+1} = d_{2k+1}, \tag{23.11}$$

where

$$\alpha_{2m+1,2k+1} = \sum_{n=1}^{\infty} \frac{1-R_n}{n} J_{2m+1}(\beta_n \alpha) J_{2k+1}(\beta_n \alpha),$$

$$a_{2m+1} = \frac{A_2(1-\nu)}{1-2\nu} A_{2m+1}.$$

To solve (23.11) we assume

$$a_{2m+1} = \sum_{k=0}^{\infty} a_{2m+1}^{(k)} d_{2k+1}, \tag{23.12}$$

where $a_{2m+1}^{(k)}$ is the solution of a system the left-hand side of which is the same as in (23.11), and the right-hand side of which has 1 at the kth position and zeros at the others.

Substituting into (23.8) now yields

$$p(t) = \frac{1-2\nu}{A_2(1-\nu)} \frac{1}{\sqrt{\alpha^2 - t^2}} \sum_{k=0}^{\infty} d_{2k+1} \left[\sum_{m=0}^{\infty} a_{2m+1}^{(k)} T_{2m+1}(t/\alpha) \right]. \tag{23.13}$$

To proceed with the analysis, the coefficients d_{2k+1} must be determined, for which purpose we turn to the local problem ($b11$) in region Ω_1 (see Fig. 23.2) and make use of the strip–surface interface condition. From the boundary conditions (19.19) and (19.20) for Ω_1 we have

$$\begin{aligned} \tau_{22}^{11} &= -c_{12}^{\mathrm{S}}, \qquad \tau_{23}^{11} = 0 \quad (\xi_2 = \pm\alpha), \\ \tau_{33}^{11} &= -c_{12}^{\mathrm{S}}, \qquad \tau_{23}^{11} = 0 \quad (z = \tfrac{1}{2} + t_0), \end{aligned} \tag{23.14}$$

where (and later) a superscript S is used to refer to the strip material.

Along the line of contact between Ω_1 and Ω_3, in accordance with (15.60), (19.21), (23.2) and (23.6), we obtain:

$$\tau_{33}^{11} = -c_{12}^{\mathrm{S}}, \qquad \tau_{23}^{11} = p^*(\xi_2) \quad (z = \tfrac{1}{2}, |\xi_2| < \alpha). \tag{23.15}$$

Furthermore, from (23.14), using the assumed small height to length ratio ($t_0 \ll 2\alpha$), we obtain:

$$\tau_{33}^{11} = -c_{12}^{\mathrm{S}}, \quad \text{in } \Omega_1, \tag{23.16}$$

which when substituted into (19.18) for τ_{33}^{11} gives

$$\frac{\partial U_3^{11}}{\partial z} = -\frac{c_{12}^S}{c_{11}^S} - \frac{1}{h_2}\frac{c_{12}^S}{c_{11}^S}\frac{\partial U_2^{11}}{\partial \xi_2},$$

and using this in turn in the expression for τ_{22}^{11} in (19.18), we have

$$\tau_{22}^{11} = -\frac{(c_{12}^S)^2}{c_{11}^S} + \frac{1}{h_2}\left(c_{11}^S - \frac{(c_{12}^S)^2}{c_{11}^S}\right)\frac{\partial U_2^{11}}{\partial \xi_2}. \tag{23.17}$$

If we average (19.16) with $i = 2$ over the thickness of the strip, we find

$$\frac{1}{h_2}\frac{\mathrm{d}\tilde{\tau}_{22}^{11}}{\partial \xi_2} + \tau_{23}^{11}\big|_{z=\frac{1}{2}+t_0} - \tau_{23}^{11}\big|_{z=\frac{1}{2}} = 0, \tag{23.18}$$

and hence, by (23.14) and (23.15),

$$\frac{\mathrm{d}\tilde{\tau}_{22}^{11}}{\mathrm{d}\xi_2} = h_2 p^*(\xi_2), \tag{23.19}$$

which when integrated yields

$$\tilde{\tau}_{22}^{11} = -c_{12}^S t_0 + h_2 \int_{-\alpha}^{\xi_2} p^*(u)\,\mathrm{d}u \tag{23.20}$$

after the use of (23.14).

On the other hand, averaging (23.17) over the thickness of the strip we have

$$\tilde{\tau}_{22}^{11} = -\frac{(c_{12}^S)^2 t_0}{c_{11}^S} + \frac{1}{h_2}\left(c_{11}^S - \frac{(c_{12}^S)^2}{c_{11}^S}\right) t_0 \frac{\mathrm{d}U_2^{11}}{\mathrm{d}\xi_2}. \tag{23.21}$$

From (23.20) and (23.21), with (23.6), (23.13) and the standard integral

$$\int_{-\alpha}^{\xi_2} \frac{T_{2m+1}(u/\alpha)\,\mathrm{d}u}{\sqrt{\alpha^2 - u^2}} = -\frac{\sqrt{1-(\xi_2/\alpha)^2}}{2m+1} U_{2m}(\xi_2/\alpha),$$

where U_{2m} is the Chebyshev polynomial of the second kind, we obtain:

$$\frac{\mathrm{d}U_2^{11}}{\mathrm{d}\xi_2} = -\frac{[c_{12}^S - (c_{12}^S)^2/c_{11}^S]h_2}{c_{11}^S - (c_{12}^S)^2/c_{11}^S} - \frac{c_{44}h_2}{(1-\nu)[c_{11}^S - (c_{12}^S)^2/c_{11}^S]t_0 A_2} \sum_{k=0}^{\infty} d_{2k+1} \sum_{m=0}^{\infty} a_{2m+1}^{(k)} \frac{U_{2m}(\xi_2/\alpha)}{2m+1}\sqrt{1-(\xi_2/\alpha)^2}. \tag{23.22}$$

This is now integrated using $U_2^{11}(0) = 0$ (which clearly follows from the symmetry of the problem), expressions for the elastic coefficients of an isotropic material and the integral

$$\int_0^{\xi_2} \sqrt{1-(\xi_2/\alpha)^2}\, U_{2m}(\xi_2/\alpha)\mathrm{d}\xi_2 = \frac{\alpha}{2m+1} T_{2m+1}(\xi_2/\alpha).$$

The result is

$$U_2^{11} = -\nu_S h_2 \xi_2 - \frac{E(1-\nu_S^2)h_2\alpha}{2E_S(1-\nu^2)t_0 A_2} \sum_{k=0}^{\infty} d_{2k+1} \sum_{m=0}^{\infty} \frac{a_{2m+1}^{(k)}}{(2m+1)^2} T_{2m+1}(\xi_2/\alpha), \tag{23.23}$$

so that using (23.2) and noting that U_2^{11} is continuous across the strip–surface interface in view of (15.60), we arrive at the following system of equations for the coefficients d_{2k+1}:

$$d_{2m+1} + \frac{E(1-\nu_S^2)h_2\alpha}{2E_S(1-\nu^2)t_0 A_2} \sum_{k=0}^{\infty} a_{2m+1}^{(k)} \frac{d_{2k+1}}{(2m+1)^2} = -\nu_S h_2 \alpha \delta_{2m+1,1}, \tag{23.24}$$

where $\delta_{2m+1,1}$ is unity for $m=0$ and zero for $m \geqslant 1$.

The local problem ($b22$) can be treated in a similar fashion, and to obtain its solution the double index $\{11\}$ must be replaced by $\{22\}$ everywhere in the above formulas. In the conditions (23.14) we write

$$\tau_{22}^{22} = -c_{11}^{S} \quad (\xi_2 = \pm\alpha)$$

in this case and, accordingly, c_{11}^S replaces c_{12}^S in (23.20) and (23.21).

One further change is that in the first term in (23.23) the factor ν_S will disappear and, accordingly, it must also be dropped in the right-hand side of the system (23.24) (which is solved for $d_{2k+1}^{(2)}$ this time rather than for $d_{2k+1} = d_{2k+1}^{(1)}$ as in the ($b11$) case). We therefore have

$$d_{2k+1}^{(1)} = \nu_S d_{2k+1}^{(2)}.$$

Having solved the local problems ($b11$) and ($b22$) we are now ready to calculate the effective elastic moduli of the shell (plate) reinforced by a regular system of thin surface strip (see Fig. 23.1). Using (19.21) we find that

$$\begin{aligned}
\langle b_{11}^{11} \rangle &= \frac{E}{1-\nu^2}\left[1 + \frac{E_S(1-\nu^2)}{E}\frac{2lt_0}{h_2} - \nu_S \frac{l}{h_2}\Sigma\right], \\
\langle b_{22}^{11} \rangle = \langle b_{11}^{22} \rangle &= \frac{\nu E}{1-\nu^2}\left(1 - \frac{1}{\nu}\frac{l}{h_2}\Sigma\right), \\
\langle b_{22}^{22} \rangle &= \frac{E}{1-\nu^2}\left(1 - \frac{1}{\nu_S}\frac{l}{h_2}\Sigma\right),
\end{aligned} \tag{23.25}$$

where we have written as a shorthand

$$\Sigma = \sum_{k=0}^{\infty} d_{2k+1} \sum_{m=0}^{\infty} \frac{a_{2m+1}^{(k)}}{(2m+1)^2}.$$

If $E_S = E$ and $\nu_S = \nu$, calculations show that the elastic moduli (23.25) satisfy the

Table 23.1 *Effective elastic moduli of shells and plates reinforced by a regular system of thin surface strips:* $\nu = 0.3$; $h_2 = 10$; $A_2 = 1$.

		$\alpha = 0.4$				$t_0 = 0.05$			
		$t_0 = 0.005$	$t_0 = 0.01$	$t_0 = 0.05$	$t_0 = 0.1$	$\alpha = 0.475$	$\alpha = 0.45$	$\alpha = 0.35$	$\alpha = 0.3$
$\dfrac{\langle b_{11}^{11} \rangle}{E/(1-\nu^2)}$	1	1.0082	1.0163	1.0811	1.1615	1.0954	1.0907	1.0708	1.0605
	2	1.0073	1.0146	1.0728	1.1456	1.0865	1.0819	1.0637	1.0546
	3	0.09%	0.17%	0.77%	1.39%	0.82%	0.81%	0.67%	0.56%
$\dfrac{\langle b_{22}^{11} \rangle}{\nu E/(1-\nu^2)}$	1	1.0097	1.0194	1.0926	1.1770	1.0998	1.0974	1.0786	1.0654
	2	1.0000	1.0000	1.0000	1.0000	1.0000	1.0000	1.0000	1.0000
	3	0.97%	1.94%	9.26%	17.7%	9.98%	9.74%	7.86%	6.54%

relations (19.28) proved in the general case for the effective moduli of homogeneous ribbed shells.

23.2. The effective elastic moduli and the local stresses

A direct comparison between (23.25) and the corresponding results of the structurally anisotropic model (Birger 1961) shows that the latter differ from the former because of the lack of terms containing Σ. The role of this correction is illustrated in Table 23.1 which shows the effective elastic moduli computed for homogeneous plates and shells for $h_2 = 10$, $\nu = 0.3$, $A_2 = 1$ and various values of α and t_0 from (23.25) (rows 1) and by the method of Birger (rows 2); the magnitude of the percentage changes is given in rows 3. It is seen that while the percentage changes for the modulus $\langle b_{11}^{11} \rangle$ are small, those for $\langle b_{22}^{11} \rangle$ are quite considerable.

The method of determining the distribution of local stresses will be illustrated by considering the tension of a homogeneous plate reinforced by a set of surface strips when one end of the strip is clamped and the other subjected to an external force of magnitude Q normal to the surface strips. The homogenized problem, i.e. that of the tension of a strip with effective moduli (23.25), is solved in an elementary fashion to give

$$N_{22} = Q, \qquad \varepsilon_{22} = Q/(\delta \langle b_{22}^{22} \rangle).$$

For the stress quantity σ_{22}, using (15.74) and the solution of the problem ($b22$), we have

$$\begin{aligned} \sigma_{22}/\sigma_{22}^{0} &= b_{22}^{22}/\langle b_{22}^{22} \rangle \\ &= -\frac{\pi}{h_2} \sum_{n=1}^{\infty} G_n \sum_{k=0}^{\infty} d_{2k+1}^{(2)} \sum_{m=0}^{\infty} (-1)^m a_{2m+1}^{(k)} J_{2m+1}(2\pi n \alpha) \cos(2\pi n y_2), \end{aligned} \tag{23.26}$$

where $\sigma_{22}^{(0)} = Q/\delta$ and G_n are known functions.

The results obtained from (23.26) for $z = 0$ (the axis of the plate) and $\alpha = 0.4$ (resp. $\alpha = 0.45$) are shown in Fig. 23.3 (resp. in Fig. 23.4) for $t_0 = 0.01$ (curves 1) and $t_0 = 0.05$

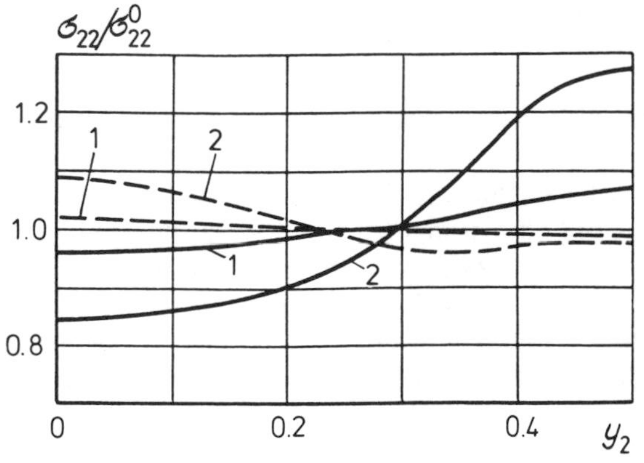

Fig. 23.3 *Stress distribution $\sigma_{22}/\sigma_{22}^{0}$ along the axis $z = 0$ for $\alpha = 0.4$, $t_0 = 0.01$ and $t_0 = 0.05$ (curves 1 and 2) and along the line $z = 0.5$ (dashed curves).*

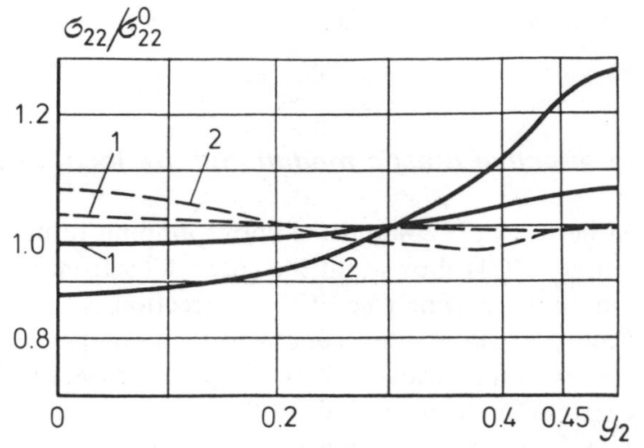

Fig. 23.4 *Stress distribution $\sigma_{22}/\sigma_{22}^{0}$ along the axis $z=0$ for $\alpha=0.45$, $t_0=0.01$ and $t_0=0.05$ (curves 1 and 2) and along the line $z=0.5$ (dashed curves).*

(curves 2). The dashed curves in both figures refer to the case $z = 0.5$. All the calculations were performed for $\nu = 0.3$ and $h_2 = 10$. Note that the line $\sigma_{22}/\sigma_{22}^{0} = 1$ corresponds to the stress distribution in the absence of the surface strips.

24. SHELLS AND PLATES WITH CORRUGATED SURFACES OF REGULAR STRUCTURE

In this section we will be concerned with different profiles of corrugated surfaces, such as trapezoidal and sinusoidal (see Figs 24.2, 24.6, 24.10, 24.11), and it is clear from the outset that since all these profiles have dimensions of the same order of magnitude, the approximate methods we have thus far developed will be of no use for the local problems that arise. The trigonometric series expansions that we employ as an alternative approach will enable us to calculate the effective moduli of the structures under study and to evaluate the stress fields in them under different types of macro deformations. It is assumed throughout that the bounding contour of the corrugated surface is nowhere perpendicular to the middle surface of the shell.

24.1. The solution in terms of trigonometric expansions

We begin by considering the local problem ($b11$) (equations (19.16), (19.18)–(19.20) with $\lambda\mu = 11$) for the unit cell shown in Fig. 24.1. We assume that the profile of the rib is symmetric with respect to the axis Oz so that $z^{+}(\xi_2)$ is an even function of ξ_2. It is also assumed that the shell material is isotropic and its properties are E and ν.

Considered together with conditions (15.57), the local problem ($b11$) belongs to the class of plane problems of the theory of elasticity, and its solution can, in this case, be represented in the form

$$U_2^{11} = \sum_{n=1}^{\infty} \{A_n^{11}\sinh(\lambda_n z) + B_n^{11}\cosh(\lambda_n z) + C_n^{11}[\cosh(\lambda_n z) + z\lambda_n \sinh(\lambda_n z)]$$

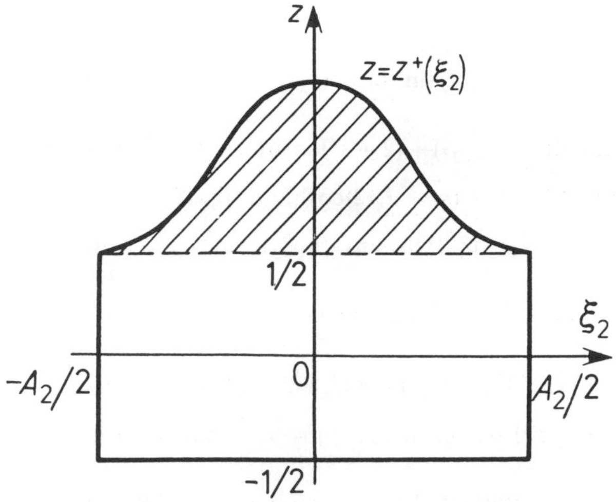

Fig. 24.1 *Unit cell for a shell with a corrugated surface in the coordinate system* ξ_1, z.

$$+ D_n^{11}[\sinh(\lambda_n z) + z\lambda_n \cosh(\lambda_n z)]\}\sin(\lambda_n h_2 \xi_2),$$

$$U_3^{11} = A_0^{11} z + \sum_{n=1}^{\infty} \{-A_n^{11}\cosh(\lambda_n z) - B_n^{11}\sinh(\lambda_n z) \tag{24.1}$$

$$+ C_n^{11}[2(1-2\nu)\sinh(\lambda_n z) - z\lambda_n\cosh(\lambda_n z)]$$

$$+ D_n^{11}[2(1-2\nu)\cosh(\lambda_n z) - z\lambda_n \sinh(\lambda_n z)]\}\cos(\lambda_n h_2 \xi_2),$$

where $\lambda_n = 2\pi n/(h_2 A_2)$.

Using (24.1) in (19.18) yields the corresponding expansions for the functions τ_{22}^{11}, τ_{23}^{11} and τ_{33}^{11} involved in the boundary conditions (19.19) and (19.20) on the contour $z = z^+(\xi_2)$ and at $z = -\frac{1}{2}$(see Fig. 24.1). From the boundary conditions at $z = -\frac{1}{2}$, we can show that

$$A_0^{11} = -\frac{\nu}{1-\nu}, \qquad A_n^{11} = C_n^{11}(sc + \lambda_n/2) - D_n^{11}(2\nu + s^2),$$

$$B_n^{11} = C_n^{11}(1 - 2\nu + s^2) + D_n^{11}(-sc + \lambda_n/2), \tag{24.2}$$

$$s = \sinh(\lambda_n/2), \qquad c = \cosh(\lambda_n/2).$$

From the boundary conditions on the contour $z = z^+(\xi_2)$, expanding once more in terms of trigonometric functions, and after some calculation, we obtain the following system of algebraic equations for detetrmining the coefficients C_n^{11} and D_n^{11}:

$$\begin{cases} \sum_{n=1}^{\infty} (\alpha_{nm} X_n + \beta_{nm} Y_n) = \alpha_m \quad (m = 1, 2, \ldots), \\ \sum_{n=1}^{\infty} (\gamma_{nm} X_n + \kappa_{nm} Y_n) = 0 \quad (m = 0, 1, 2, \ldots), \end{cases} \tag{24.3}$$

$$\lambda_n C_n^{11} = \frac{\nu}{1-\nu} X_n, \qquad \lambda_n D_n^{11} = \frac{\nu}{1-\nu} Y_n, \tag{24.4}$$

where

$$\alpha_{nm} = \frac{4}{A_2}\int_0^{A_2/2} \{-[(sc+\lambda_n/2)\sinh(\lambda_n z^+(\xi_2))$$
$$+ z^+(\xi_2)\lambda_n \sinh(\lambda_n z^+(\xi_2)) + (2+s^2)\cosh(\lambda_n z^+(\xi_2))](z^+(\xi_2))'\cos(\lambda_n h_2 \xi_2)$$
$$+ [c^2 \sinh(\lambda_n z^+(\xi_2)) + (sc+\lambda_n/2)\cosh(\lambda_n z^+(\xi_2))$$
$$+ z^+(\xi_2)\lambda_n \cosh(\lambda_n z^+(\xi_2))]h_2 \sin(\lambda_n h_2 \xi_2)\}\sin(\lambda_m h_2 \xi_2)\,d\xi_2, \tag{24.5}$$

$$\beta_{nm} = \frac{4}{A_2}\int_0^{A_2/2} \{-[(1-s^2)\sinh(\lambda_n z^+) + (\lambda_n/2 - sc)\cosh(\lambda_n z^+)$$
$$+ z^+\lambda_n \cosh(\lambda_n z^+)](z^+(\xi_2))'\cos(\lambda_n h_2 \xi_2) + [(\lambda_n/2 - sc)\sinh(\lambda_n z^+)$$
$$+ z^+\lambda_n \sinh(\lambda_n z^+) - s^2\cosh(\lambda_n z^+)]h_2 \sin(\lambda_n h_2 \xi_2)\}\sin(\lambda_m h_2 \xi_2) d\xi_2,$$

$$\gamma_{nm} = \frac{4}{A_2}\int_0^{A_2/2} \{-[c^2 \sinh(\lambda_n z^+) + (sc+\lambda_n/2)\cosh(\lambda_n z^+)$$
$$+ z^+\lambda_n \cosh(\lambda_n z^+)](z^+(\xi_2))'\sin(\lambda_n h_2 \xi_2) - [(\lambda_n/2 + sc)\sinh(\lambda_n z^+)$$
$$+ z^+\lambda_n \sinh(\lambda_n z^+) + s^2\cosh(\lambda_n z^+)]h_2 \cos(\lambda_n h_2 \xi_2)\}\cos(\lambda_m h_2 \xi_2)\,d\xi_2,$$

$$\kappa_{nm} = \frac{4}{A_2}\int_0^{A_2/2} \{-[(\lambda_n/2 - sc)\sinh(\lambda_n z^+) + z^+\lambda_n \sinh(\lambda_n z^+)$$
$$- s^2\cosh(\lambda_n z^+)](z^+(\xi_2))'\sin(\lambda_n h_2 \xi_2) + [c^2\sinh(\lambda_n z^+) + (sc - \lambda_n/2)\cosh(\lambda_n z^+)$$
$$- z^+\lambda_n \cosh(\lambda_n z^+)]h_2 \cos(\lambda_n h_2 \xi_2)\}\cos(\lambda_m h_2 \xi_2)\,d\xi_2,$$

$$\alpha_m = \frac{4}{A_2}\int_0^{A_2/2} (z^+(\xi_2))'\sin(\lambda_m h_2 \xi_2)\,d\xi_2, \tag{24.6}$$

where a prime indicates an ordinary derivative with respect to ξ_2.

Using (19.18) and (19.21) along with (24.1) and (24.2), the functions we are interested in are found to be related by

$$b_{11}^{11} = \nu(\tau_{22}^{11} + \tau_{33}^{11}) + c_{11}, \qquad b_{22}^{11} = \tau_{22}^{11} + c_{12},$$
$$b_{33}^{11} = \tau_{33}^{11} + c_{12}, \qquad b_{23}^{11} = \tau_{23}^{11},$$

$$\tau_{22}^{11} = -\frac{c_{12}^2}{c_{11}} + (c_{11} - c_{12})\sum_{n=1}^{\infty} \{C_n^{11}[(sc+\lambda_n/2)\sinh(\lambda_n z)$$
$$+ z\lambda_n \sinh(\lambda_n z) + (2+s^2)\cosh(\lambda_n z)] + D_n^{11}[(1-s^2)\sinh(\lambda_n z)$$
$$+ (\lambda_n/2 - sc)\cosh(\lambda_n z) + z\lambda_n\cosh(\lambda_n z)]\}\lambda_n \cos(\lambda_n h_2 \xi_2),$$

$$\tau_{33}^{11} = -c_{12} + (c_{11} - c_{12})\sum_{n=1}^{\infty} \{-C_n^{11}[(\lambda_n/2 + sc)\sinh(\lambda_n z) \tag{24.7}$$
$$+ z\lambda_n \sinh(\lambda_n z) + s^2\cosh(\lambda_n z)]$$
$$+ D_n^{11}[c^2\sinh(\lambda_n z) + (sc - \lambda_n/2)\cosh(\lambda_n z) - z\lambda_n\cosh(\lambda_n z)]\}\lambda_n \cos(\lambda_n h_2 \xi_2),$$

$$\tau_{23}^{11} = (c_{11} - c_{12}) \sum_{n=1}^{\infty} \{C_n^{11}[c^2 \sinh(\lambda_n z) + (sc + \lambda_n/2)\cosh(\lambda_n z)$$
$$+ z\lambda_n \cosh(\lambda_n z)] + D_n^{11}[(\lambda_n/2 - sc)\sinh(\lambda_n z)$$
$$+ z\lambda_n \sinh(\lambda_n z) - s^2 \cosh(\lambda_n z)]\}\lambda_n \sin(\lambda_n h_2 \xi_2).$$

The solution of the local problem ($b22$) (equations (19.16) and (19.18)–(19.20) with $\lambda\mu = 22$) may be obtained along lines similar to those for ($b11$), with the superscript pair $\{11\}$ replaced by $\{22\}$ in (24.1) and (24.2). From (24.3), the coefficients C_n^{22} and D_n^{22} are found to be

$$\lambda_n C_n^{22} = \frac{1}{1-v} X_n, \qquad \lambda_n D_n^{22} = \frac{1}{1-v} Y_n. \tag{24.8}$$

Now if the expansions employed for the solution of the ($b11$) problem are compared with those for the ($b22$) problem then, with the aid of (24.3) and (24.8), the relations

$$\begin{aligned} b_{11}^{11} &= v b_{11}^{22} + E, & b_{22}^{11} &= v b_{22}^{22}, \\ b_{33}^{11} &= v b_{33}^{22}, & b_{23}^{11} &= v b_{23}^{22}. \end{aligned} \tag{24.9}$$

are obtained, which supplement the general conditions (19.32) and (19.33) valid for any homogeneous ribbed shell and plate made of an orthotropic material. Because of the relationships established in (24.9) between the functions b_{ij}^{22} and b_{ij}^{11}, it suffices to solve only the local problem ($b11$) for the determination of these functions.

We consider next the local problems (b^*11) and (b^*22) set by equations (19.22), (19.24)–(19.26) with, respectively, $\lambda\mu = 11$ and $\lambda\mu = 22$ on the unit cell shown in Fig. 24.1. These problems also belong to the class of plane elastic problems and are solved in the same manner as problems ($b11$) and ($b22$).

The exact analytical solutions of the problem (b^*11) is of the form

$$V_2^{11} = \sum_{n=1}^{\infty} \{\bar{C}_n^{11}[(\lambda_n/2 + sc)\sinh(\lambda_n z) + \lambda_n z \sinh(\lambda_n z)$$
$$+ (s^2 + 2 - 2v)\cosh(\lambda_n z)] + \bar{D}_n^{11}[(1 - 2v - s^2)\sinh(\lambda_n z)$$
$$+ (\lambda_n/2 - sc)\cosh(\lambda_n z) + z\lambda_n \cosh(\lambda_n z)]\} \sin(\lambda_n h_2 \xi_2),$$

$$V_3^{11} = -\frac{v}{1-v}\frac{z^2}{2} + \sum_{n=1}^{\infty} \{\bar{C}_n^{11}[(1 - 2v - s^2)\sinh(\lambda_n z) \tag{24.10}$$
$$- (\lambda_n/2 + sc)\cosh(\lambda_n z) - z\lambda_n \cosh(\lambda_n z)]$$
$$+ \bar{D}_n^{11}[(sc - \lambda_n/2)\sinh(\lambda_n z) - z\lambda_n \sinh(\lambda_n z)$$
$$+ (2 - 2v + s^2)\cosh(\lambda_n z)]\} \cos(\lambda_n h_2 \xi_2),$$

and for the functions b_{ij}^{*11} we have

$$\begin{aligned} b_{11}^{*11} &= v(\bar{\tau}_{22}^{11} + \bar{\tau}_{33}^{11}) + zc_{11}, & b_{22}^{*11} &= \bar{\tau}_{22}^{11} + zc_{12}, \\ b_{33}^{*11} &= \bar{\tau}_{33}^{11} + zc_{12}, & b_{23}^{*11} &= \bar{\tau}_{23}^{11}, \end{aligned}$$

$$\bar{\tau}_{22}^{11} = -\frac{c_{12}^2}{c_{11}} z + (c_{11} - c_{12}) \sum_{n=1}^{\infty} \{\bar{C}_n^{11}[(sc + \lambda_n/2)\sinh(\lambda_n z)$$
$$+ z\lambda_n \sinh(\lambda_n z) + (2 + s^2)\cosh(\lambda_n z)] + \bar{D}_n^{11}[(1 - s^2)\sinh(\lambda_n z)$$
$$+ (\lambda_n/2 - sc)\cosh(\lambda_n z) + z\lambda_n \cosh(\lambda_n z)]\} \lambda_n \cos(\lambda_n h_2 \xi_2),$$

$$\bar{\tau}_{33}^{11} = -c_{12}z + (c_{11} - c_{12}) \sum_{n=1}^{\infty} \{-\bar{C}_n^{11}[(\lambda_n/2 + sc)\sinh(\lambda_n z) \tag{24.11}$$
$$+ z\lambda_n \sinh(\lambda_n z) + s^2 \cosh(\lambda_n z)] + \bar{D}_n^{11}[c^2 \sinh(\lambda_n z)$$
$$+ (sc - \lambda_n/2)\cosh(\lambda_n z) - z\lambda_n \cosh(\lambda_n z)]\} \lambda_n \cos(\lambda_n h_2 \xi_2),$$

$$\bar{\tau}_{23}^{11} = (c_{11} - c_{12}) \sum_{n=1}^{\infty} \{\bar{C}_n^{11}[c^2 \sinh(\lambda_n z) + (sc + \lambda_n/2)\cosh(\lambda_n z)$$
$$+ z\lambda_n \cosh(\lambda_n z)] + \bar{D}_n^{11}[(\lambda_n/2 - sc)\sinh(\lambda_n z) + z\lambda_n \sinh(\lambda_n z)$$
$$- s^2 \cosh(\lambda_n z)]\} \lambda_n \sin(\lambda_n h_2 \xi_2).$$

The coefficients $\bar{C}_n^{11}$ and $\bar{D}_n^{11}$ in the expansions (24.10) and (24.11) are determined from the equations

$$\lambda_n \bar{C}_n^{11} = \frac{v}{1 - v} \bar{X}_n, \qquad \lambda_n \bar{D}_n^{11} = \frac{v}{1 - v} \bar{Y}_n, \tag{24.12}$$

where $\bar{X}_n$ and $\bar{Y}_n$ are the solutions of the algebraic system

$$\begin{aligned} &\sum_{n=1}^{\infty} (\alpha_{nm} \bar{X}_n + \beta_{nm} \bar{Y}_n) = \beta_m \quad (m = 1, 2, \ldots), \\ &\sum_{n=1}^{\infty} (\gamma_{nm} \bar{X}_n + \kappa_{nm} \bar{Y}_n) = 0 \quad (m = 0, 1, 2, \ldots), \end{aligned} \tag{24.13}$$

in which the coefficients are defined by (24.5) and the right-hand side of the first equation is

$$\beta_m = \frac{4}{A_2} \int_0^{A_2/2} z^+(\xi_2)(z^+(\xi_2))' \sin(\lambda_m h_2 \xi_2)\, d\xi_2. \tag{24.14}$$

The solutions of the local problem (b^*22) are obtained from (24.10) merely by replacing the superscript pair $\{11\}$ by $\{22\}$ and, similarly to the group (b) local problems, it is readily shown that

$$\bar{C}_n^{11} = v\bar{C}_n^{22}, \qquad \bar{D}_n^{11} = v\bar{D}_n^{22}. \tag{24.15}$$

Using this and comparing the series representations for the solutions of the (b^*11) and (b^*22) problems, we arrive at the following relations analogous to (24.9) and complementary to (19.34) and (19.35) (which express, recall, the general properties of homogeneous ribbed shells):

$$\begin{aligned} b_{11}^{*11} &= vb_{11}^{*22} + zE, \qquad & b_{22}^{*11} &= vb_{22}^{*22}, \\ b_{33}^{*11} &= vb_{33}^{*22}, \qquad & b_{23}^{*11} &= vb_{23}^{*22}. \end{aligned} \tag{24.16}$$

By means of (24.15) and (24.16), we can express the solutions of the problem (b^*22) through the corresponding solutions of the problem (b^*11).

The local problem ($b12$) we consider next is set on the unit cell of Fig. 24.1 by equations (19.17)–(19.20) with $\lambda\mu = 12$. The problem belongs to the class of antiplane elastic problems and its solution may be given in the form

$$U_1^{12} = \sum_{n=1}^{\infty} B_n^{12}[t\sinh(\lambda_n z) + \cosh(\lambda_n z)]\sin(\lambda_n h_2 \xi_2),$$
$$t = \tanh(\lambda_n/2), \tag{24.17}$$

where the coefficients B_n^{12} are found from the following system of algebraic equations:

$$\sum_{n=1}^{\infty} (\lambda_n B_n^{12})\theta_{nm} = \alpha_m \quad (m = 1, 2, \dots), \tag{24.18}$$

$$\theta_{nm} = \frac{4}{A_2}\int_0^{A_2/2} \{-[\cosh(\lambda_n z^+) + t\sinh(\lambda_n z^+)](z^+(\xi_2))' \cos(\lambda_n h_2 \xi_2)$$
$$+ [\sinh(\lambda_n z^+) + t\cosh(\lambda_n z^+)] h_2 \sin(\lambda_n h_2 \xi_2)\} \sin(\lambda_m h_2 \xi_2)\,\mathrm{d}\xi_2, \tag{24.19}$$

with α_m defined in (24.6).

Having solved the problem ($b12$), the functions

$$b_{12}^{12} = G + G\sum_{n=1}^{\infty} \lambda_n B_n^{12}[t\sinh(\lambda_n z) + \cosh(\lambda_n z)]\cos(\lambda_n h_2 \xi_2)$$
$$b_{13}^{12} = G\sum_{n=1}^{\infty} \lambda_n B_n^{12}[t\cosh(\lambda_n z) + \sinh(\lambda_n z)]\sin(\lambda_n h_2 \xi_2) \tag{24.20}$$

are determined.

The solution of the remaining local problem (b^*12), posed by equations (19.23)–(19.26) with $\lambda\mu = 12$, is obtained in the same way as for the problem ($b12$) and may be represented in the form

$$V_1^{12} = \sum_{n=1}^{\infty} \bar{B}_n^{12}[t\sinh(\lambda_n z) + \cosh(\lambda_n z)]\sin(\lambda_n h_2 \xi_2),$$
$$b_{12}^{*12} = zG + G\sum_{n=1}^{\infty} \lambda_n \bar{B}_n^{12}[t\sinh(\lambda_n z) + \cosh(\lambda_n z)]\cos(\lambda_n h_2 \xi_2), \tag{24.21}$$
$$b_{13}^{*12} = G\sum_{n=1}^{\infty} \lambda_n \bar{B}_n^{12}[t\cosh(\lambda_n z) + \sinh(\lambda_n z)]\sin(\lambda_n h_2 \xi_2),$$

where the coefficients $\bar{B}_n^{12}$ are determined from the following system of algebraic equations:

$$\sum_{n=1}^{\infty} (\lambda_n \bar{B}_n^{12})\theta_{nm} = \beta_m \quad (m = 1, 2, \dots), \tag{24.22}$$

in which θ_{nm} are defined by (24.19) and β_m by (24.14).

24.2. The effective elastic moduli and the local stresses

The relevant local problems having been solved, we can now proceed to the numerical analysis of the effective moduli and the local stress distributions in plate and shell structures with corrugated surfaces of various types. The question that immediately presents itself is: Given the desired accuracy of the calculations, where should the (theoretically infinite) system of equations (24.3), (24.13), (24.18), (24.22) be truncated? In order to get a feeling for the role of the size of the system, results obtained from 5×5 and 8×8 systems were compared. It was found that while the effective moduli are virtually insensitive to the order of the system, the changes in local stress distribution are quite substantial, especially in the vicinity of the edge of the rib (up to 20% for b_{22}^{22}); however, they become progressively less away from the edge, and for the no stress component exceed 1 or 2% at the centre of the composite shell. It was decided therefore that a 20×20 system should provide sufficient computational accuracy. In what follows, the results obtained for a few specific profiles of the reinforcing element are discussed.

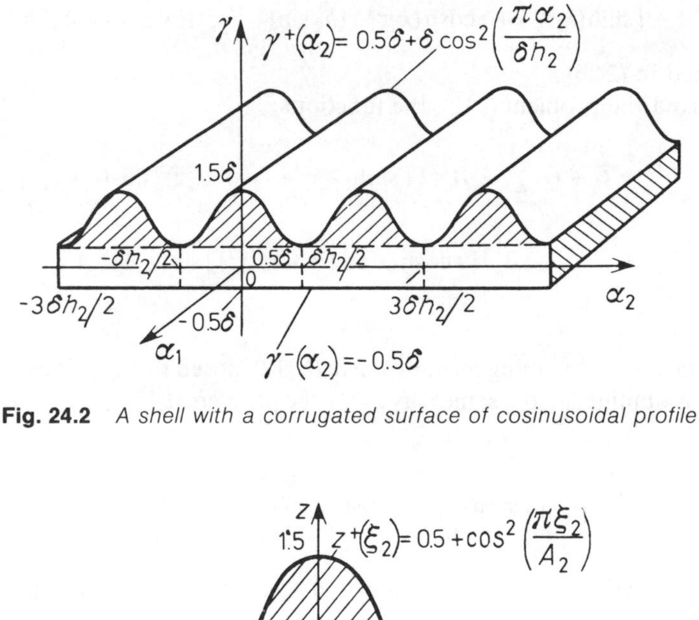

Fig. 24.2 *A shell with a corrugated surface of cosinusoidal profile*

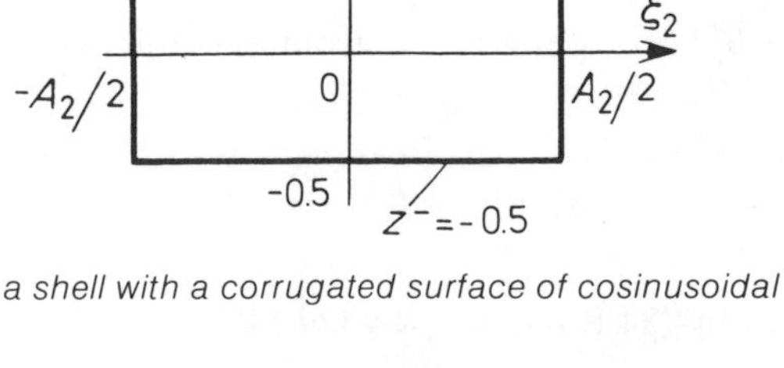

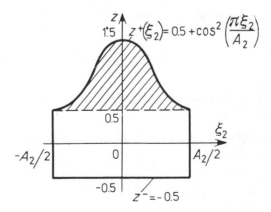

Fig. 24.3 *Unit cell for a shell with a corrugated surface of cosinusoidal profile, in the coordinate system ξ_2, z.*

For the case of a cosinusoidal profile, see Fig. 24.2, the unit cell referred to the coordinate system ξ_2, z is shown in Fig. 24.3 and the effective moduli calculated by the above scheme for $\nu = 0.3$ and $A_2 = 1$ are summarized in Table 24.1. The table also shows the corresponding results of the structurally anisotropic model (Korolev 1971), both in absolute terms and in terms of percentage changes. These latter are seen to be quite considerable for some effective moduli.

The local distributions of the functions $b_{\alpha\beta}^{\lambda\mu}$ and $b_{\alpha\beta}^{*\lambda\mu}$ over a number of important cross-sections of the unit cell of Fig. 24.3 with $h_2 = 3\pi$, $\nu = 0.3$ and $A_2 = 1$ are shown in Fig. 24.4 (for $\xi_2 = 0$) and in Fig. 24.5 (for $\xi_2 = 0.5$). By means of (15.74), these functions determine the local distribution of stresses at various macro deformations of the shell.

Fig. 24.6 shows a shell-like structure reinforced by ribs of trapezoidal profile, and Fig. 24.7 shows its unit cell referred to the coordinates ξ_2, z. The shape of the rib is determined by the parameters H, a and b $(0 \leqslant a < b < 0.5)$. The dimensions of the rib, as referred to the coordinate system α_2 and γ, are δH, $\delta a h_2$ and $\delta b h_2$, the quantity δh_2 representing the distance between two neighbouring ribs.

In rows I of Table 24.2 we show the effective stiffness properties of such structures obtained for $\nu = 0.3$ and $A_2 = 1$, and for various values of the parameters H, a, b and h_2. Rows II and III of the table present, respectively, the values computed from structurally anisotropic theory (Korolev 1971) and the differences between the results given in the first two rows. Again, the percentage changes prove to be quite large for many stiffness moduli.

For case II of Table 24.2, the local distributions of the functions $b_{\alpha\beta}^{\lambda\mu}$ and $b_{\alpha\beta}^{*\lambda\mu}$ over

Table 24.1 *Effective stiffness moduli of shells and plates with corrugated surfaces of a cosinusoidal profile (Fig. 24.2):* $\nu = 0.3$, $A_2 = 1$.

Effective stiffness moduli	Considered cases I	II	III	Structurally anisotropic model	Percentage changes
	$h_2 = 5$	$h_2 = 2\pi$	$h_2 = 3\pi$		
$\langle b_{11}^{11} \rangle / E$	1.618	1.619	1.616	1.599	1.1–1.3
$\langle b_{22}^{22} \rangle / E$	1.311	1.322	1.289	1.099	17–20
$\langle b_{22}^{11} \rangle / E$	0.393	0.397	0.387	0.330	17–20
$\langle b_{12}^{12} \rangle / G$	1.345	1.368	1.391	1.000	35–39
$\langle b_{11}^{*11} \rangle / E$	0.448	0.451	0.449	0.438	2–3
$\langle b_{22}^{*22} \rangle / E$	0.117	0.150	0.128	0	—
$\langle b_{22}^{*11} \rangle / E$	0.035	0.045	0.038	0	—
$\langle b_{12}^{*12} \rangle / G$	0.283	0.306	0.330	0.219	29–51
$\langle z b_{11}^{*11} \rangle / E$	0.515	0.518	0.517	0.508	1–2
$\langle z b_{22}^{*22} \rangle / E$	0.167	0.200	0.189	0.092	82–117
$\langle z b_{22}^{*11} \rangle / E$	0.050	0.060	0.057	0.028	82–117
$\langle z b_{12}^{*12} \rangle / G$	0.344	0.366	0.391	—	—

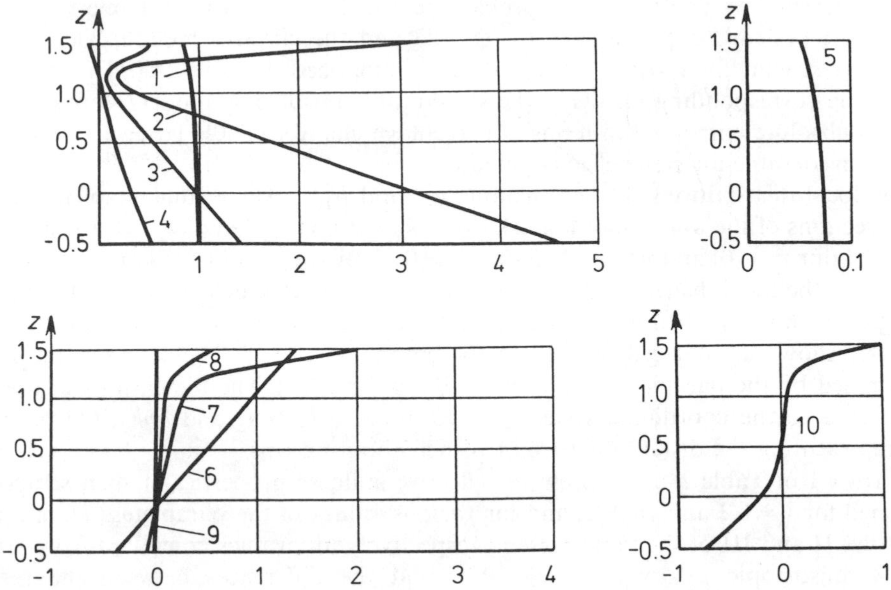

Fig. 24.4 *Local stress distribution over the cross-sections of the unit cell of Fig. 24.3*, for $h_1 = 3\pi$, $\nu = 0.3$, $A_2 = 1$. *Denotation of the curves:*

$$1{:}\ b_{11}^{11}\left(\frac{E}{1-\nu^2}\right)^{-1};\quad 2{:}\ b_{22}^{22}\left(\frac{E}{1-\nu^2}\right)^{-1};\quad 3{:}\ b_{11}^{22}\left(\frac{\nu E}{1-\nu^2}\right)^{-1};$$

$$4{:}\ b_{22}^{11}\left(\frac{\nu E}{1-\nu^2}\right)^{-1};\quad 5{:}\ b_{12}^{12}\left(\frac{E}{2(1+\nu)}\right)^{-1};$$

$$6{:}\ b_{11}^{*11}\left(\frac{E}{1-\nu^2}\right)^{-1};\quad 7{:}\ b_{22}^{*22}\left(\frac{E}{1-\nu^2}\right)^{-1};\quad 8{:}\ b_{11}^{*22}\left(\frac{\nu E}{1-\nu^2}\right)^{-1};$$

$$9{:}\ b_{22}^{*11}\left(\frac{\nu E}{1-\nu^2}\right)^{-1};\quad 10{:}\ b_{12}^{*12}\left(\frac{E}{2(1+\nu)}\right)^{-1}.$$

some of the important cross-sections of the unit cell of Fig. 24.7 are shown in Figs 24.8 and 24.9 for $\xi_2 = 0$ and $\xi_2 = 0.5$, respectively.

It should be noted that at small a/b ratios (high trapezoids) or in the limiting case when $a = 0$ (triangle), the violent oscillations and singularities that appear in solutions in the vicinity of ribs edges complicates the numerical work considerably in terms of both increased grid density and, as a consequence, increased computer time.

It is also of interest to estimate the effect that fillets between the ribs and the carrying surface have on the effective properties of, and local stress fields in, the shell. We consider as an example the trapezoidal profile of Figs 24.6 and 24.7, choosing case III of Table 24.2 as a reference case and introducing a rounded-off fillet as shown in Fig. 24.10. The values of the effective moduli calculated for $\nu = 0.3$ and $A_2 = 1$ are found to be

$$\langle b_{11}^{11}\rangle/E = 1.651;\qquad \langle b_{11}^{*11}\rangle/E = 0.723;\qquad \langle zb_{11}^{*11}\rangle/E = 1.220,$$

$$\langle b_{12}^{12}\rangle/G = 1.874;\qquad \langle b_{12}^{*12}\rangle/G = 1.054;\qquad \langle zb_{12}^{*12}\rangle/G = 0.800,$$

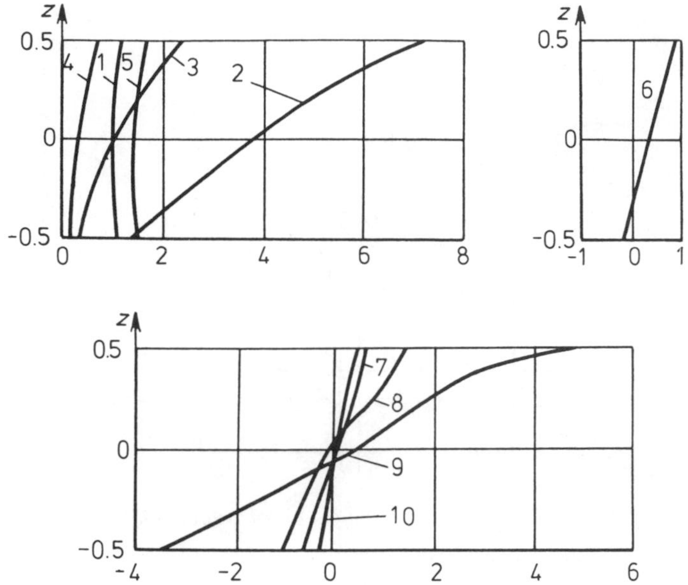

Fig. 24.5 *Local stress distributions over the cross-section $\xi_2 = 0.5$ of the unit cell of Fig. 24.3, for $h_2 = 3\pi$, $\nu = 0.3$, $A_2 = 1$. See the denotation of the curves in Fig. 24.4.*

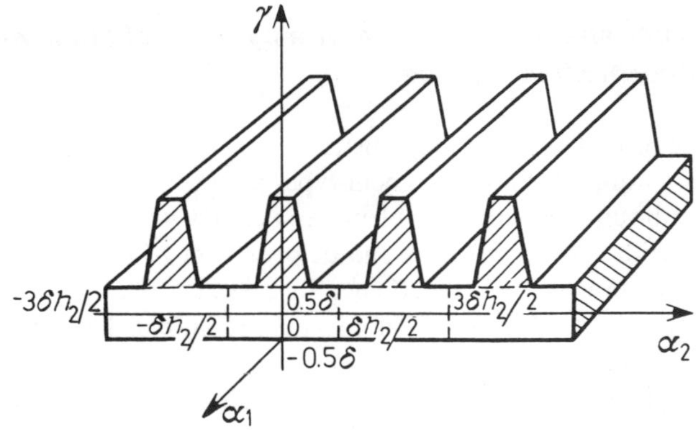

Fig. 24.6 *Shell with strengthening ribs of trapezoidal profile.*

showing a 5–16% decrease in $\langle b_{11}^{11} \rangle$, $\langle b_{11}^{*11} \rangle$ and $\langle zb_{11}^{*11} \rangle$ and a 22–40% increase in $\langle b_{12}^{12} \rangle$, $\langle b_{12}^{*12} \rangle$ and $\langle zb_{12}^{*12} \rangle$. The local distributions of the functions $b_{\alpha\beta}^{\lambda\mu}$ and $b_{\alpha\beta}^{*\lambda\mu}$ over the contour of the unit cell of Fig. 24.10 is shown in Table 24.3.

As a final example we consider a shell (or plate) with a system of parallel grooves of circular cross-section, see Fig. 24.11. The values of the effective moduli of such structures were computed for the case in which $\nu = 0.3$ and $A_2 = 1$ and are listed in Table 24.4 (column 1), where the corresponding results obtained from the structurally anisotropic model of Korolev (1971) are also shown (column 2). The values of $b_{\alpha\beta}^{\lambda\mu}$ and $b_{\alpha\beta}^{*\lambda\mu}$ at some points on the external contour of the unit cell of Fig. 24.11 are listed in Table 24.5.

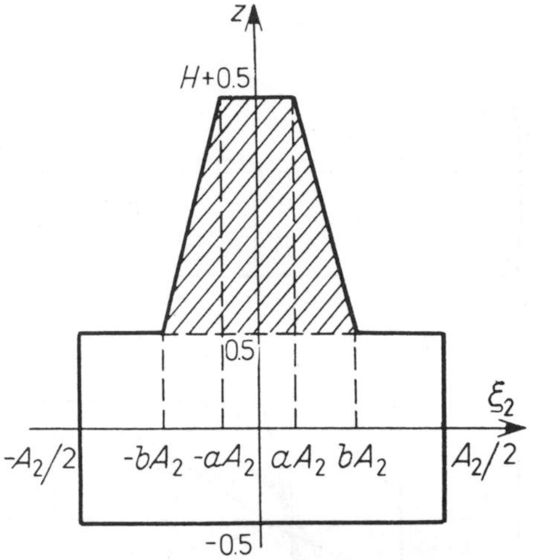

Fig. 24.7 *Unit cell for a shell with strengthening ribs of trapezoidal profile in the coordinate system ξ_2, z.*

24.3. Local stress distribution in a cylindrical shell with a corrugated surface

The effective elastic moduli and the local distributions of the functions $b^{\lambda\mu}_{\alpha\beta}$ and $b^{*\lambda\mu}_{\alpha\beta}$ we obtained above for a number of reinforcement types may be utilized for the consideration of various types of boundary value problems associated with shell-like structures. As an example, we consider the problem of the deformation of an axially free cylindrical shell the outer surface of which is strengthened by a system of stringers, while the inner surface, which is smooth, is subjected to an internal pressure q. The radius of the middle surface of the shell is taken to be R.

If the constitutive relations of the homogenized shell, (15.88), are substituted into the equations of motion of a cylindrical shell, (15.95), we find after some manipulation that the deflection w is determined by the equation

$$\frac{d^4 w}{d\alpha_1^4} + 2\beta_2 \frac{d^2 w}{d\alpha_1^2} + \beta_0 w = q^*, \tag{24.23}$$

where the relations (15.94) have been used and where

$$q^* = q/D^*, \qquad \beta_2 = C^*/(RD^*), \qquad \beta_0 = B^*/(R^2 D^*),$$

$$D^* = \delta^3[\langle z b^{*11}_{11}\rangle - (\langle b^{*11}_{11}\rangle)^2/\langle b^{11}_{11}\rangle],$$

$$C^* = \delta^2[\langle b^{11}_{22}\rangle\langle b^{*11}_{11}\rangle/\langle b^{11}_{11}\rangle - \langle b^{*11}_{22}\rangle],$$

$$B^* = \delta[\langle b^{22}_{22}\rangle - (\langle b^{11}_{22}\rangle)^2/\langle b^{11}_{11}\rangle].$$

Table 24.2 *Effective stiffness moduli of shells and plates reinforced by ribs of trapezoidal profile (Fig. 24.6);* $\nu = 0.3$, $A_2 = 1$.

Effective stiffness moduli		Considered cases I: $H = 0.5$, $h_2 = 1$, $a = 0.25$, $b = 0.45$	II: $H = 1.5$, $h_2 = 1$, $a = 0.1$, $b = 0.2$	III: $H = 2$, $h_2 = 3$, $a = 0.1$, $b = 1/6$
$\langle b_{11}^{11} \rangle / E$	I	1.4835	1.5923	1.7440
	II	1.4489	1.5489	1.6319
	III	2.4%	2.8%	6.9%
$\langle b_{22}^{22} \rangle / E$	I	1.4833	1.5811	2.3444
	II	1.0989	1.0989	1.0989
	III	35%	43.9%	113.3%
$\langle b_{22}^{11} \rangle / E$	I	0.4450	0.4743	0.7033
	II	0.3297	0.3297	0.3297
$\langle b_{12}^{12} \rangle / G$	I	1.344	1.190	1.539
	II	1.000	1.000	1.000
	III	34.4%	19%	53.9%
$\langle b_{11}^{*11} \rangle / E$	I	0.2802	0.5758	0.8626
	II	0.254	0.525	0.755
	III	10.3%	9.7%	14.3%
$\langle b_{22}^{*22} \rangle / E$	I	0.2911	0.5644	1.1956
	II	0	0	0
$\langle b_{22}^{*11} \rangle / E$	I	0.0873	0.1693	0.3587
	II	0	0	0
$\langle b_{12}^{*12} \rangle / G$	I	0.250	0.246	0.752
$\langle zb_{11}^{*11} \rangle / E$	I	0.3038	0.8524	1.4668
	II	0.2836	0.7854	1.3356
	III	7.1%	8.5%	9.8%
$\langle zb_{22}^{*22} \rangle / E$	I	0.3167	0.8367	1.550
	II	0.0916	0.0916	0.0916
$\langle zb_{22}^{*11} \rangle / E$	I	0.095	0.251	0.465
	II	0.0275	0.0275	0.0275
$\langle zb_{12}^{*12} \rangle / G$	I	0.2697	0.2688	0.6083

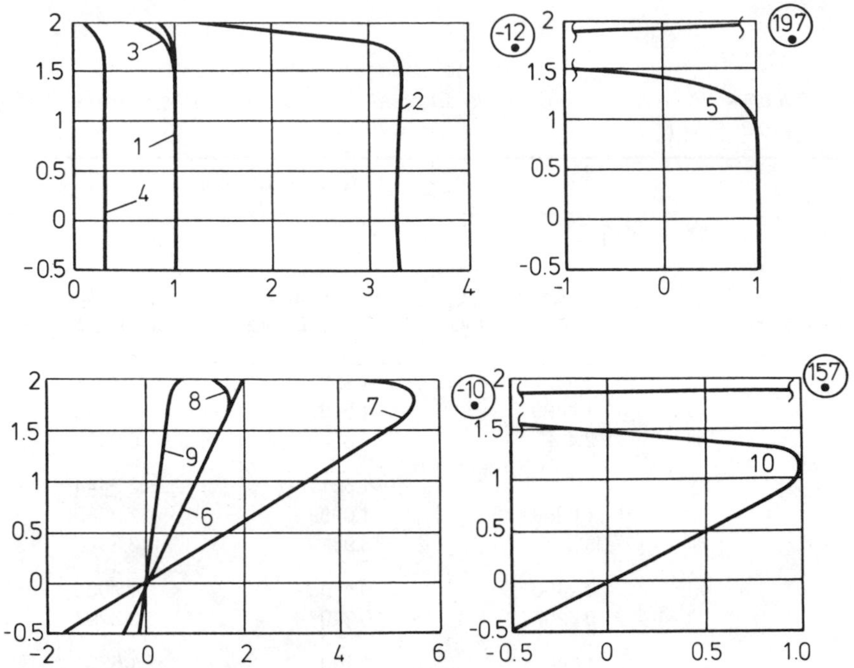

Fig. 24.8 *Local stress distributions over the cross-section $\xi_2 = 0$ of the unit cell of Fig. 24.7, for $\nu = 0.3$, $A_2 = 1$, $H = 1.5$, $h_2 = 1$, $a = 0.1$, $b = 0.2$. See the denotation of the curves in Fig. 24.4.*

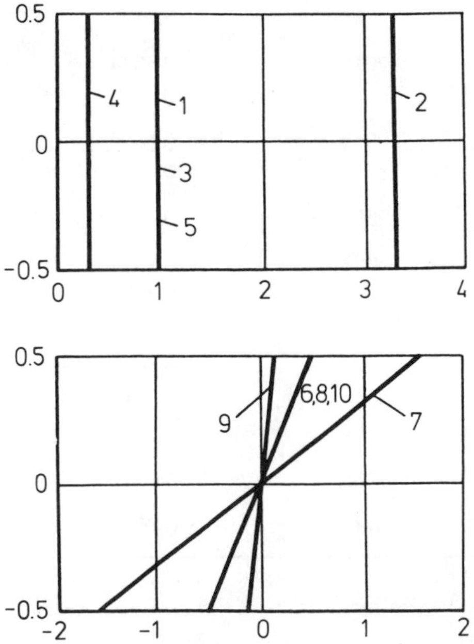

Fig. 24.9 *Local stress distributions over the cross-section $\xi_2 = 0.5$ of the unit cell of Fig. 24.7, for $\nu = 0.3$, $A_2 = 1$, $H = 1.5$, $h_2 = 1$, $a = 0.1$, $b = 0.2$. See the denotation of the curves in Fig. 24.4*

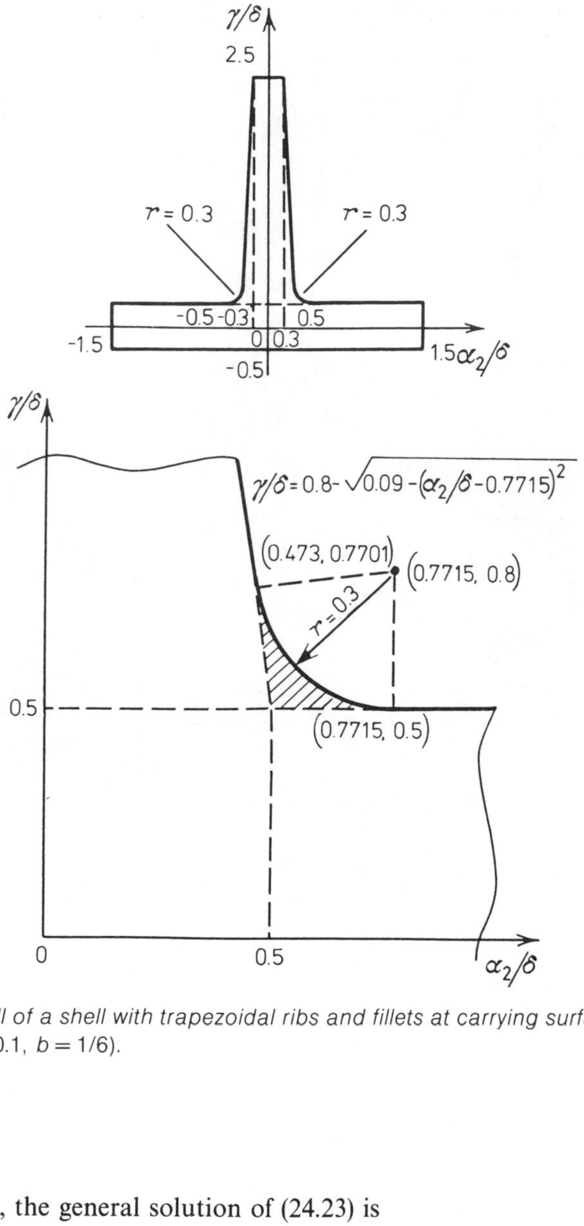

Fig. 24.10 *Unit cell of a shell with trapezoidal ribs and fillets at carrying surfaces–ribs junctions* ($H=2$, $h_2=3$, $a=0.1$, $b=1/6$).

For $q^* = \text{const.}$, the general solution of (24.23) is

$$w = e^{\lambda_1 \alpha_1}[C_1 \cos(\lambda_2 \alpha_1) + C_2 \sin(\lambda_2 \alpha_1)] + e^{-\lambda_1 \alpha_1}[C_3 \cos(\lambda_2 \alpha_1) + C_4 \sin(\lambda_2 \alpha_1)] + q^* \beta_0^{-1},$$

where the constants C_i, $i = 1, 2, 3, 4$, are to be determined from the end restraint conditions and clearly describe the boundary effects occurring in the shell. If the shell is long, then we have the momentless solution

$$w = q^* \beta_0^{-1} \tag{24.24}$$

Table 24.3 *Local distributions of the functions $b_{\alpha\beta}^{\lambda\mu}$ and $b_{\alpha\beta}^{*\lambda\mu}$ over the contour of the unit cell of Fig. 24.10.*

Points on the external contour of the unit cell	$b_{11}^{11} \Big/ \dfrac{E}{1-\nu^2}$	$b_{22}^{11} \Big/ \dfrac{\nu E}{1-\nu^2}$	$b_{11}^{*11} \Big/ \dfrac{E}{1-\nu^2}$	$b_{22}^{*11} \Big/ \dfrac{\nu E}{1-\nu^2}$	b_{12}^{12}/G	b_{12}^{*12}/G
$\xi_2 = 0$ $z = 2.5$ $(\alpha_2 = 0)$	0.185	−5.70	1.12	−26.0	187.0	−10.0
$\xi_2 = 0.0625$ $z = 2.5$ $(\alpha_2 = 0.1875\delta)$	0.147	−4.20	1.22	−22.0	184.0	−0.84
$\xi_2 = 0.1250$ $z = 1.75$ $(\alpha_2 = 0.375\delta)$	1.02	0.572	1.64	0.131	−3.0	−2.1
$\xi_2 = 0.1875$ $z = 0.5849$ $(\alpha_2 = 0.5625\delta)$	1.00	0.942	0.580	0.575	0.97	0.521
$\xi_2 = 0.25$ $z = 0.5008$ $(\alpha_2 = 0.75\delta)$	1.00	0.982	0.501	0.493	1.03	0.52
$\xi_2 = 0.3125$ $z = 0.5$ $(\alpha_2 = 0.9375\delta)$	1.00	1.02	0.504	0.492	1.08	0.583
$\xi_2 = 0.3750$ $z = 0.5$ $(\alpha_2 = 1.125\delta)$	0.999	1.06	0.507	0.496	1.10	0.626
$\xi_2 = 0.4375$ $z = 0.5$ $(\alpha_2 = 1.3125\delta)$	0.998	1.10	0.509	0.500	1.10	0.649
$\xi_2 = 0.5$ $z = 0.5$ $(\alpha_2 = 1.5\delta)$	0.998	1.11	0.510	0.502	1.10	0.656

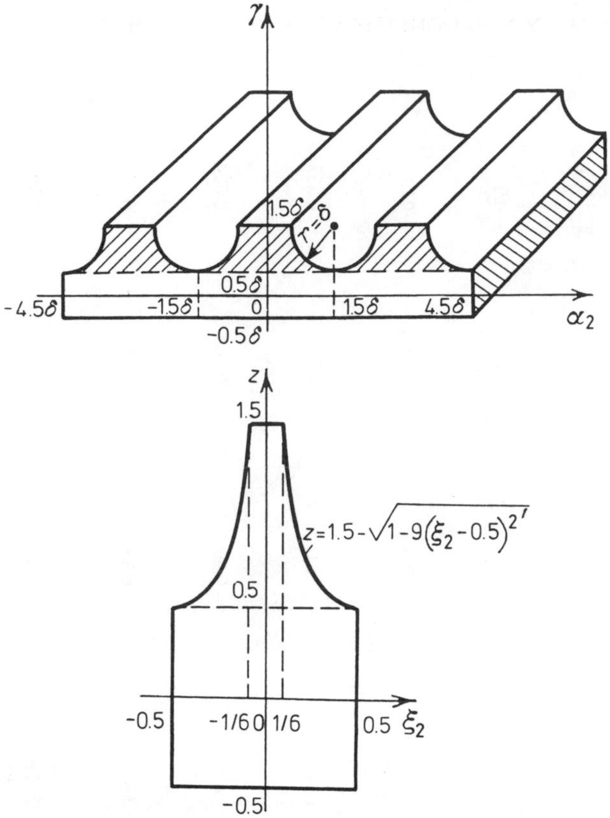

Fig. 24.11 *A shell with parallel grooves of circular cross-section; unit cell in the coordinates ξ_2, z.*

Table 24.4 *Effective stiffness moduli of shell or plate with a system of parallel grooves of circular cross-section (Fig. 24.11); $\nu = 0.3$, $A_2 = 1$.*

Effective stiffness moduli	Solutions of local local problems	Structurally anisotropic model	Percentage changes
$\langle b_{11}^{11} \rangle / E$	1.638	1.575	4
$\langle b_{22}^{22} \rangle / E$	1.796	1.099	63.3
$\langle b_{22}^{11} \rangle / E$	0.539	0.330	63.3
$\langle b_{12}^{12} \rangle / G$	1.266	1.000	26.6
$\langle b_{11}^{*11} \rangle / E$	0.510	0.437	16.7
$\langle b_{22}^{*22} \rangle / E$	0.811	0	—
$\langle b_{22}^{*11} \rangle / E$	0.243	0	—
$\langle b_{12}^{*12} \rangle / G$	0.235	0.217	8.3
$\langle z b_{11}^{*11} \rangle / E$	0.612	0.533	14.8
$\langle z b_{22}^{*22} \rangle / E$	0.974	0.092	—
$\langle z b_{22}^{*11} \rangle / E$	0.292	0.028	—
$\langle z b_{12}^{*12} \rangle / G$	0.124	—	—

Table 24.5 *Local distributions of the functions* $\overset{\lambda\mu}{b}_{\alpha\beta}$ *and* $\overset{*\lambda\mu}{b}_{\alpha\beta}$ *over the contour of the unit cell of Fig. 24.11.*

Points on the external contour of the unit cell	$b_{11}^{11}\Big/\dfrac{E}{1-\nu^2}$	$b_{22}^{11}\Big/\dfrac{\nu E}{1-\nu^2}$	$b_{11}^{*11}\Big/\dfrac{E}{1-\nu^2}$	$b_{22}^{*11}\Big/\dfrac{\nu E}{1-\nu^2}$	b_{12}^{12}/G	b_{12}^{*12}/G
$\xi_2 = 0$ $z = 1.5$ $(\alpha_2 = 0)$	1.39	−8.3	2.1	−12.0	−0.91	−1.0
$\xi_2 = 0.0625$ $z = 1.5$ $(\alpha_2 = 0.1875\delta)$	1.28	−12.0	1.95	−19.0	−0.8	−0.9
$\xi_2 = 0.1250$ $z = 1.5$ $(\alpha_2 = 0.375\delta)$	1.03	−33.0	1.59	−50.0	−0.73	−0.71
$\xi_2 = 0.1875$ $z = 1.152$ $(\alpha_2 = 0.5625\delta$	0.961	11.8	1.11	18.0	0.701	0.297
$\xi_2 = 0.25$ $z = 0.8386$ $(\alpha_2 = 0.75\delta)$	0.98	−0.24	0.814	−0.99	0.143	0.024
$\xi_2 = 0.3125$ $z = 0.6732$ $(\alpha_2 = 0.9375\delta)$	0.988	−0.051	0.652	−0.93	0.732	0.547
$\xi_2 = 0.3750$ $z = 0.573$ $(\alpha_2 = 1.125\delta)$	0.991	0.306	0.554	−0.53	1.33	1.06
$\xi_2 = 0.4375$ $z = 0.5177$ $(\alpha_2 = 1.3125\delta)$	0.993	0.495	0.499	−0.33	1.73	1.40
$\xi_2 = 0.5$ $z = 0.5$ $(\alpha_2 = 1.5\delta)$	0.994	0.551	0.481	−0.27	1.87	1.52

at points remote from the end, and the corresponding strains are given by

$$\varepsilon_{11} = -\frac{\langle b_{22}^{11}\rangle}{\langle b_{11}^{11}\rangle}\frac{q^*}{R\beta_0}, \qquad \varepsilon_{22} = \frac{q^*}{R\beta_0}, \qquad \tau_{11} = \tau_{22} = 0. \tag{24.25}$$

The stress component σ_{22}, the most important one in this case, is, from (15.74),

$$\sigma_{22} = b_{22}^{11}\varepsilon_{11} + b_{22}^{22}\varepsilon_{22} \tag{24.26}$$

or, substituting from (24.25) and making use of (24.9),

$$\sigma_{22} = \frac{qR}{\delta}\tilde{b}_{22}^{22}\left(1 - \frac{\nu\langle\tilde{b}_{22}^{11}\rangle}{\langle\tilde{b}_{11}^{11}\rangle}\right)\left[\langle\tilde{b}_{22}^{22}\rangle - \frac{(\langle\tilde{b}_{22}^{11}\rangle)^2}{\langle\tilde{b}_{11}^{11}\rangle}\right]^{-1}, \tag{24.27}$$

where a tilde denotes normalization to $E/(1-\nu^2)$ of the quantity so marked.

For the trapezoidal stringer profile, case II of Table 24.2, (24.27) yields

$$\sigma_{22} = \frac{qR}{\delta}0.695\tilde{b}_{22}^{22} \tag{24.28}$$

so that the distribution of the quantity $\sigma_{22}\delta/(0.695qR)$ coincides with that of $\tilde{b}_{22}^{22}$ as shown in Figs 24.8 and 24.9 for $\xi_2 = 0$ and $\xi_2 = 0.5$, respectively. In particular, it follows from (24.28) that the value which σ_{22} assumes between the stringers is 2.3 times qR/δ, i.e. the σ_{22} value in a stringerless shell of thickness δ (see Fig. 24.9).

For the cosinusoidal stringer profile (case $h_2 = 3\pi$ in Table 24.1), (24.27) gives

$$\sigma_{22} = \frac{qR}{\delta}0.853\tilde{b}_{22}^{22}. \tag{24.29}$$

The local distributions of the functions $\tilde{b}_{22}^{22}$ are shown for this case in Figs 24.4 and 24.5 for the cross-sections $\xi_2 = 0$ and $\xi_2 = 0.5$, respectively. Analysis of these distributions shows that the use of cosinusoidal stringers results in an even greater concentration of the stress component σ_{22}. For example, at a point on the surface which is equidistant from the two neighbouring stringers (and whose coordinates are $\xi_2 = 0.5$, $(A_2 = 1)$, $z = 0.5$ in Fig. 24.3) we obtain $\sigma_{22} = 6.1\, qR/\delta$ using (24.29) and referring to Fig. 24.5.

7 Network and Framework Reinforced Shells and Plates with Regular Structure

Plate and shell structures provided with spatially periodic network- or framework-type reinforcements have found numerous and varied applications in many branches of modern technology. If the network in question is dense enough, it has turned out, as indeed one would expect, that the mechanical behaviour of such a system can be well predicted by means of a continuum model; the actual network is then visualized as a continuous shell (or plate) with elastic constitutive relations of some special form. Pshenichnov (1982) derived such forms for a number of specific network geometries by considering stress–strain and moment–strain relations in bars connecting the sites of a network. It is the purpose of this section to discuss an alternative method for the derivation of the constitutive relations for network-type shell structures.

The method was developed in papers by Kalamkarov (1987a, 1988b, 1988c, 1989) and is essentially based on the general homogenized shell model described in Chapter 5. A network shell is considered in this approach as a non-homogeneous layer which has zero material properties in the regions of perforation and whose overall (or effective) properties are estimated from the solution of the appropriate local problems set on the unit cell of the layer. Our goal in this section is to derive working formulas for the entire complex of the (effective) elastic, thermoelastic and heat conduction properties of a periodic network reinforced shell formed by several families of bars of elliptic (in particular, circular) cross-section. For illustrative purposes we derive constitutive relations for rectangular, rhombic and triangular network configurations, and we also compare our stiffness moduli results with those of Pshenichnov and obtain formulae for the elastic moduli of composite shells with a high stiffness framework-type reinforcement.

25. EFFECTIVE ELASTIC MODULI OF A NETWORK REINFORCED SHELL

Consider a network reinforced shell of regular structure formed by N families of mutually parallel bars $S_1^\delta, S_2^\delta, \dots, S_N^\delta$. Let $S^\delta = \cup_{i=1}^N S_i^\delta \cap \Omega_\delta$, where Ω_δ is the unit cell of the shell, and let S_i and S be the images of (respectively) S_i^δ and S^δ in the coordinate system ξ_1, ξ_2, z. It is assumed that the bars are made of a homogeneous isotropic material, so that everywhere in S the elastic properties c_{ijmn} are defined by (15.91); at points outside the region S we have $c_{ijmn} = 0$.

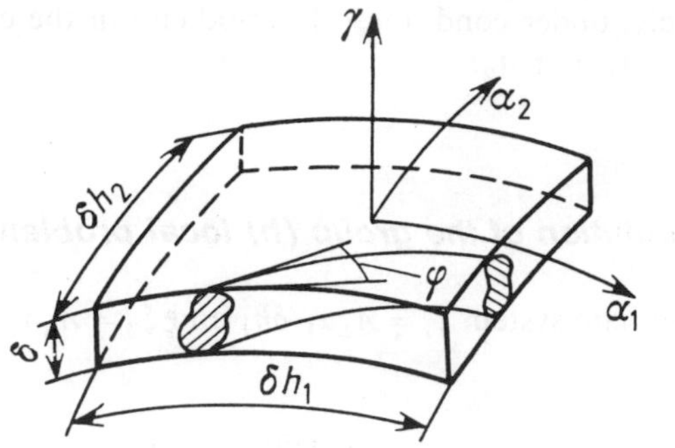

Fig. 25.1 *Bar S_j^δ within the unit cell Ω_δ of a network shell.*

We first consider the local elastic problem for an individual bar S_j^δ. Within the unit cell of the structure, this bar is a curved cylinder making an angle φ with the coordinate line α_1 (see Fig. 25.1). The coordinates of the middle cross-section of the cylinder will be denoted as δC_1 and δC_2.

The local problems of types (b) and (b^*) (that is, equations (15.55)–(15.61)) can be written in this case as

$$\frac{1}{h_\beta}\frac{\partial}{\partial \xi_\beta} b_{i\beta}^{\lambda\mu} + \frac{\partial}{\partial z} b_{i3}^{\lambda\mu} = 0,$$

$$\left.\left(\frac{1}{h_\beta} n_\beta b_{i\beta}^{\lambda\mu} + n_3 b_{i3}^{\lambda\mu}\right)\right|_{\partial S_j} = 0, \tag{25.1}$$

and

$$\frac{1}{h_\beta}\frac{\partial}{\partial \xi_\beta} b_{i\beta}^{*\lambda\mu} + \frac{\partial}{\partial z} b_{i3}^{*\lambda\mu} = 0,$$

$$\left.\left(\frac{1}{h_\beta} n_\beta b_{i3}^{*\lambda\mu} + n_3 b_{i3}^{*\lambda\mu}\right)\right|_{\partial S_j} = 0, \tag{25.2}$$

where n_i denotes the unit normal to the lateral side ∂S_j of the bar and is related to the coordinate system ξ_1, ξ_2, z.

The functions $b_{ij}^{\lambda\mu}$ and $b_{ij}^{*\lambda\mu}$ are related to $U_i^{\lambda\mu}$ and $V_i^{\lambda\mu}$ by equations (15.41) and (15.53), which, using (15.33), can be rewritten as

$$b_{ij}^{\lambda\mu} = \frac{1}{h_\beta} c_{ijm\beta}\frac{\partial U_m^{\lambda\mu}}{\partial \xi_\beta} + c_{ijm3}\frac{\partial U_m^{\lambda\mu}}{\partial z} + c_{ij\lambda\mu}, \tag{25.3}$$

$$b_{ij}^{*\lambda\mu} = \frac{1}{h_\beta} c_{ijm\beta}\frac{\partial V_m^{\lambda\mu}}{\partial \xi_\beta} + c_{ijm3}\frac{\partial V_m}{\partial z} + z c_{ij\lambda\mu}. \tag{25.4}$$

The functions $U_m^{\lambda\mu}$ and $V_m^{\lambda\mu}$ are obtained by solving the problems (25.1), (25.3) and

(25.2), (25.4), respectively, under conditions of periodicity in the coordinates ξ_1 and ξ_2 with respective periods A_1 and A_2.

25.1. The solution of the group (b) local problems

In relation to the coordinate system $\xi_1 = A_1\alpha_1/\delta h_1$ and $\xi_2 = A_2\alpha_2/\delta h_2$, the bar S_j makes an angle

$$\varphi' = \arctan\left(\frac{A_2 h_1}{A_1 h_2}\tan\varphi\right) \tag{25.5}$$

with the coordinate line ξ_1, and the centre of its middle cross-section has the coordinates

$$C_1' = \frac{A_1 C_1}{h_1}, \quad C_2' = \frac{A_2 C_2}{h_2}.$$

Shifting by the vector (C_1', C_2') and rotating through an angle φ' we now transform from ξ_1 and ξ_2 to the coordinates η_1 and η_2:

$$\begin{aligned} \eta_1 &= (\xi_1 - C_1')\cos\varphi' + (\xi_2 - C_2')\sin\varphi', \\ \eta_2 &= -(\xi_1 - C_1')\sin\varphi' + (\xi_2 - C_2')\cos\varphi', \end{aligned} \tag{25.6}$$

in which the bar under consideration is parallel to the coordinate line η_1 and has its middle cross-section centre coincident with the origin.

The periodic solutions of the group (*b*) local problems (posed by equations (25.1) and (25.3)) will only depend on the coordinates η_2 and z so that, using (25.6) and recalling the isotropy assumption, we have from (25.3):

$$\begin{aligned}
b_{11}^{\lambda\mu} &= -\frac{c_{11}}{h_1}\sin\varphi'\frac{\partial U_1^{\lambda\mu}}{\partial\eta_2} + \frac{c_{12}}{h_2}\cos\varphi'\frac{\partial U_2^{\lambda\mu}}{\partial\eta_2} + c_{12}\frac{\partial U_3^{\lambda\mu}}{\partial z} + c_{11\lambda\mu}, \\
b_{22}^{\lambda\mu} &= -\frac{c_{12}}{h_1}\sin\varphi'\frac{\partial U_1^{\lambda\mu}}{\partial\eta_2} + \frac{c_{11}}{h_2}\cos\varphi'\frac{\partial U_2^{\lambda\mu}}{\partial\eta_2} + c_{12}\frac{\partial U_3^{\lambda\mu}}{\partial z} + c_{22\lambda\mu}, \\
b_{33}^{\lambda\mu} &= -\frac{c_{12}}{h_1}\sin\varphi'\frac{\partial U_1^{\lambda\mu}}{\partial\eta_2} + \frac{c_{12}}{h_2}\cos\varphi'\frac{\partial U_2^{\lambda\mu}}{\partial\eta_2} + c_{11}\frac{\partial U_3^{\lambda\mu}}{\partial z} + c_{33\lambda\mu}, \\
b_{12}^{\lambda\mu} &= \frac{c_{44}}{h_2}\cos\varphi'\frac{\partial U_1^{\lambda\mu}}{\partial\eta_2} - \frac{c_{44}}{h_1}\sin\varphi'\frac{\partial U_2^{\lambda\mu}}{\partial\eta_2} + c_{12\lambda\mu}, \\
b_{13}^{\lambda\mu} &= c_{44}\frac{\partial U_1^{\lambda\mu}}{\partial z} - \frac{c_{44}}{h_1}\sin\varphi'\frac{\partial U_3^{\lambda\mu}}{\partial\eta_2}, \\
b_{23}^{\lambda\mu} &= c_{44}\frac{\partial U_2^{\lambda\mu}}{\partial z} + \frac{c_{44}}{h_2}\cos\varphi'\frac{\partial U_3^{\lambda\mu}}{\partial\eta_2}.
\end{aligned} \tag{25.7}$$

We next turn to (25.1), which can be rewritten in terms of the coordinates η_2 and z as

$$-\frac{\sin\varphi'}{h_1}\frac{\partial}{\partial\eta_2}b_{i1}^{\lambda\mu}+\frac{\cos\varphi'}{h_2}\frac{\partial}{\partial\eta_2}b_{i2}^{\lambda\mu}+\frac{\partial}{\partial z}b_{i3}^{\lambda\mu}=0,$$

$$\left[n_2'\left(-\frac{\sin\varphi'}{h_1}b_{i1}^{\lambda\mu}+\frac{\cos\varphi'}{h_2}b_{i2}^{\lambda\mu}\right)+n_3'b_{i3}^{\lambda\mu}\right]\Bigg|_{\partial S_j'}=0, \tag{25.8}$$

where n_2' and n_3', the components of the unit normal to the lateral surface $\partial S_j'$ of the bar S_j, are related to the coordinate system η_1, η_2, z (note that $n_1'=0$).

Substituting from (25.7) into (25.8) now results in a boundary value problem for the functions $U_i^{\lambda\mu}(\eta_2,z)$, and it can be shown that the solutions of this problem are linear in η_2 and z whatever the shape of the cross-section of the bar. To see this, let us try to solve (25.7) and (25.8) under the assumption that the functions $U_1^{\lambda\mu}$ and $U_2^{\lambda\mu}$ are linear in η_2 and independent of z, while $U_3^{\lambda\mu}$ is, on the other hand, linear in z and independent of η_2. Then, from (25.7),

$$b_{13}^{\lambda\mu}=b_{23}^{\lambda\mu}=0. \tag{25.9}$$

The problem (25.8) will be solved if

$$-\frac{\sin\varphi'}{h_1}b_{11}^{\lambda\mu}+\frac{\cos\varphi'}{h_2}b_{12}^{\lambda\mu}=0,$$

$$-\frac{\sin\varphi'}{h_1}b_{12}^{\lambda\mu}+\frac{\cos\varphi'}{h_2}b_{22}^{\lambda\mu}=0, \qquad b_{33}^{\lambda\mu}=0. \tag{25.10}$$

Substitution of the appropriate formulas from (25.7) into (25.10) yields a system of three linear equations with constant coefficients for three unknowns, $\mathrm{d}U_1^{\lambda\mu}/\mathrm{d}\eta_2$, $\mathrm{d}U_2^{\lambda\mu}/\mathrm{d}\eta_2$ and $\mathrm{d}U_3^{\lambda\mu}/\mathrm{d}z$ (which are in fact constant in view of the linearity of the required solutions in both η_2 and z). This system neither depends on nor contradicts relations (25.9), and having been solved, produces the above three derivatives and hence, by (25.7), the required functions $b_{ij}^{\lambda\mu}$. We have thus proved our statement that the problem (b) has a solution which is linear in η_2 and z. It should be admitted that the above method of solution is discouragingly cumbersome the way we have described it, but we can simplify our task by noting that it is actually the functions $b_{ij}^{\lambda\mu}$ that we need. Accordingly, we express the derivatives $\mathrm{d}U_1^{\lambda\mu}/\mathrm{d}\eta_2$, $\mathrm{d}U_2^{\lambda\mu}/\mathrm{d}\eta_2$ and $\mathrm{d}U_3^{\lambda\mu}/\mathrm{d}z$ in terms of the functions $b_{ij}^{\lambda\mu}$ (by solving (25.7) for $b_{11}^{\lambda\mu}$, $b_{22}^{\lambda\mu}$ and $b_{12}^{\lambda\mu}$) and substitute the results into the third of equations (25.10) to obtain:

$$\frac{\sin\varphi'}{h_1}\left(\frac{c_{11}^2}{c_{12}}-c_{12}\right)\left[\frac{\sin\varphi'}{h_1}\frac{(b_{11}^{\lambda\mu}-b_{22}^{\lambda\mu}+c_{22\lambda\mu}-c_{11\lambda\mu})}{(c_{12}-c_{11})}+\frac{\cos\varphi'}{h_2}\frac{(b_{12}^{\lambda\mu}-c_{12\lambda\mu})}{c_{44}}\right]$$
$$+\frac{\cos\varphi'}{h_2}(c_{12}-c_{11})\left[-\frac{\sin\varphi'}{h_1}\frac{(b_{12}^{\lambda\mu}-c_{12\lambda\mu})}{c_{44}}+\frac{\cos\varphi'}{h_2}\frac{(b_{11}^{\lambda\mu}-b_{22}^{\lambda\mu}+c_{22\lambda\mu}-c_{11\lambda\mu})}{(c_{12}-c_{11})}\right]$$
$$+\left(\frac{\sin^2\varphi'}{h_1^2}+\frac{\cos^2\varphi'}{h_2^2}\right)\left(\frac{c_{11}}{c_{12}}b_{11}^{\lambda\mu}-\frac{c_{11}}{c_{12}}c_{11\lambda\mu}+c_{33\lambda\mu}\right)=0. \tag{25.11}$$

These, together with the first two of equations (25.10), form a system of three simultaneous algebraic equations for the functions $b_{11}^{\lambda\mu}$, $b_{22}^{\lambda\mu}$ and $b_{12}^{\lambda\mu}$. Having solved the system and using the expressions for the elastic coefficients of an isotropic material, we find

$$
\begin{aligned}
b_{11}^{11} &= E\frac{\cos^2\varphi'}{h_2^4}\left(\frac{\sin^2\varphi'}{h_1^2}+\frac{\cos^2\varphi'}{h_2^2}\right)^{-2},\\
b_{11}^{12} &= E\frac{\sin\varphi'\cos^3\varphi'}{h_1h_2^3}\left(\frac{\sin^2\varphi'}{h_1^2}+\frac{\cos^2\varphi'}{h_2^2}\right)^{-2},\\
b_{11}^{22} &= E\frac{\sin^2\varphi'\cos^2\varphi'}{h_1^2h_2^2}\left(\frac{\sin^2\varphi'}{h_1^2}+\frac{\cos^2\varphi'}{h_2^2}\right)^{-2},\\
b_{12}^{\lambda\mu} &= \tan\varphi'\frac{h_2}{h_1}b_{11}^{\lambda\mu}, \qquad b_{22}^{\lambda\mu} = \tan^2\varphi'\frac{h_2^2}{h_1^2}b_{11}^{\lambda\mu}.
\end{aligned}
\tag{25.12}
$$

These expressions are constants and hence independent of the choice of the coordinate system. The final solution of the local problem (*b*) is obtained by substituting (25.5) into (25.12). The result is (Kalamkarov 1987a)

$$
\begin{aligned}
&b_{11}^{11} = E\frac{A_1^4}{B^4}c^4, \qquad b_{11}^{12} = E\frac{A_1^3A_2}{B^4}c^3s, \qquad b_{22}^{12} = E\frac{A_1A_2^3}{B^4}cs^3,\\
&b_{11}^{22} = b_{12}^{12} = E\frac{A_1^2A_2^2}{B^4}c^2s^2, \qquad b_{22}^{22} = E\frac{A_2^4}{B^4}s^4, \qquad b_{3i}^{\lambda\mu} = 0,
\end{aligned}
\tag{25.13}
$$

where we have introduced the notation

$$c = \cos\varphi, \qquad s = \sin\varphi, \qquad B^2 = A_1^2c^2 + A_2^2s^2. \tag{25.14}$$

Taking into account the symmetry properties (15.75) and (15.76), equations (25.13) exhaust all possible combinations of indices on the functions *b*.

25.2. The solution of the group (b^*) local problems

The group (b^*) local problems for the bar S_j are expressed by equations (25.2) and (25.4).

Working in coordinates η_1, η_2, z, and noting that the functions sought for are independent of η_1, it follows from (25.4) and (25.6) that, for an isotropic material,

$$
\begin{aligned}
b_{11}^{*\lambda\mu} &= -\frac{c_{11}}{h_1}\sin\varphi'\frac{\partial V_1^{\lambda\mu}}{\partial\eta_2} + \frac{c_{12}}{h_2}\cos\varphi'\frac{\partial V_2^{\lambda\mu}}{\partial\eta_2} + c_{12}\frac{\partial V_3^{\lambda\mu}}{\partial z} + zc_{11\lambda\mu},\\
b_{22}^{*\lambda\mu} &= -\frac{c_{12}}{h_1}\sin\varphi'\frac{\partial V_1^{\lambda\mu}}{\partial\eta_2} + \frac{c_{11}}{h_2}\cos\varphi'\frac{\partial V_2^{\lambda\mu}}{\partial\eta_2} + c_{12}\frac{\partial V_3^{\lambda\mu}}{\partial z} + zc_{22\lambda\mu},\\
b_{33}^{*\lambda\mu} &= -\frac{c_{12}}{h_1}\sin\varphi'\frac{\partial V_1^{\lambda\mu}}{\partial\eta_2} + \frac{c_{12}}{h_2}\cos\varphi'\frac{\partial V_2^{\lambda\mu}}{\partial\eta_2} + c_{11}\frac{\partial V_3^{\lambda\mu}}{\partial z} + zc_{33\lambda\mu},
\end{aligned}
\tag{25.15}
$$

$$b_{12}^{*\lambda\mu} = \frac{c_{44}}{h_2}\cos\varphi'\frac{\partial V_1^{\lambda\mu}}{\partial\eta_2} - \frac{c_{44}}{h_1}\sin\varphi'\frac{\partial V_2^{\lambda\mu}}{\partial\eta_2} + zc_{12\lambda\mu},$$

$$b_{13}^{*\lambda\mu} = c_{44}\frac{\partial V_1^{\lambda\mu}}{\partial z} - \frac{c_{44}}{h_1}\sin\varphi'\frac{\partial V_3^{\lambda\mu}}{\partial\eta_2},$$

$$b_{23}^{*\lambda\mu} = c_{44}\frac{\partial V_2^{\lambda\mu}}{\partial z} + \frac{c_{44}}{h_2}\cos\varphi'\frac{\partial V_3^{\lambda\mu}}{\partial\eta_2}.$$

Written in terms of coordinates η_2 and z, equations (25.2) take the same form as (25.8):

$$-\frac{\sin\varphi'}{h_1}\frac{\partial}{\partial\eta_2}b_{i1}^{*\lambda\mu} + \frac{\cos\varphi'}{h_2}\frac{\partial}{\partial\eta_2}b_{i2}^{*\lambda\mu} + \frac{\partial}{\partial z}b_{i3}^{*\lambda\mu} = 0,$$

$$\left[n_2'\left(-\frac{\sin\varphi'}{h_1}b_{i1}^{*\lambda\mu} + \frac{\cos\varphi'}{h_2}b_{i2}^{*\lambda\mu}\right) + n_3'b_{i3}^{*\lambda\mu}\right]\Bigg|_{\partial S_j'} = 0. \quad (25.16)$$

Because of the presence of the coordinate z in equations (25.15), the solution of the group (b^*) local problems clearly requires a knowledge of the specific shape of the cross-section of the bar S_j, and in what follows an important practical example of an elliptic cross-section will be considered. We note, first of all, that the cross-section in question changes its shape on transforming from the original coordinate system α_1, α_2, γ to coordinates ξ_1, ξ_2, z or η_1, η_2, z. If, for example, the value of the eccentricity of the cross-section was e in the original system, it becomes

$$e' = \left[1 - \frac{A_1^2A_2^2(1-e^2)}{A_1^2h_2^2c^2 + A_2^2h_1^2s^2}\right]^{1/2} \quad (25.17)$$

in coordinate systems ξ_1, ξ_2, z and η_1, η_2, z.

In the case of an elliptic cross-section with an eccentricity e' (see Fig. 25.2), the components n_2' and n_3' of the normal vector are given by

$$n_2' = \eta_2[1-(e')^2]^{-1}, \quad n_3' = z, \quad (25.18)$$

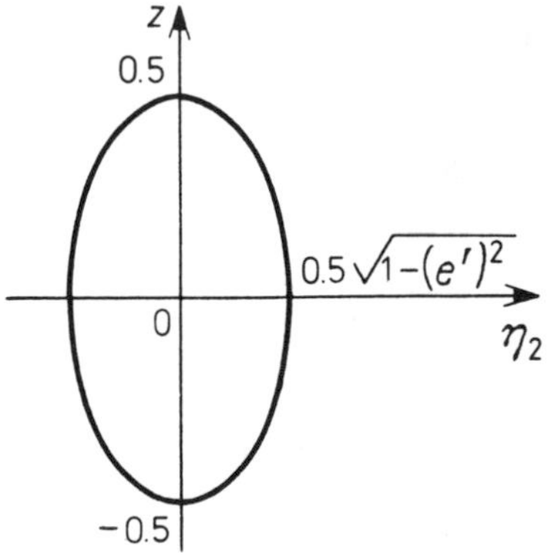

Fig. 25.2 *Elliptic cross-section of a bar in coordinates η_2, z.*

and it should be noted that, since the condition (25.16) on the contour $\partial S'_j$ is homogeneous, there is no need to normalize this vector.

With this result in mind, we now show that the solution of the problem (25.15), (25.16) may be taken in the form

$$V_i^{\lambda\mu} = W_{i1}^{\lambda\mu}\eta_2 z + W_{i2}^{\lambda\mu}\frac{\eta_2^2}{2} + W_{i3}^{\lambda\mu}\frac{z^2}{2}, \tag{25.19}$$

where $W_{ij}^{\lambda\mu}$ are constant coefficients. To this end, substitute (25.18) and (25.19) into (25.15) and (25.16) and equate to zero the coefficients of various combinations of η_2 and z in the resulting expressions. We then have

$$W_{12}^{\lambda\mu} = W_{13}^{\lambda\mu} = W_{22}^{\lambda\mu} = W_{23}^{\lambda\mu} = W_{31}^{\lambda\mu} = 0, \tag{25.20}$$

which with (25.19) indicates that

$$\begin{gathered} V_1^{\lambda\mu} = W_{11}^{\lambda\mu}\eta_2 z, \qquad V_2^{\lambda\mu} = W_{21}^{\lambda\mu}\eta_2 z, \\ V_3^{\lambda\mu} = W_{32}^{\lambda\mu}\frac{\eta_2^2}{2} + W_{33}^{\lambda\mu}\frac{z^2}{2}. \end{gathered} \tag{25.21}$$

The remaining four coefficients that are involved in this equation must satisfy the following system of four linear algebraic equations:

$$\begin{aligned} &-\frac{\sin\varphi'}{h_1}W_{11}^{\lambda\mu} + \frac{\cos\varphi'}{h_2}W_{21}^{\lambda\mu} + \left(\frac{\sin^2\varphi'}{h_1^2} + \frac{\cos^2\varphi'}{h_2^2}\right)W_{32}^{\lambda\mu} = 0, \\ &-c_{12}\frac{\sin\varphi'}{h_1}W_{11}^{\lambda\mu} + c_{12}\frac{\cos\varphi'}{h_2}W_{21}^{\lambda\mu} + c_{11}W_{33}^{\lambda\mu} + c_{33\lambda\mu} = 0, \\ &\left(c_{11}\frac{\sin^2\varphi'}{h_1^2} + c_{44}\frac{\cos^2\varphi'}{h_2^2}\right)W_{11}^{\lambda\mu} - (c_{12}+c_{44})\frac{\sin\varphi'\cos\varphi'}{h_1 h_2}W_{21}^{\lambda\mu} - c_{12}\frac{\sin\varphi'}{h_1}W_{33}^{\lambda\mu} \\ &\quad + c_{12\lambda\mu}\frac{\cos\varphi'}{h_2} - c_{11\lambda\mu}\frac{\sin\varphi'}{h_1} + c_{44}[1-(e')^2]\left(W_{11}^{\lambda\mu} - \frac{\sin\varphi'}{h_1}W_{32}^{\lambda\mu}\right) = 0, \\ &-(c_{12}+c_{44})\frac{\sin\varphi'\cos\varphi'}{h_1 h_2}W_{11}^{\lambda\mu} + \left(c_{44}\frac{\sin^2\varphi'}{h_1^2} + c_{11}\frac{\cos^2\varphi'}{h_2^2}\right)W_{21}^{\lambda\mu} \\ &\quad + c_{12}\frac{\cos\varphi'}{h_2}W_{33}^{\lambda\mu} + c_{22\lambda\mu}\frac{\cos\varphi'}{h_2} - c_{12\lambda\mu}\frac{\sin\varphi'}{h_1} \\ &\quad + c_{44}[1-(e')^2]\left(W_{21}^{\lambda\mu} + \frac{\cos\varphi'}{h_2}W_{32}^{\lambda\mu}\right) = 0. \end{aligned} \tag{25.22}$$

Having solved this system, all the functions $b_{ij}^{*\lambda\mu}$ can be obtained from (25.15) and (25.16), which completes the proof that the solution of the local problem (b^*) may indeed be represented in the form of (25.19) or, more precisely, (25.21). It is advisable, however, by analogy with the group (b) problems, to derive a system to directly determine the functions $b_{ij}^{*\lambda\mu}$ (which are actually what we need) by expressing them in terms of the coefficients in equations (25.21).

Substituting (25.21) into (25.15) yields

$$\begin{aligned}
&b_{11}^{*\lambda\mu} = zB_{11}^{\lambda\mu},\\
&B_{11}^{\lambda\mu} = -c_{11}\frac{\sin\varphi'}{h_1}W_{11}^{\lambda\mu} + c_{12}\frac{\cos\varphi'}{h_2}W_{21}^{\lambda\mu} + c_{12}W_{33}^{\lambda\mu} + c_{11\lambda\mu},\\
&b_{22}^{*\lambda\mu} = zB_{22}^{\lambda\mu},\\
&B_{22}^{\lambda\mu} = -c_{12}\frac{\sin\varphi'}{h_1}W_{11}^{\lambda\mu} + c_{11}\frac{\cos\varphi'}{h_2}W_{21}^{\lambda\mu} + c_{12}W_{33}^{\lambda\mu} + c_{22\lambda\mu},\\
&b_{33}^{*\lambda\mu} = zB_{33}^{\lambda\mu},\\
&B_{33}^{\lambda\mu} = -c_{12}\frac{\sin\varphi'}{h_1}W_{11}^{\lambda\mu} + c_{12}\frac{\cos\varphi'}{h_2}W_{21}^{\lambda\mu} + c_{11}W_{33}^{\lambda\mu} + c_{33\lambda\mu},\\
&b_{12}^{*\lambda\mu} = zB_{12}^{\lambda\mu}, \qquad B_{12}^{\lambda\mu} = c_{44}\frac{\cos\varphi'}{h_2}W_{11}^{\lambda\mu} - c_{44}\frac{\sin\varphi'}{h_1}W_{21}^{\lambda\mu} + c_{12\lambda\mu},\\
&b_{13}^{*\lambda\mu} = \eta_2 B_{13}^{\lambda\mu}, \qquad B_{13}^{\lambda\mu} = c_{44}W_{11}^{\lambda\mu} - c_{44}\frac{\sin\varphi'}{h_1}W_{32}^{\lambda\mu},\\
&b_{23}^{*\lambda\mu} = \eta_2 B_{23}^{\lambda\mu}, \qquad B_{23}^{\lambda\mu} = c_{44}W_{21}^{\lambda\mu} + c_{44}\frac{\cos\varphi'}{h_2}W_{32}^{\lambda\mu},
\end{aligned} \tag{25.23}$$

and using these in (25.22) we find

$$\begin{aligned}
&-\frac{\sin\varphi'}{h_1}B_{13}^{\lambda\mu} + \frac{\cos\varphi'}{h_2}B_{23}^{\lambda\mu} = 0, \qquad B_{33}^{\lambda\mu} = 0,\\
&-\frac{\sin\varphi'}{h_1}B_{11}^{\lambda\mu} + \frac{\cos\varphi'}{h_2}B_{12}^{\lambda\mu} + [1-(e')^2]\,B_{13}^{\lambda\mu} = 0,\\
&-\frac{\sin\varphi'}{h_1}B_{12}^{\lambda\mu} + \frac{\cos\varphi'}{h_2}B_{22}^{\lambda\mu} + [1-(e')^2]\,B_{23}^{\lambda\mu} = 0.
\end{aligned} \tag{25.24}$$

From expressions for $B_{11}^{\lambda\mu}$, $B_{22}^{\lambda\mu}$ and $B_{12}^{\lambda\mu}$ in (25.23) we obtain:

$$\begin{aligned}
W_{11}^{\lambda\mu} &= \left(\frac{\sin\varphi'}{h_1}\,\frac{B_{11}^{\lambda\mu} - B_{22}^{\lambda\mu} + c_{22\lambda\mu} - c_{11\lambda\mu}}{c_{12} - c_{11}} + \frac{\cos\varphi'}{h_2}\,\frac{B_{12}^{\lambda\mu} - c_{12\lambda\mu}}{c_{44}}\right)\\
&\quad\times\left(\frac{\sin^2\varphi'}{h_1^2} + \frac{\cos^2\varphi'}{h_2^2}\right)^{-1},\\
W_{21}^{\lambda\mu} &= \left(-\frac{\sin\varphi'}{h_1}\,\frac{B_{12}^{\lambda\mu} - c_{12\lambda\mu}}{c_{44}} + \frac{\cos\varphi'}{h_2}\,\frac{B_{11}^{\lambda\mu} - B_{22}^{\lambda\mu} + c_{22\lambda\mu} - c_{11\lambda\mu}}{c_{12} - c_{11}}\right)\left(\frac{\sin^2\varphi'}{h_1^2} + \frac{\cos^2\varphi'}{h_2^2}\right)^{-1},\\
W_{33}^{\lambda\mu} &= \frac{B_{11}^{\lambda\mu} - c_{11\lambda\mu}}{c_{12}} + \frac{c_{11}}{c_{12}}\,\frac{\sin\varphi'}{h_1}W_{11}^{\lambda\mu} - \frac{\cos\varphi'}{h_2}W_{21}^{\lambda\mu}.
\end{aligned} \tag{25.25}$$

If this is now substituted into the expression for $B_{33}^{\lambda\mu}$ in (25.23) and the result is equated to zero in accordance with the second of equations (25.24), an equation for determining the quantities $B_{11}^{\lambda\mu}$, $B_{22}^{\lambda\mu}$ and $B_{12}^{\lambda\mu}$ is obtained, which coincides with (25.11) except for the fact that $B_{\alpha\beta}^{\lambda\mu}$ replaces $b_{\alpha\beta}^{\lambda\mu}$ throughout.

Elimination of $W_{32}^{\lambda\mu}$ from the first of (25.24) and the expressions for $B_{13}^{\lambda\mu}$ and $B_{23}^{\lambda\mu}$ in (25.23) yields

$$B_{13}^{\lambda\mu} = c_{44}\frac{\cos\varphi'}{h_2}\left(\frac{\sin\varphi'}{h_1}W_{21}^{\lambda\mu} + \frac{\cos\varphi'}{h_2}W_{11}^{\lambda\mu}\right)\left(\frac{\cos^2\varphi'}{h_2^2} + \frac{\sin^2\varphi'}{h_1^2}\right)^{-1}. \tag{25.26}$$

Substituting this into the third of (25.24) and using (25.25) gives the second equation for determining $B_{11}^{\lambda\mu}$, $B_{22}^{\lambda\mu}$ and $B_{12}^{\lambda\mu}$:

$$-\frac{\sin\varphi'}{h_1}B_{11}^{\lambda\mu} + \frac{\cos\varphi'}{h_2}B_{12}^{\lambda\mu} + \frac{\cos\varphi'}{h_2}[1-(e')^2]\left[\left(\frac{\cos^2\varphi'}{h_2^2} - \frac{\sin^2\varphi'}{h_1^2}\right)(B_{12}^{\lambda\mu} - c_{12\lambda\mu})\right.$$
$$\left. - \frac{\sin\varphi'\cos\varphi'}{h_1 h_2}(B_{11}^{\lambda\mu} - B_{22}^{\lambda\mu} + c_{22\lambda\mu} - c_{11\lambda\mu})\right]\left(\frac{\sin^2\varphi'}{h_1^2} + \frac{\cos^2\varphi'}{h_2^2}\right)^{-2} = 0. \tag{25.27}$$

The third equation can be obtained by combining the first, third and fourth of equations (25.24) to give

$$\frac{\sin^2\varphi'}{h_1^2}B_{11}^{\lambda\mu} - \frac{2\sin\varphi'\cos\varphi'}{h_1 h_2}B_{12}^{\lambda\mu} + \frac{\cos^2\varphi'}{h_2^2}B_{22}^{\lambda\mu} = 0. \tag{25.28}$$

The linear algebraic system for the coefficients $B_{11}^{\lambda\mu}$, $B_{22}^{\lambda\mu}$ and $B_{12}^{\lambda\mu}$ thus includes equation (25.11) with $b_{\alpha\beta}^{\lambda\mu}$ replaced by $B_{\alpha\beta}^{\lambda\mu}$, and equations (25.27) and (25.28). Having solved the system and using (25.5), (25.17), (25.23) and the properties of the elastic moduli of an isotropic material, the solution of the group (b^*) local problems may finally be written as (Kalamkarov 1987a)

$$b_{\alpha\beta}^{*\lambda\mu} = zB_{\alpha\beta}^{\lambda\mu}, \qquad b_{33}^{*\lambda\mu} = 0, \tag{25.29}$$

$$\begin{aligned}
B_{11}^{11} &= \frac{E}{1+v}\frac{A_1^4}{B^4}c^2[2A_2^4 s^2(1-e^2)\Delta + c^2(1+v)],\\
B_{11}^{12} = B_{12}^{11} &= \frac{E}{1+v}\frac{A_1^3 A_2}{B^4}cs[A_2^2(A_2^2 s^2 - A_1^2 c^2)(1-e^2)\Delta + c^2(1+v)],\\
B_{11}^{22} = B_{22}^{11} &= \frac{E}{1+v}\frac{A_1^2 A_2^2}{B^4}c^2 s^2[-2A_1^2 A_2^2(1-e^2)\Delta + 1 + v],\\
B_{12}^{12} &= \frac{E}{1+v}\frac{A_1^2 A_2^2}{2B^4}[(A_1^2 c^2 - A_2^2 s^2)(1-e^2)\Delta + 2c^2 s^2(1+v)],\\
B_{12}^{22} = B_{22}^{12} &= \frac{E}{1+v}\frac{A_1 A_2^3}{B^4}cs[A_1^2(A_1^2 c^2 - A_2^2 s^2)(1-e^2)\Delta + s^2(1+v)],\\
B_{22}^{22} &= \frac{E}{1+v}\frac{A_2^4}{B^4}s^2[2A_1^4 c^2(1-e^2)\Delta + s^2(1+v)],
\end{aligned} \tag{25.30}$$

where, in addition to the previous notation, we have introduced

$$\Delta = [B^2 + A_1^2 A_2^2(1 - e^2)]^{-1}. \tag{25.31}$$

It is useful to note that although the functions $b_{13}^{*\lambda\mu}$ and $b_{23}^{*\lambda\mu}$ are different from zero and can in principle be determined from equations (25.23), (25.24) and (25.30), they do vanish after the averaging procedure (which is in accordance with the general formulas (15.62)) and therefore contribute nothing to the elastic relations of the problem.

We note also, if without a proof, that formulae (25.30) for B_{11}^{11} (with $\varphi = 0$) and for B_{22}^{22} (with $\varphi = \pi/2$) hold for any shape of the cross-section of the bar.

25.3. Effective moduli

Expressions (25.13) and (25.29), (25.30) for the functions $b_{\alpha\beta}^{\lambda\mu}$ and $b_{\alpha\beta}^{*\lambda\mu}$ make it possible to derive formulas for the effective elastic moduli of a network shell. Let $\delta^3 V$ be the volume of one bar within the unit cell of the shell as related to the coordinate system α_1, α_2, γ. For a bar with an elliptic cross-section, noting that the $b_{\alpha\beta}^{\lambda\mu}$ are constant and the $b_{\alpha\beta}^{*\lambda\mu}$ proportional to z, we have

$$\langle b_{\alpha\beta}^{\lambda\mu} \rangle = \frac{V}{h_1 h_2} b_{\alpha\beta}^{\lambda\mu}, \qquad \langle z b_{\alpha\beta}^{\lambda\mu} \rangle = \langle b_{\alpha\beta}^{*\lambda\mu} \rangle = 0,$$
$$\langle z b_{\alpha\beta}^{*\lambda\mu} \rangle = \frac{V}{16 h_1 h_2} B_{\alpha\beta}^{\lambda\mu}. \tag{25.32}$$

Note that the vanishing of the skew-symmetric coefficients $\langle z b_{\alpha\beta}^{\lambda\mu} \rangle$ and $\langle b_{\alpha\beta}^{*\lambda\mu} \rangle$ is due to the symmetry of the elliptic cross-section of the bar with respect to the shell middle surface.

In the case of a regular network with bars of elliptic cross-section, equations (25.13), (25.30) and (25.32) enable us, whatever the combination of the bars, to derive explicit expressions for the leading-order terms of the effective moduli of the homogenized plate (or shell) structure. The way of doing this is by first calculating the elastic moduli of each individual bar and subsequently summing over the bars. This approach suffers, of course, from neglecting the interaction that exists between the bars at their intersections, but if each of the bars is much smaller in size than the unit cell as a whole, the error in estimating the averaged moduli clearly must be minor in view of the fact that the stress field arising near an intersection is sharply confined to this vicinity and its contribution therefore will be negligible after the averaging procedure has been carried out. A mathematical justification for this kind of argument, in the form of the so-called principle of the split homogenized operator, has been provided by Bakhvalov and Panasenko (1984) in their work on the homogenization of periodic framework-type structures.

In conformity with the above, we find that

$$\langle b_{\alpha\beta}^{\lambda\mu} \rangle = \sum_{j=1}^{N} E_j b_j \gamma_j, \qquad \langle z b_{\alpha\beta}^{*\lambda\mu} \rangle = \sum_{j=1}^{N} E_j b_j \left(1 + \frac{c_j}{1 + \nu_j}\right) \frac{\gamma_j}{16} \tag{25.33}$$

for the case when the network shell with a regular structure is formed by N families of mutually parallel bars of elliptic cross-section. The parameters b_j and c_j in (25.33) depend on the choice of the fourfold index combination $\alpha\beta\lambda\mu$ and are determined by the following equations:

$$\begin{aligned}
\alpha\beta\lambda\mu = 1111{:}\ & b_j = A_1^4 B_j^{-4}\cos^4\varphi_j,\\
& c_j = 2A_2^4\tan^2\varphi_j(1-e_j^2)\Delta_j,\\
\alpha\beta\lambda\mu = 2222{:}\ & b_j = A_2^4 B_j^{-4}\sin^4\varphi_j,\\
& c_j = 2A_1^4\cot^2\varphi_j(1-e_j^2)\Delta_j,\\
\alpha\beta\lambda\mu = 1212{:}\ & b_j = A_1^2 A_2^2 B_j^{-4}\cos^2\varphi_j\sin^2\varphi_j\\
& c_j = \tfrac{1}{2}(A_1^2\cot^2\varphi_j + A_2^4\tan^2\varphi_j - 2A_1^2A_2^2)(1-e_j^2)\Delta_j,\\
\alpha\beta\lambda\mu = 1122, 2211{:}\ & b_j = A_1^2A_2^2B_j^{-4}\cos^2\varphi_j\sin^2\varphi_j,\\
& c_j = -2A_1^2A_2^2(1-e_j^2)\Delta_j,\\
\alpha\beta\lambda\mu = 1112, 1211{:}\ & b_j = A_1^3A_2B_j^{-4}\cos^3\varphi_j\sin\varphi_j,\\
& c_j = A_2^2(A_2^2\tan^2\varphi_j - A_1^2)(1-e_j^2)\Delta_j,\\
\alpha\beta\lambda\mu = 1222, 2212{:}\ & b_j = A_1A_2^3B_j^{-4}\cos\varphi_j\sin^3\varphi_j,\\
& c_j = A_1^2(A_1^2\cot^2\varphi_j - A_2^2)(1-e_j^2)\Delta_j.
\end{aligned} \tag{25.34}$$

The notation used in (25.33) and (25.34) is

$$\begin{aligned}
B_j^2 &= A_1^2\cos^2\varphi_j + A_2^2\sin\varphi_j,\\
\Delta_j &= [B_j^2 + A_1^2A_2^2(1-e_j^2)]^{-1}.
\end{aligned} \tag{25.35}$$

φ_j is the angle the coordinate line α_1 makes with the bars of the jth family; $\gamma_j = V_j(h_1h_2)^{-1}$ is the volumetric content of the bars of the jth family in the unit cell; e_j and E_j, ν_j are, respectively, the eccentricity and material constants of the jth family bars.

26. EFFECTIVE THERMOELASTIC PROPERTIES OF NETWORK REINFORCED SHELLS

In this section we have to consider the local thermoelastic problems (s) and (s^*), posed by equations (17.9), (17.11) and (17.14)–(17.16).

For an individual bar S_j^δ (see Fig. 25.1) this pair of problems is the same as the pair (25.1), (25.2) but with $b_{ik}^{\lambda\mu}$ and $b_{ik}^{*\lambda\mu}$ replaced by s_{ik} and s_{ik}^*, respectively. Note also that, instead of relations (25.3) and (25.4), equations (17.9) should be used, along with (15.33), (17.2) and (17.5).

The detailed solution of these problems was given by Kalamkarov (1988c) and will not be reproduced here. For the functions s_{ik} and s_{ik}^*, in the notation of (25.14), we have

$$\begin{aligned}
&\delta s_{11} = \alpha^{\mathrm{T}} E A_1^2 B^{-2} c^2, \qquad \delta s_{22} = \alpha^{\mathrm{T}} E A_2^2 B^{-2} s^2,\\
&\delta s_{12} = \delta s_{21} = \alpha^{\mathrm{T}} E A_1 A_2 B^{-2} cs, \qquad s_{3k} = 0,\\
&s_{ik}^* = z s_{ik}.
\end{aligned} \tag{26.1}$$

These equations hold for any shape of the cross-section of the bar S_j. For the special case of an elliptic cross-section, applying the averaging procedure, it is found that

$$\langle s_{\alpha\beta} \rangle = \frac{V}{h_1 h_2} s_{\alpha\beta}, \qquad \langle z s_{\alpha\beta} \rangle = \langle s^*_{\alpha\beta} \rangle = 0,$$
$$\langle z s^*_{\alpha\beta} \rangle = \frac{V}{16 h_1 h_2} s_{\alpha\beta}, \tag{26.2}$$

Now if we sum over all N bars of the network and make use of (26.1) and (26.2) we obtain the following expressions for the effective thermoelastic properties of network reinforced shells with bars of elliptic cross-section:

$$\delta \langle s_{\alpha\beta} \rangle = \sum_{j=1}^{N} \alpha_j^{\mathrm{T}} E_j s_j \gamma_j, \qquad \delta \langle z s^*_{\alpha\beta} \rangle = \frac{s_{\alpha\beta}}{16}, \tag{26.3}$$

where α_j^{T} is the coefficient of linear expansion of the material of the jth family bar, and the parameters s_j are defined by

$$\alpha\beta = 11: s_j = A_1^2 B_j^{-2} \cos^2 \varphi_j,$$
$$\alpha\beta = 22: s_j = A_2^2 B_j^{-2} \sin^2 \varphi_j, \tag{26.4}$$
$$\alpha\beta = 12, 21: s_j = A_1 A_2 B_j^{-2} \cos \varphi_j \sin \varphi_j.$$

Alternatively, we can deduce relations (26.2)–(26.4) from the general equations (17.17) making use of the solutions (25.13), (25.29) and (25.30) of the local problems (b) and (b^*). It is perhaps useful to note that while the group (b^*) solution, as given by (25.29) and (25.30), was obtained specifically for bars of elliptic cross-section, the solutions (26.1) for the local problems (s) and (s^*) are valid for arbitrary cross-section geometries.

27. HEAT CONDUCTION OF NETWORK REINFORCED PLATES AND SHELLS

In this section we discuss the problem of determining the effective heat conduction properties involved in the heat conduction relations of the homogenized shell, equations (16.35); we have, that is, to consider—in the context of network reinforced plates and shells with a regular structure—the local heat conduction problems (l) and (l^*) posed by equations (16.24), (16.26)–(16.31).

For an individual bar S_j^δ, see Fig. 25.1, these problems may be expressed in the form

$$\frac{1}{h_\mu} \frac{\partial l_{\mu\beta}}{\partial \xi_\mu} + \frac{\partial l_{3\beta}}{\partial z} = 0,$$
$$\left. \left(\frac{1}{h_\mu} n_\mu l_{\mu\beta} + n_3 l_{3\beta} \right) \right|_{\partial S_j} = 0, \quad l_{i\beta} \leftrightarrow l^*_{i\beta}, \tag{27.1}$$

if we introduce the definitions

$$l_{i\beta}=\frac{\lambda_{i\mu}}{h_\mu}\frac{\partial W_\beta}{\partial \xi_\mu}+\lambda_{i3}\frac{\partial W_\beta}{\partial z}+\lambda_{i\beta},$$
$$l^*_{i\beta}=\frac{\lambda_{i\mu}}{h_\mu}\frac{\partial W^*_\beta}{\partial \xi_\mu}+\lambda_{i3}\frac{\partial W^*_\beta}{\partial z}+z\lambda_{i\beta}, \tag{27.2}$$

and denote by n_i the components of the vector normal to the lateral surface ∂S_j.

The functions W_β and W^*_β are determined from the solution of the problems (27.1), (27.2) using the periodicity in the coordinates ξ_1 and ξ_2 with respective periods A_1 and A_2.

To solve these problems it is again convenient to change to the coordinate system η_1, η_2, z defined by (25.6) and to utilize the independence of the solutions of the coordinate η_1. The local problem (l) for the functions W_β is solved exactly whatever the shape of the cross-section of the bar S_j. We omit all the intermediate steps of the solution and write down at once the final results for the functions $l_{i\beta}$:

$$l_{11}=\lambda A_1^2B^{-2}c^2, \qquad l_{22}=\lambda A_2^2B^{-2}s^2,$$
$$l_{12}=l_{21}=\lambda A_1A_2B^{-2}cs, \; l_{31}=l_{32}=0, \tag{27.3}$$

using the notation of (25.14) and denoting by λ the heat conductivity of the bar material.

The solution of the local problem (l^*) for the functions W^*_β requires, as in the case of the group (b^*) problems, that the shape of the bar cross-section be prescribed, and is carried out analytically in the special case of an elliptic cross-section. Denoting by e the eccentricity of the cross-section in the coordinate system α_1, α_2, γ, the functions $l^*_{\alpha\beta}$ are found to be given by

$$l^*_{11}=z\lambda A_1^2[c^2+A_2^2(1-e^2)]\Delta,$$
$$l^*_{22}=z\lambda A_2^2[s^2+A_1^2(1-e^2)]\Delta, \tag{27.4}$$
$$l^*_{12}=l^*_{21}=z\lambda A_1A_2cs\Delta,$$

in the notation of (25.31). We note also that, again in analogy with the group (b^*) problems, the functions $l^*_{3\beta}$ are different from zero but vanish after averaging and do not therefore contribute (16.35).

For a bar with an elliptic cross-section, using (27.3), (27.4), and averaging we have

$$\langle l_{\alpha\beta}\rangle=\frac{V_1}{h_1h_2}l_{\alpha\beta}, \qquad \langle zl_{\alpha\beta}\rangle=\langle l^*_{\alpha\beta}\rangle=0,$$
$$\langle zl^*_{\alpha\beta}\rangle=\frac{V}{16h_1h_2}\frac{l^*_{\alpha\beta}}{z}. \tag{27.5}$$

Summing over all N bars and using (27.3)–(27.5) we obtain the following expressions for the effective heat conduction properties of a network reinforced shell with bars of elliptic cross-section:

$$\langle l_{\alpha\beta}\rangle = \sum_{j=1}^{N} \lambda_j s_j \gamma_j, \qquad \langle z l^*_{\alpha\beta}\rangle = \sum_{j=1}^{N} \lambda_j l_j \gamma_j/16. \tag{27.6}$$

Here λ_j is the thermal conductivity of the bar material of the jth family, the parameters s_j are defined by (26.4) and

$$\begin{aligned} \alpha\beta = 11: \quad & l_j = A_1^2[\cos^2\varphi_j + A_2^2(1-e_j^2)]\,\Delta_j, \\ \alpha\beta = 22: \quad & l_j = A_2^2[\sin^2\varphi_j + A_1^2(1-e_j^2)]\,\Delta_j, \\ \alpha\beta = 12, 21: & \, l_j = A_1 A_2 \cos\varphi_j \sin\varphi_j \Delta_j. \end{aligned} \tag{27.7}$$

We now proceed to find the coefficients occurring in equations (16.32) and (16.34) for the heat conduction of the homogenized shell. In the case of an elliptic cross-section,

$$\begin{aligned} \langle c_v\rangle = \frac{V}{h_1 h_2} c_v, \qquad & \langle z c_v\rangle = 0, \qquad \langle \lambda_{33}\rangle = \frac{V}{h_1 h_2}\lambda, \\ \langle z\lambda_{33}\rangle = 0, \qquad & \langle z^2 c_v\rangle = \frac{V}{16 h_1 h_2} c_v \end{aligned} \tag{27.8}$$

The coefficients $\mathfrak{J}_m$ ($m = 0, 1, 2$), R_m and Z_m ($m = 0, 1$) are calculated from (16.33) and (15.21). For a bar having the form of an elliptic cylinder of length δl, with e the eccentricity of the (elliptic) cross-section and φ the angle between the cross-section and the coordinate line α_1 (cf. Fig. 25.1), assuming that $\alpha_S^{\pm}$ and $g_S^{\pm}$ are independent of the 'rapid' coordinates y_1 and y_2, we find

$$\begin{aligned} \mathfrak{J}_0 &= (\alpha_S^+ + \alpha_S^-) l B E(\kappa)(h_1 h_2 A_1 A_2)^{-1}, \\ \mathfrak{J}_1 &= (\alpha_S^+ - \alpha_S^-) l \sqrt{1-e^2}\, a(2 h_1 h_2)^{-1}, \\ \mathfrak{J}_2 &= (\alpha_S^+ + \alpha_S^-) l B[(2-\kappa^{-2})E(\kappa) + (\kappa^{-2}-1)K(\kappa)](12 h_1 h_2 A_1 A_2)^{-1}, \\ R_0 &= (g_S^+ + g_S^-) l B E(\kappa)(h_1 h_2 A_1 A_2)^{-1}, \\ R_1 &= (g_S^+ - g_S^-) l \sqrt{1-e^2}\, a(2 h_1 h_2)^{-1}, \\ Z_0 &= \pi\lambda l \sqrt{1-e^2}\,(k_1 A_2^2 s^2 + k_2 A_1^2 c^2)\varepsilon(2 h_1 h_2 B^2)^{-1}, \\ Z_1 &= 0. \end{aligned} \tag{27.9}$$

Here we use the notation of (25.14) and also denote

$$\begin{aligned} \kappa &= [1-(1-e^2)A_1^2 A_2^2 B^{-2}]^{1/2}, \\ a &= \frac{1}{2\sqrt{1-\kappa^2}} + \frac{\sqrt{1-\kappa^2}}{4\kappa}\ln\frac{1+\kappa}{1-\kappa}, \\ \varepsilon &= (1-\kappa^2/2-\sqrt{1-\kappa^2})\kappa^{-4}. \end{aligned} \tag{27.10}$$

The quantities $K(\kappa)$ and $E(\kappa)$ in (27.9) are respectively the complete elliptic integrals of the first and second kinds. Note that for $\kappa = 0$, $K(0) = E(0) = \pi/2$, $a = 1$ and $\varepsilon = \frac{1}{8}$.

In the special case of $e = 0$ (circular cross-section) and $A_1 = A_2 = 1$, we have $B = 1$, $\kappa = 0$ and, from (27.9) and (27.10), we see that

$$\begin{aligned}
&\mathfrak{J}_0 = (\alpha_S^+ + \alpha_S^-)S_l(2h_1h_2)^{-1},\\
&\mathfrak{J}_1 = (\alpha_S^+ - \alpha_S^-)S_l(2\pi h_1h_2)^{-1},\\
&\mathfrak{J}_2 = (\alpha_S^- + \alpha_S^+)S_l(24h_1h_2)^{-1}, \qquad R_0 = (g_S^+ + g_S^-)S_l(2h_1h_2)^{-1},\\
&R_1 = (g_S^+ - g_S^-)S_l(2\pi h_1h_2)^{-1},\\
&Z_0 = \lambda S_l(k_1s^2 + k_2c^2)(16h_1h_2)^{-1},
\end{aligned} \tag{27.11}$$

where $\delta^2 S_l$ is the area of the lateral surface of the bar in the coordinate system $\alpha_1, \alpha_2, \gamma$ (for a bar of circular cross-section, it is obvious that $S_l = \pi l$).

The effective coefficients entering into equations (16.32) and (16.34) can now be found by first calculating the corresponding quantities from relations (27.8)–(27.11) for each individual bar and then summing over all the bars. The error in estimating the coefficients by the above scheme will not be appreciable if the volumes of the individual bars are much smaller than the volume of the entire unit cell.

28. CONSTITUTIVE EQUATIONS FOR NETWORK REINFORCED PLATES AND SHELLS OF RECTANGULAR, RHOMBIC AND TRIANGULAR STRUCTURE

Network reinforced thin-walled systems of rectangular, rhombic or triangular configurations are important structural materials, adaptable to many uses. In this section we employ the general relations (25.33), (26.3) and (27.6) to derive the constitutive relations and effective properties for each of these reinforcement types.

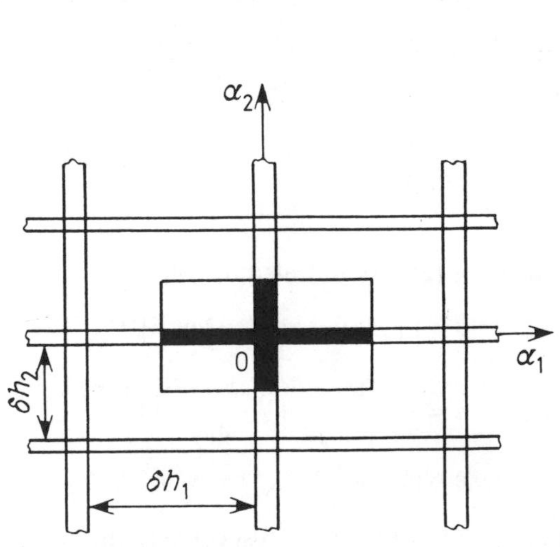

Fig. 28.1 *Rectangular network.*

28.1 Rectangular network

Let the reinforcing network of the plate (or shell) be formed by two mutually perpendicular families of bars of elliptic cross-section, the unit cell of the structure being as shown in Fig. 28.1. Using the subscript $\beta = 1,2$ to refer to a family, we denote by $E_\beta, \nu_\beta, \lambda_\beta, \alpha_\beta^{\mathrm{T}}, e_\beta$ and F_β the material and geometrical parameters of the families ($\delta^2 F_\beta$ representing the cross-sectional area in the coordinate system $\alpha_1, \alpha_2, \gamma$) and by δh_1 and δh_2 the inter-bar spacings. From this point on we formally consider the system $\alpha_1, \alpha_2, \gamma$ as Cartesian when denoting the geometrical characteristics of the reinforcing network.

The constitutive relations of the homogenized shell (cf. (17.23)) in this case may be represented in the form

$$
\begin{aligned}
&N_1 = \delta E_1 \frac{F_1}{h_2} \varepsilon_1 - \delta \alpha_1^{\mathrm{T}} E_1 \frac{F_1}{h_2} \theta_1^{(0)}, \qquad N_2 = \delta E_2 \frac{F_2}{h_1} \varepsilon_2 - \delta \alpha_2^{\mathrm{T}} E_2 \frac{F_2}{h_1} \theta_1^{(0)}, \\
&N_{12} = 0, \qquad M_1 = \delta^3 E_1 \frac{F_1}{16 h_2} \kappa_1 - \delta^2 \alpha_1^{\mathrm{T}} E_1 \frac{F_2}{16 h_2} \theta_2^{(0)}, \\
&M_2 = \delta^3 E_2 \frac{F_2}{16 h_1} \kappa_2 - \delta^2 \alpha_2^{\mathrm{T}} E_2 \frac{F_2}{16 h_1} \theta_2^{(0)}, \\
&M_{12} = \delta^3 \left[\frac{E_1}{1+\nu_1} \frac{F_1}{16 h_2} \frac{A_2^2(1-e_1^2)}{1 + A_2^2(1-e_1^2)} + \frac{E_2}{1+\nu_2} \frac{F_2}{16 h_1} \frac{A_1^2(1-e_2^2)}{1 + A_1^2(1-e_2^2)} \right] \tau,
\end{aligned}
\tag{28.1}
$$

and the homogenized shell heat conduction relations (cf.(16.35)) become

$$
\begin{aligned}
&\langle q_1^{(0)} \rangle = -\lambda_1 \frac{F_1}{h_2 A_1} \frac{\partial \theta_1^{(0)}}{\partial \alpha_1}, \qquad \langle q_2^{(0)} \rangle = -\lambda_2 \frac{F_2}{h_1 A_2} \frac{\partial \theta_1^{(0)}}{\partial \alpha_2}, \\
&\langle z q_1^{(0)} \rangle = -\frac{1}{16 A_1} \left[\lambda_1 \frac{F_1}{h_2} + \lambda_2 \frac{F_2}{h_1} \frac{A_1^2(1-e_2^2)}{1 + A_1^2(1-e_2^2)} \right] \frac{\partial \theta_2^{(0)}}{\partial \alpha_1}, \\
&\langle z q_2^{(0)} \rangle = -\frac{1}{16 A_2} \left[\lambda_2 \frac{F_2}{h_1} + \lambda_1 \frac{F_1}{h_2} \frac{A_2^2(1-e_1^2)}{1 + A_2^2(1-e_1^2)} \right] \frac{\partial \theta_2^{(0)}}{\partial \alpha_2}.
\end{aligned}
\tag{28.2}
$$

28.2 Rhombic network

The problem considered is shown in Fig. 28.2. The shell is composed of two families of mutually parallel bars of the same material and of the same elliptic cross-section, with eccentricity e and cross-sectional area $\delta^2 F$. The distance between two neighbouring bars of the same family is δa, and the angle formed by bars of different families is 2φ. The constitutive equations of the homogenized shell, (17.23) and (16.35), take the form

$$
\begin{aligned}
&N_1 = \delta 2 E \frac{F}{a} A_1^2 B^{-4} c^2 (A_1^2 c^2 \varepsilon_1 + A_2^2 s^2 \varepsilon_2) - \delta 2 \alpha^{\mathrm{T}} E \frac{F}{a} A_1^2 B^{-2} c^2 \theta_1^{(0)}, \\
&N_2 = N_1 A_2^2 s^2 (A_1^2 c^2)^{-1}, \qquad N_{12} = \delta 2 E \frac{F}{a} A_1^2 A_2^2 B^{-4} c^2 s^2 \omega,
\end{aligned}
$$

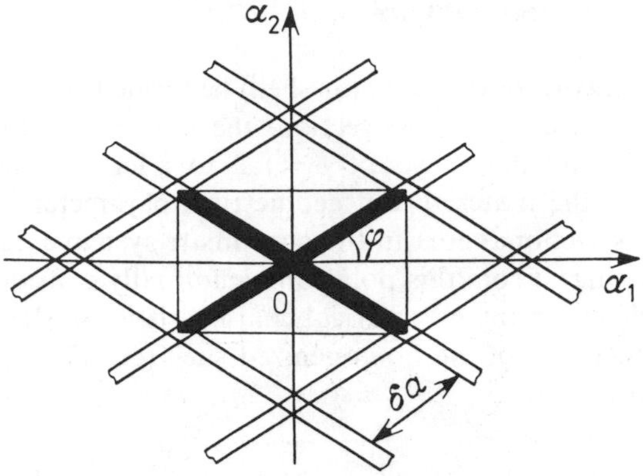

Fig. 28.2 *Rhombic network.*

$$M_1 = \delta^3 \frac{E}{(1+\nu)} \frac{F}{8a} \frac{A_1^2 c^2}{B^4} \{A_1^2[2A_2^4 s^2(1-e^2)\Delta + c^2(1+\nu)]\kappa_1$$

$$+ A_2^2 s^2[-2A_1^2 A_2^2(1-e^2)\Delta + 1 + \nu]\kappa_2\} - \delta^2 \alpha^{\mathrm{T}} E \frac{F}{8a} \frac{A_1^2}{B^2} c^2 \theta_2^{(0)}, \tag{28.3}$$

$$M_2 = \delta^3 \frac{E}{(1+\nu)} \frac{F}{8a} \frac{A_2^2 s^2}{B^4} \{A_1^2 c^2[-2A_1^2 A_2^2(1-e^2)\Delta + 1 + \nu]\kappa_1$$

$$+ A_2^2[2A_1^4 c^2(1-e^2)\Delta + s^2(1+\nu)]\kappa_2\} - \delta^2 \alpha^{\mathrm{T}} E \frac{F}{8a} \frac{A_2^2}{B^2} s^2 \theta_2^{(0)},$$

$$M_{12} = \delta^3 \frac{E}{(1+\nu)} \frac{F}{8a} \frac{A_1^2 A_2^2}{B^4} [(A_1^2 c^2 - A_2^2 s^2)^2 (1-e^2)\Delta + 2c^2 s^2(1+\nu)]\tau,$$

$$\langle q_1^{(0)} \rangle = -\lambda \frac{2F}{a} \frac{A_1}{B^2} c^2 \frac{\partial \theta_1^{(0)}}{\partial \alpha_1}, \qquad \langle q_2^{(0)} \rangle = -\lambda \frac{2F}{a} \frac{A_2}{B^2} s^2 \frac{\partial \theta_1^{(0)}}{\partial \alpha_2},$$

$$\langle z q_1^{(0)} \rangle = -\lambda \frac{F}{8a} A_1 [c^2 + A_2^2(1-e^2)] \Delta \frac{\partial \theta_2^{(0)}}{\partial \alpha_1}, \tag{28.4}$$

$$\langle z q_2^{(0)} \rangle = -\lambda \frac{F}{8a} A_2 [s^2 + A_1^2(1-e^2)] \Delta \frac{\partial \theta_2^{(0)}}{\partial \alpha_2},$$

where we have used the notations described in connection with (25.14) and (25.31).

28.3. Network formed by equilateral triangles

The network reinforced shell consists of three families of mutually parallel bars which intersect to form equilateral triangles, as shown in Fig. 28.3. The cross-sections of all bars are circles of diameter δ (that is, $e = 0$, $F = \pi/4$), and the neighbouring

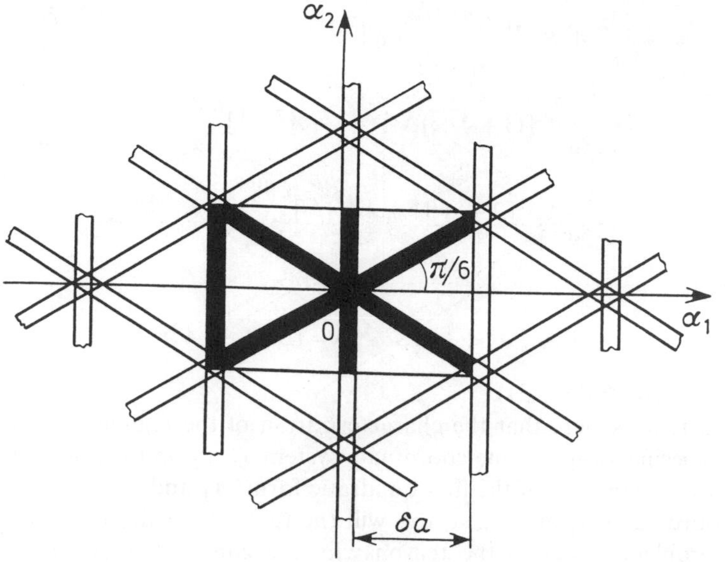

Fig. 28.3 *Network formed by equilateral triangles.*

bars of the same family are a distance δa apart. Assuming that all the bars are made of the same material, the constitutive relations of the homogenized shell, equations (17.23) and (16.35), may be written in the following manner:

$$N_1 = \delta E \frac{3\pi}{32a} \frac{A_1^2}{B^4}(3A_1^2\varepsilon_1 + A_2^2\varepsilon_2) - \delta\alpha^{\mathrm{T}} E \frac{3\pi}{8a} \frac{A_1^2}{B^2} \theta_1^{(0)},$$

$$N_2 = \delta E \frac{\pi}{32a} B^{-4}[3A_1^2A_2^2\varepsilon_1 + (8B^4 + A_2^4)\varepsilon_2] - \delta\alpha^{\mathrm{T}} E \frac{\pi}{8a}\left(2 + \frac{A_2^2}{B^2}\right)\theta_1^{(0)},$$

$$N_{12} = \delta E \frac{3\pi}{32a} \frac{A_1^2 A_2^2}{B^4}\omega,$$

$$M_1 = \delta^3 \frac{E}{(1+\nu)} \frac{3\pi}{512a} \frac{A_1^2}{B^4}\{A_1^2[2A_2^4\Delta + 3(1+\nu)]\kappa_1$$

$$+ A_2^2[-2A_1^2A_2^2\Delta + 1 + \nu]\kappa_2\} - \delta^2\alpha^{\mathrm{T}} E \frac{3\pi}{128a} \frac{A_1^2}{B^2}\theta_2^{(0)}, \tag{28.5}$$

$$M_2 = \delta^3 \frac{E}{(1+\nu)} \frac{\pi}{512a} B^{-4}\{3A_1^2A_2^2[-2A_1^2A_2^2\Delta + 1 + \nu]\kappa_1$$

$$+ [6A_1^4A_2^4\Delta + (A_2^4 + 8B^4)(1+\nu)]\kappa_2\} - \delta^2\alpha^{\mathrm{T}} E \frac{\pi}{128a}\left(2 + \frac{A_2^2}{B^2}\right)\theta_2^{(0)},$$

$$M_{12} = \delta^3 \frac{E}{(1+\nu)} \frac{\pi}{512a} \frac{A_1^2A_2^2}{B^4}\{[3A_1^2 - A_2^2)^2\Delta + 6(1+\nu)$$

$$+ 8A_2^{-2}(1 + A_1^2)^{-1}B^4\}\tau,$$

$$\langle q_1^{(0)} \rangle = -\lambda \frac{3\pi}{8a} A_1 B^{-2} \frac{\partial \theta_1^{(0)}}{\partial \alpha_1}, \qquad \langle q_2^{(0)} \rangle = -\lambda \frac{\pi}{8a} A_2^{-1} \left(2 + \frac{A_2^2}{B^2}\right) \frac{\partial \theta_1^{(0)}}{\partial \alpha_2},$$

$$\langle z q_1^{(0)} \rangle = -\lambda \frac{\pi}{128a} A_1 [(3 + 4A_2^2)\Delta + 2(1 + A_1^2)^{-1}] \frac{\partial \theta_2^{(0)}}{\partial \alpha_1}, \tag{28.6}$$

$$\langle z q_2^{(0)} \rangle = -\lambda \frac{\pi}{128a} A_2^{-1} [2 + A_2^2(1 + 4A_1^2)\Delta] \frac{\partial \theta_2^{(0)}}{\partial \alpha_2},$$

where we write

$$B^2 = \tfrac{3}{4} A_1^2 + \tfrac{1}{4} A_2^2, \qquad \Delta = (B^2 + A_1^2 A_2^2)^{-1},$$

as we did in (25.14) and (25.31).

We should remark here that the characterization of the regular structure of a shell requires the specification of some coordinate system α_1, α_2 on the middle surface of the shell. Since the coefficients of the first quadratic form, A_1 and A_2, generally depend on the 'slow' coordinates α_1 and α_2, so too will the real (physical) dimensions of the unit cell of the problem, which is the reason why the constitutive relations contain the coefficients $A_1(\alpha_1, \alpha_2)$ and $A_2(\alpha_1, \alpha_2)$ and the homogenized shell is, in general, quasi-homogeneous. In the special case of network reinforced plates, the coordinate system α_1, α_2 may be a Cartesian one, and hence $A_1 \equiv 1$ and $A_2 \equiv 1$ in all the results obtained. Omitting the effect of temperature on the state of stress, the elastic relations (28.1), (28.3) and (28.5) coincide in this case with similar results obtained in Pshenichnov (1982) by a different method; see also Parton and Kalamkarov (1988a).

It should be also noted that the above constitutive relations are invariant with respect to transformation of the coordinate system α_1, α_2 provided the physical dimensions of the unit cell remain unchanged. A change in the functions A_1 and A_2 due to transformation is compensated in all the above formulas by corresponding changes in the values of the parameters $h_1, h_2, \alpha, \varphi, e$ and F.

29. COMPOSITE SHELLS WITH HIGH-STIFFNESS FRAMEWORK-TYPE REINFORCEMENT

The term high-stiffness composite material usually refers to polymer matrix fibre reinforced composites in which the Young's modulus of the fibre phase, E_F, is much larger that of the matrix phase, E_M. Accordingly, the mechanical behaviour of the composite will be predicted with an error of the order of only $E_M/E_F \ll 1$ if we assume that, for comparable fibre and matrix percentage contents, the role of the matrix is negligible and the stressed state of the composite is determined by the deformation of the fibre system alone. The state of stress in the matrix itself will then be found from the problem set in the region occupied by the matrix, under appropriately formulated fibre–matrix interface conditions (see Annin *et al.* 1990).

We thus consider an angle-ply composite shell formed by N layers of parallel fibres, as shown in Fig. 29.1, and we assume, in accordance with the above, that the fibre material is much stiffer than that of the matrix. In this case the solution of the local elastic problems (b) and (b*) (equations (25.1), (25.3) and (25.2), (25.4), respectively) are

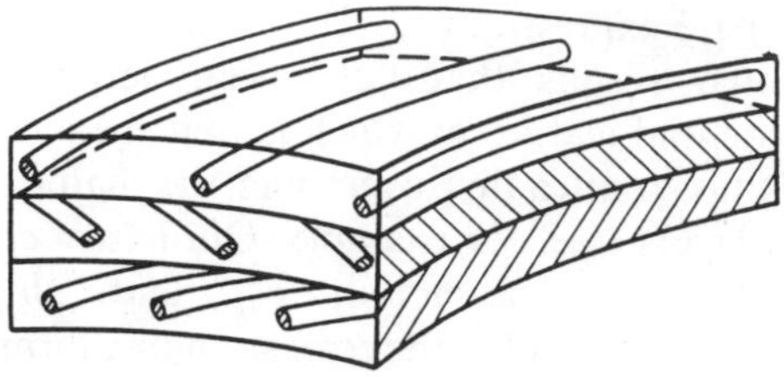

Fig. 29.1 *Two-way reinforced composite shell.*

much simplified owing to the decoupling of the regions occupied by the fibres and the matrix, and we will employ this fact when solving these problems for a fibre of the jth layer of the system (the layer making an angle φ_j with the coordinate line α_1, the departure of the axis of a fibre from the shell middle surface, $\gamma = 0$ will be denoted by δa_j in coordinates $\alpha_1, \alpha_2, \gamma$). In Section 25, exact solutions of the local problems were found for the case in which $a_j = 0$ and under the assumption (introduced specifically for group (b^*) problems) that the fibres were elliptic in cross-section. To extend these solutions to the case $a_j \neq 0$, we modify the problems (b) and (b^*) by everywhere replacing $z' + a_j$. An analysis of the relations so obtained shows that, as far as the local problem (b) is concerned, the solution remains the same as for $a_j = 0$ and is therefore given by equations (25.13). With regard to the local problem (b^*) for the case $a_j \neq 0$, it proves possible to show that its solutions are related to those for $a_j = 0$ by the equations

$$b^{*\lambda\mu}_{\alpha\beta} = z' B^{\lambda\mu}_{\alpha\beta} + a_j b^{\lambda\mu}_{\alpha\beta}, \tag{29.1}$$

where $b^{\lambda\mu}_{\alpha\beta}$ and $B^{\lambda\mu}_{\alpha\beta}$ are given respectively by (25.13) and (25.30), with the replacements $c \to \cos\varphi_j$, $s \to \sin\varphi_j$, $B \to B_j$, $e \to e_j$ and $\Delta \to \Delta_j$ in the notation of (25.14), (25.31) and (25.35). Note that equation (29.1), like (25.29), holds for the case of an elliptic cross-section of a fibre, e_j being the eccentricity of the ellipse.

Having solved the local problems, we average the functions $b^{\lambda\mu}_{\alpha\beta}$ and $b^{*\lambda\mu}_{\alpha\beta}$ over the fibre volume and then sum over all N layers to obtain the following expressions for the effective elastic moduli of disconnected framework formed by fibres of elliptic cross-section:

$$\begin{aligned} \langle b^{\lambda\mu}_{\alpha\beta} \rangle &= \sum_{j=1}^{N} E_j b_j \gamma_j, \\ \langle z b^{\lambda\mu}_{\alpha\beta} \rangle = \langle b^{*\lambda\mu}_{\alpha\beta} \rangle &= \sum_{j=1}^{N} E_j b_j a_j \gamma_j, \\ \langle z b^{*\lambda\mu}_{\alpha\beta} \rangle &= \sum_{j=1}^{N} E_j b_j \left[a_j^2 + \frac{1}{16}\left(1 + \frac{c_j}{1+\nu_j}\right)\right]\gamma_j. \end{aligned} \tag{29.2}$$

Here b_j and c_j are determined from (25.34) for each combination $\alpha\beta\lambda\mu$; E_j and ν_j are the material properties of the fibres in the jth layer; and γ_j is the volumetric fibre content in the jth layer. Note that if we set $a_j = 0 (j = 1, 2, \ldots, N)$ in equations (29.2), these latter reduce to formulae (25.33) for the effective elastic moduli of a network reinforced shell.

It is of interest to compare expressions (29.2) for high-stiffness framework-type shell reinforcements with similar results that have been derived from the structurally anisotropic model, the essential feature of which is that the average over the thickness of a multilayered shell is taken after first averaging the material characteristics of the constituent (orthotropic) layers (see, for example, Obraztsov *et al.* 1977. Tamuzs and Protasov 1986, Vasil'ev 1988). For the moduli $\langle b_{\alpha\beta}^{\lambda\mu} \rangle$ and $\langle zb_{\alpha\beta}^{\lambda\mu} \rangle$ it is found that the expressions given by (29.2) are identical to the corresponding formulae for the generalized properties of a multilayered shell working in a tension–compression regime, provided the contribution of the matrix to the reduced properties of the orthotropic layers is negligible. The flexural and torsional stiffnesses, $\langle zb_{\alpha\beta}^{*\lambda\mu} \rangle$, do differ from the corresponding results of the structurally anistropic model and may be converted to these latter by setting $e_j = 1$, $j = 1, 2, \ldots, N$ (which means neglecting the shape of the cross-section) and replacing by 12 the factor 16 arising in the denominator through the moment of inertia of the elliptic fibre cross-section.

To obtain an estimate of the magnitude of the corrections, consider a three-layered angle-ply shell with reinforcement angles $\varphi_1 = \pi/4$, $\varphi_2 = 0$ and $\varphi_3 = -\pi/4$ in which the fibres in all three layers are made of the same material (with isotropic properties E and ν) and have the same circular cross-section. We assume that $\gamma_1 = \gamma_2 = \gamma_3 = \gamma_0$, $a_1 = 1$, $a_2 = 0$ and $a_3 = -1$. From (29.2), the non-zero flexural and torsional moduli of the shell are then given by

$$\begin{aligned} \langle zb_{11}^{*11} \rangle &= E[0.594 + 0.0313(1+\nu)^{-1}]\gamma_0, \\ \langle zb_{11}^{*22} \rangle &= E[0.5 + 0.0313\,\nu(1+\nu)^{-1}]\gamma_0, \\ \langle zb_{22}^{*22} \rangle &= E[0.531 + 0.0313(1+\nu)^{-1}]\gamma_0, \\ \langle zb_{12}^{*12} \rangle &= E[0.531 + 0.0156(1+\nu)^{-1}]\gamma_0, \end{aligned} \tag{29.3}$$

where, setting $\nu = 0.25$ for specificity, the quantities in brackets are found to be equal to 0.619, 0.506, 0.556 and 0.543, respectively. The corresponding expressions resulting from the structurally anisotropic approach are

$$\begin{aligned} \langle zb_{11}^{*11} \rangle &= 0.625\,E\gamma_0, \qquad & \langle zb_{11}^{*22} \rangle &= 0.542\,E\gamma_0, \\ \langle zb_{22}^{*22} \rangle &= 0.542\,E\gamma_0, \qquad & \langle zb_{12}^{*12} \rangle &= 0.542\,E\gamma_0, \end{aligned} \tag{29.4}$$

the maximum percentage change in the values of the quantities in brackets being 7% for the elastic modulus $\langle zb_{11}^{*22} \rangle$.

8 The Fundamental Solution of the Periodic Elasticity Problem

As noted earlier, the local problems formulated in Section 15 are doubly periodic three-dimensional problems of the theory of elasticity which are set on the unit cell of the structure under study and the solutions of which are assumed to satisfy the prescribed boundary conditions at the surfaces $z = z^+$, $z = z^-$ and periodicity requirements on the coordinates ξ_1 and ξ_2, with respective periods A_1 and A_2. Obviously, problems of this level of complexity are generally amenable only to a numerical solution, for example by the Boundary Element (BE) method. The application of the BE method to three-dimensional elasticity problems depends on the use of the fundamental solution of the equations of the theory of elasticity for a concentrated force acting in a unbounded elastic medium. For the case of a homogeneous isotropic medium such a solution was obtained by Lord Kelvin (see, for example, Love 1944) and has been employed in combination with the BE method by a number of workers (see Banerjee and Butterfield 1981, Crouch and Starfield 1983), but as far as the local problems of interest here are concerned, unfortunately this approach is all but inadequate, because of the periodicity of the stress and strain fields in the coordinates ξ_1 and ξ_2. It therefore becomes necessary to construct a periodic fundamental solution of the three-dimensional elasticity problem which would enable a modification of the BE algorithm into a form suitable for the solution of the local problems relevant to the study of composites (Kalamkarov *et al.* 1989).

In what follows we confine ourselves to the case of an isotropic medium and set $A_1 = A_2$ and $h_1 = h_2 = h$.

30. THE DERIVATION OF A DOUBLY PERIODIC FUNDAMENTAL SOLUTION OF THE THREE-DIMENSIONAL ELASTICITY PROBLEM

In accordance with the way in which the local problems (19.6), (19.7) and (19.11), (19.12) were formulated, by the fundamental doubly periodic solution we mean the solution of the system

$$\frac{1}{h}\frac{\partial \tau_{i\alpha}}{\partial \xi_\alpha} + \frac{\partial \tau_{i3}}{\partial z} = -P_i\delta(\xi_1)\delta(\xi_2)\delta(z) \quad (i = 1, 2, 3),$$

$$\tau_{\alpha\beta} = \frac{c_{44}}{h}\left(\frac{\partial U_\alpha}{\partial \xi_\beta} + \frac{\partial U_\beta}{\partial \xi_\alpha}\right) + c_{12}\delta_{\alpha\beta}\left(\frac{\partial U_\gamma}{h\partial \xi_\gamma} + \frac{\partial U_3}{\partial z}\right),$$

$$\tau_{33} = 2c_{44}\frac{\partial U_3}{\partial z} + c_{12}\left(\frac{1}{h}\frac{\partial U_\gamma}{\partial \xi_\gamma} + \frac{\partial U_3}{\partial z}\right), \tag{30.1}$$

$$\tau_{3\alpha} = c_{44}\left(\frac{1}{h}\frac{\partial U_3}{\partial \xi_\alpha} + \frac{\partial U_\alpha}{\partial z}\right)$$

in the region $|\xi_\alpha| < A/2$ and $|z| < \infty$, with functions U_1, U_2 and U_3 periodic in ξ_1 and ξ_2 with period A. The quantities P_i in the first of equations (30.1) are the components of the (concentrated) force applied to the origin. As usual, Greek indices range from 1 to 2.

We represent the solution of the system as a sum of three solutions, each corresponding to one of the three components of the force. Setting $P_2 = P_3 = 0$ first and noting that

$$\delta(\xi_\alpha) = \frac{1}{A} + \frac{2}{A}\sum_{n=1}^{\infty} \cos(\beta_n \xi_\alpha) \quad (\beta_n = 2\pi n/A),$$

we assume the solution of (30.1) to be of the form

$$U_1^{(1)} = \sum_{n=0}^{\infty}\sum_{m=0}^{\infty} U_{1nm}^{(1)}(z)\cos(\beta_n\xi_1)\cos(\beta_m\xi_2),$$

$$U_2^{(1)} = \sum_{n=1}^{\infty}\sum_{m=1}^{\infty} U_{2nm}^{(1)}(z)\sin(\beta_n\xi_1)\sin(\beta_m\xi_2), \tag{30.2}$$

$$U_3^{(1)} = \sum_{n=1}^{\infty}\sum_{m=0}^{\infty} U_{3nm}^{(1)}(z)\sin(\beta_n\xi_1)\cos(\beta_m\xi_2),$$

and substitute into (30.1) to obtain the following system of ordinary differential equations:

$$c_{44}\frac{d^2U_{100}^{(1)}}{dz^2} = -\frac{P_1}{A^2}\delta(z), \qquad c_{44}\left(\frac{d^2U_{10m}^{(1)}}{dz^2} - \lambda_m^2 U_{10m}^{(1)}\right) = -\frac{2P_1}{A^2}\delta(z),$$

$$c_{44}\frac{d^2U_{1n0}^{(1)}}{dz^2} - c_{11}\lambda_n^2 U_{1n0}^{(1)} + (c_{44}+c_{12})\lambda_n\frac{dU_{3n0}^{(1)}}{dz} = -\frac{2P_1}{A^2}\delta(z),$$

$$-(c_{44}+c_{12})\lambda_n\frac{dU_{1n0}^{(1)}}{dz} + c_{11}\frac{d^2U_{3n0}^{(1)}}{dz^2} - c_{44}\lambda_n^2 U_{3n0}^{(1)} = 0,$$

$$c_{44}\frac{d^2U_{1nm}^{(1)}}{dz^2} - (c_{11}\lambda_n^2 + c_{44}\lambda_m^2)U_{1nm}^{(1)} + (c_{44}+c_{12})\lambda_n\lambda_m U_{2nm}^{(1)}$$

$$+(c_{44}+c_{12})\lambda_n\frac{dU_{3nm}^{(1)}}{dz} = -\frac{4P_1}{A^2}\delta(z), \tag{30.3}$$

$$(c_{44}+c_{12})\lambda_n\lambda_m U_{1nm}^{(1)} + c_{44}\frac{d^2U_{2nm}^{(1)}}{dz^2} - (c_{44}\lambda_n^2 + c_{11}\lambda_m^2)U_{2nm}^{(1)} - (c_{44}+c_{12})\lambda_m\frac{dU_{3nm}^{(1)}}{dz} = 0,$$

$$-(c_{44}+c_{12})\lambda_n\frac{dU_{1nm}^{(1)}}{dz} + (c_{44}+c_{12})\lambda_m\frac{dU_{2nm}^{(1)}}{dz} + c_{11}\frac{d^2U_{3nm}}{dz^2} - c_{44}(\lambda_n^2 + \lambda_m^2)U_{3nm}^{(1)} = 0,$$

$$\lambda_n = \beta_n/h = 2\pi n/(hA).$$

Recalling that

$$\delta(z) = \frac{1}{\pi}\int_0^\infty \cos(z\eta)\,\mathrm{d}\eta,$$

we represent the solution to (30.3) in the form

$$\begin{aligned}
\begin{Bmatrix} U_{1n0}^{(1)} \\ U_{10m}^{(1)} \end{Bmatrix} &= \frac{2}{\pi}\int_0^\infty \begin{Bmatrix} \tilde{U}_{1n0}^{(1)}(\eta) \\ \tilde{U}_{10m}^{(1)}(\eta) \end{Bmatrix} \cos(z\eta)\,\mathrm{d}\eta, \\
U_{3n0}^{(1)} &= \frac{2}{\pi}\int_0^\infty \tilde{U}_{3n0}^{(1)}(\eta)\sin(z\eta)\,\mathrm{d}\eta, \\
\begin{Bmatrix} U_{1nm}^{(1)} \\ U_{2nm}^{(1)} \end{Bmatrix} &= \frac{2}{\pi}\int_0^\infty \begin{Bmatrix} \tilde{U}_{1nm}^{(1)}(\eta) \\ \tilde{U}_{2nm}^{(1)}(\eta) \end{Bmatrix} \cos(z\eta)\,\mathrm{d}\eta, \\
U_{3nm}^{(1)} &= \frac{2}{\pi}\int_0^\infty \tilde{U}_{3nm}^{(1)}(\eta)\sin(z\eta)\,\mathrm{d}\eta, \quad (n, m = 1, 2, \ldots),
\end{aligned} \tag{30.4}$$

which when substituted into (30.3) yields a system of algebraic equations for $\tilde{U}_{1nm}^{(1)}$, $\tilde{U}_{2nm}^{(1)}$ and $\tilde{U}_{3nm}^{(1)}$. The solution of (30.3) is obtained by solving the system and subsequently evaluating the integrals involved in (30.4).

Returning to (30.2), the fundamental solution of the problem, in terms of the unitary function $\eta(z)$, is

$$\begin{aligned}
U_1^{(1)} = &-\frac{P_1}{A^2 c_{44}} z\eta(z) + \frac{P_1}{A^2}\frac{(c_{11}+c_{44})}{2c_{11}c_{44}} \sum_{n=1}^\infty \frac{e^{-|z|\lambda_n}}{\lambda_n}\cos(\beta_n\xi_1) \\
&- \frac{P_1}{A^2}\frac{(c_{11}-c_{44})}{2c_{11}c_{44}} |z| \sum_{n=1}^\infty e^{-|z|\lambda_n}\cos(\beta_n\xi_1) + \frac{P_1}{A^2 c_{44}} \sum_{m=1}^\infty \frac{e^{-|z|\lambda_m}}{\lambda_m}\cos(\beta_m\xi_2) \\
&+ \frac{2P_1}{A^2 c_{44}} \sum_{n=1}^\infty\sum_{m=1}^\infty \frac{e^{-|z|\sqrt{\lambda_n^2+\lambda_m^2}}}{\sqrt{\lambda_n^2+\lambda_m^2}} \cos(\beta_n\xi_1)\cos(\beta_m\xi_2) \\
&- \frac{P_1(c_{11}-c_{44})}{A^2 c_{11}c_{44}} \sum_{n=1}^\infty\sum_{m=1}^\infty [(\lambda_n^2+\lambda_m^2)^{-3/2} \\
&+ |z|(\lambda_n^2+\lambda_m^2)^{-1}]\,\lambda_n^2 e^{-\sqrt{\lambda_n^2+\lambda_m^2}|z|}\cos(\beta_n\xi_1)\cos(\beta_m\xi_2),
\end{aligned} \tag{30.5}$$

$$\begin{aligned}
U_2^{(1)} = &\frac{P_1(c_{11}-c_{44})}{A^2 c_{11}c_{44}} \sum_{n=1}^\infty\sum_{m=1}^\infty [(\lambda_n^2+\lambda_m^2)^{-3/2} \\
&+ |z|(\lambda_n^2+\lambda_m^2)^{-1}]\,\lambda_n\lambda_m e^{-\sqrt{\lambda_n^2+\lambda_m^2}|z|}\sin(\beta_n\xi_1)\sin(\beta_m\xi_2),
\end{aligned} \tag{30.6}$$

$$\begin{aligned}
U_3^{(1)} = &\frac{P_1(c_{11}-c_{44})}{A^2 2c_{11}c_{44}} z \sum_{n=1}^\infty e^{-|z|\lambda_n}\sin(\beta_n\xi_1) \\
&+ \frac{P_1(c_{11}-c_{44})}{A^2 c_{11}c_{44}} z \sum_{n=1}^\infty\sum_{m=1}^\infty (\lambda_n^2+\lambda_m^2)^{-1/2} e^{-\sqrt{\lambda_n^2+\lambda_m^2}|z|\lambda_n}\sin(\beta_n\xi_1)\cos(\beta_m\xi_2),
\end{aligned} \tag{30.7}$$

31. TRANSFORMATION OF THE DOUBLY PERIODIC FUNDAMENTAL SOLUTION OF THE ELASTICITY PROBLEM

It is essential for our following discussion to transform equations (30.5)–(30.7) in such a way as to separate singular terms from regular ones. We note first of all that the summation of the single series in these equations can be effected with the aid of the following well-known formulae:

$$\sum_{n=1}^{\infty} \mathrm{e}^{-ny}\cos(n\varphi) = -\frac{1}{2} + \frac{\sinh y}{2(\cosh y - \cos\varphi)},$$

$$\sum_{n=1}^{\infty} \mathrm{e}^{-ny}\sin(n\varphi) = \frac{\sin\varphi}{2(\cosh y - \cos\varphi)},$$

$$\sum_{n=1}^{\infty} \frac{1}{n}\mathrm{e}^{-ny}\cos(n\varphi) = \frac{y}{2} - \tfrac{1}{2}\ln(2|\cosh y - \cos\varphi|) \quad (y > 0, |\varphi| < \pi),$$

given, for example, in Whittaker and Watson (1927).

To transform the double sums in (30.5)–(30.7) use will be made of two representations of the Jacobi theta function ϑ (also given in Whittaker and Watson 1927):

$$\vartheta_3(\xi|\mathrm{i}\tau) = 1 + 2\sum_{n=1}^{\infty} \mathrm{e}^{-\pi n^2\tau}\cos(2n\xi),$$

$$\vartheta_3(\xi|\mathrm{i}\tau) = \mathrm{e}^{-\xi^2/\pi\tau}/\sqrt{\tau} + \frac{1}{\sqrt{\tau}}\sum_{n=1}^{\infty}[\mathrm{e}^{-(\xi+\pi n)^2/\pi\tau} + \mathrm{e}^{-(\xi-\pi\tau)^2/\pi\tau}].$$

Multiplying the equation

$$1 + 2\sum_{n=1}^{\infty} \mathrm{e}^{-\pi n^2\tau}\cos(2n\xi) = \frac{1}{\sqrt{\tau}}\mathrm{e}^{-\xi^2/\pi\tau} + \frac{1}{\sqrt{\tau}}\sum_{n=1}^{\infty}[\mathrm{e}^{-(\xi+\pi n)^2/\pi\tau} + \mathrm{e}^{-(\xi-\pi n)^2/\pi\tau}]$$

by $(1/\sqrt{\tau})\exp(-z^2/\pi\tau)$, taking the Laplace transform with respect to τ and noting that (Bateman and Erdélyi 1954)

$$\mathrm{e}^{-a/(4\tau)}/\sqrt{\tau} \doteqdot (\sqrt{\pi}/\sqrt{p})\,\mathrm{e}^{-\sqrt{ap}}, \qquad \mathrm{e}^{-a/(4\tau)}/\tau \doteqdot 2K_0(\sqrt{ap}),$$

$$\frac{1}{\sqrt{\tau}}\mathrm{e}^{-a/4\tau-\pi n^2\tau} \doteqdot \frac{\sqrt{\pi}}{\sqrt{p+\pi n^2}}\mathrm{e}^{-\sqrt{a(p+\pi n^2)}},$$

where p is the transform parameter and $K_0(x)$ the MacDonald function, we obtain:

$$\sqrt{\pi/p}\,\mathrm{e}^{-2|z|\sqrt{p/\pi}} + 2\sqrt{\pi}\sum_{n=1}^{\infty}\frac{\mathrm{e}^{-2|z|\sqrt{p/\pi+n^2}}}{\sqrt{p+\pi n^2}}\cos(2n\xi) = 2K_0(2\sqrt{z^2+\xi^2}\sqrt{p/\pi})$$
$$+ 2\sum_{n=1}^{\infty}[K_0(2\sqrt{z^2+(\xi+\pi n)^2}\sqrt{p/\pi}) + K_0(2\sqrt{z^2+(\xi-\pi n)^2}\sqrt{p/\pi})]. \qquad (31.1)$$

Setting $p=\pi k^2$ in (31.1) yields

$$\sum_{n=1}^{\infty}\frac{e^{-2|z|\sqrt{k^2+n^2}}}{\sqrt{k^2+n^2}}\cos(2n\xi)=\frac{1}{2k}e^{-2|z|k}+K_0(2k\sqrt{z^2+\xi^2})$$

$$+\sum_{n=1}^{\infty}[K_0(2k\sqrt{z^2+(\xi+\pi n)^2})+K_0(2k\sqrt{z^2+(\xi-\pi n)^2}]. \qquad (31.2)$$

Multiplying (31.2) by $\cos(2k\eta)$, summing over k and using the following well-known expansion (Prudnikov *et al.* 1983):

$$\sum_{k=1}^{\infty}K_0(kx)\cos(ka)=\frac{\pi}{2}(x^2+a^2)^{-1/2}+\frac{1}{2}\left(C+\ln\frac{x}{4\pi}\right)$$

$$+\frac{\pi}{2}\sum_{k=1}^{\infty}\left[\frac{1}{\sqrt{(2\pi k-a)^2+x^2}}-\frac{1}{2\pi k}\right]+\frac{\pi}{2}\sum_{k=1}^{\infty}\left[\frac{1}{\sqrt{(2\pi k+a)^2+x^2}}-\frac{1}{2\pi k}\right], \qquad (31.3)$$

where $x>0$ and C is Euler's constant, the required representation of one of the double sums of interest is

$$\sum_{k=1}^{\infty}\sum_{n=1}^{\infty}\frac{e^{-2|z|\sqrt{k^2+n^2}}}{\sqrt{k^2+n^2}}\cos(2n\xi)\cos(2k\eta)=\tfrac{1}{2}\ln\left[\frac{1}{2\pi}\sqrt{2|\cosh(2z)-\cos(2\eta)|(z^2+\xi^2)}\right]$$

$$+\frac{\pi}{4}(z^2+\xi^2+\eta^2)^{-1/2}+\Phi_0(\xi,\eta,z)\quad(|\xi|<\pi/2,|\eta|<\pi/2), \qquad (31.4)$$

where $\Phi_0(\xi,\eta,z)$ is a regular term expressible in the form

$$\Phi_0(\xi,\eta,z)=-\tfrac{1}{2}(|z|-C)+\frac{\pi}{4}\sum_{k=1}^{\infty}\left[\frac{1}{\sqrt{(\pi k-\eta)^2+z^2+\xi^2}}-\frac{1}{\pi k}\right]$$

$$+\frac{\pi}{4}\sum_{k=1}^{\infty}\left[\frac{1}{\sqrt{(\pi k+\eta)^2+z^2+\xi^2}}-\frac{1}{\pi k}\right]$$

$$+\sum_{k=1}^{\infty}\sum_{n=1}^{\infty}[K_0(2k\sqrt{z^2+(\pi n+\eta)^2})+K_0(2k\sqrt{z^2+(\pi n-\eta)^2})]\cos(2k\eta). \qquad (31.5)$$

The second double sum representation for (30.5) is found by differentiating (31.2) with respect to k, multiplying by $k\cos(2k\eta)$ and finally summing over k with the aid of the expansion

$$x\sum_{k=1}^{\infty}kK_1(kx)\cos(ka)=\frac{\pi}{2}x^2(x^2+a^2)^{-3/2}$$

$$-\frac{1}{2}+\frac{\pi}{2}\sum_{k=1}^{\infty}\left[\frac{x^2}{((2\pi k-a)^2+x^2)^{3/2}}+\frac{x^2}{((2\pi k+a)^2+x^2)^{3/2}}\right],$$

which follows from (31.3). The result is

$$\sum_{k=1}^{\infty}\sum_{n=1}^{\infty}[(k^2+n^2)^{-3/2}+2|z|(k^2+n^2)^{-1}]\,e^{-2|z|\sqrt{n^2+k^2}}\,k^2\cos(2n\xi)\cos(2k\eta)$$

$$=\frac{\pi}{4}(z^2+\eta^2)(z^2+\xi^2+\eta^2)^{-3/2}+\tfrac{1}{2}\ln\left(\sqrt{2|\cosh(2z)-\cos(2\xi)|}\right)$$

$$-\frac{1}{2}\frac{z\sinh(2z)}{\cosh(2z)-\cos(2\xi)}+\Psi_0(\xi,\eta,z), \tag{31.6}$$

which the regular term $\Psi_0(\xi,\eta,z)$ is of the form

$$\Psi_0(\xi,\eta,z)=-\frac{1}{2}+\frac{\pi}{4}\sum_{k=1}^{\infty}\left[\frac{z^2+\eta^2}{((\pi k-\xi)^2+z^2+\eta^2)^{3/2}}+\frac{z^2+\eta^2}{((\pi k+\xi)^2+z^2+\eta^2)^{3/2}}\right]$$

$$+\sum_{k=1}^{\infty}\sum_{n=1}^{\infty}[2k\sqrt{z^2+(\eta+\pi n)^2}\,K_1(2k\sqrt{z^2+(\eta+\pi n)^2})$$

$$+2k\sqrt{z^2+(\eta-\pi n)^2}\,K_1(2k\sqrt{z^2+(\eta-\pi n)^2})]. \tag{31.7}$$

To separate singular terms in the double sum in (30.6), we differentiate (31.2) with respect to k and ξ, in that order, and then multiply the result by $\sin(2k\eta)$ and sum over k. Making use of the expansion

$$\sum_{k=1}^{\infty}kK_0(kx)\sin(ka)=\frac{\pi}{2}a(x^2+a^2)^{-3/2}$$

$$+\frac{\pi}{2}\sum_{k=1}^{\infty}\left[\frac{a-2\pi k}{((a-2\pi k)^2+x^2)^{3/2}}+\frac{a+2\pi k}{((a+2\pi k)^2+x^2)^{3/2}}\right],$$

we find

$$\sum_{k=1}^{\infty}\sum_{n=1}^{\infty}[(k^2+n^2)^{-3/2}+2z(k^2+n^2)^{-1}]\,e^{-2z\sqrt{k^2+n^2}}\,kn\sin(2n\xi)\sin(2k\eta)$$

$$=\frac{\pi}{4}\xi\eta(z^2+\xi^2+\eta^2)^{-3/2}+F_0(\xi,\eta,z), \tag{31.8}$$

where

$$F_0(\xi,\eta,z)=\frac{\pi}{4}\eta\sum_{k=1}^{\infty}\left[\frac{\xi-\pi k}{((\xi-\pi k)^2+z^2+\eta^2)^{3/2}}+\frac{\xi+\pi k}{((\xi+\pi k)^2+z^2+\eta^2)^{3/2}}\right]$$

$$+2\sum_{k=1}^{\infty}\sum_{n=1}^{\infty}[(\eta+\pi n)K_0(2k\sqrt{z^2+(\eta+\pi n)^2})$$

$$+(\eta-\pi n)K_0(2k\sqrt{z^2+(\eta-\pi n)^2})]k\sin(2k\xi). \tag{31.9}$$

Using (31.4)–(31.9) we may rearrange (30.5)–(30.7) to

$$U_1^{(1)}=\frac{P_1h^2}{16\pi G(1-\nu)}[(3-4\nu)(z^2+h^2\xi_1^2+h^2\xi_2^2)^{-1/2}+h^2\xi_1^2(z^2+h^2\xi_1^2+h^2\xi_2^2)^{-3/2}]$$

$$+\frac{P_1}{A^2G}\left[-z\eta(z)+2\Phi_3(\xi_1,\xi_2,z)+\frac{1}{2(1-\nu)}\Phi_1(\xi_1,\xi_2,z)\right],$$

$$U_2^{(1)}=\frac{P_1h^2}{16\pi G(1-\nu)}h^2\xi_1\xi_2(z^2+h^2\xi_1^2+h^2\xi_2^2)^{-3/2}+\frac{P_1}{2A^2G(1-\nu)}\Phi_2(\xi_1,\xi_2,z), \quad (31.10)$$

$$U_3^{(1)}=\frac{P_1h^2}{16\pi G(1-\nu)}zh\xi_1(z^2+h^2\xi_1^2+h^2\xi_2^2)^{-3/2}$$

$$-\frac{P_1}{2A^2G(1-\nu)}\frac{z}{h}\frac{\partial}{\partial\xi_1}\Phi_3(\xi_2,\xi_1,z),$$

where

$$\Phi_1(\xi_1,\xi_2,z)=\frac{Ah}{4\pi}-\frac{Ah}{8\pi}\sum_{k=1}^{\infty}\left[\frac{A^{-2}h^{-2}(z^2+h^2\xi_2^2)}{((k^2-\xi_1/A)^2+A^{-2}h^{-2}(z^2+h^2\xi_2^2))^{3/2}}\right.$$

$$\left.+\frac{A^{-2}h^{-2}(z^2+h^2\xi_2^2)}{((k+\xi_1/A)^2+A^{-2}h^{-2}(z^2+h^2\xi_2^2))^{3/2}}\right]-Ah\sum_{k=1}^{\infty}\sum_{n=1}^{\infty}$$

$$\times[\sqrt{z^2A^{-2}h^{-2}+(\xi_2/A+m)^2}K_1(2\pi k\sqrt{z^2A^{-2}h^{-2}+(\xi_2/A+m)^2})$$

$$+\sqrt{z^2A^{-2}h^{-2}+(\xi_2/A-m)^2}K_1(2\pi k\sqrt{z^2A^{-2}h^{-2}+(\xi_2/A-m)^2})]$$

$$\times k\cos(\beta_k\xi_1), \quad (31.11)$$

$$\Phi_2(\xi_1,\xi_2,z)=\frac{h\xi_2}{8\pi}\sum_{k=1}^{\infty}\left[\frac{\xi_1/A-k}{((k-\xi_1/A)^2+A^{-2}h^{-2}(z^2+h^2\xi_2^2))^{3/2}}\right.$$

$$\left.+\frac{\xi_1/A+k}{((k+\xi_1/A)^2+A^{-2}h^{-2}(z^2+h^2\xi_2^2))^{3/2}}\right]$$

$$+Ah\sum_{k=1}^{\infty}\sum_{m=1}^{\infty}[(\xi_2/A+m)K_0(2\pi k\sqrt{z^2A^{-2}h^{-2}+(\xi_2/A+m)^2})$$

$$+(\xi_2/A-m)K_0(2\pi k\sqrt{z^2A^{-2}h^{-2}+(\xi_2/(A-m)^2})]k\sin(\beta_k\xi_1), \quad (31.12)$$

$$\Phi_3(\xi_1,\xi_2,z)=\frac{Ah}{4\pi}C-\frac{Ah}{4\pi}\left[\ln\sqrt{2\left|\cosh\left(\frac{2\pi z}{Ah}\right)-\cos\left(\frac{2\pi\xi_1}{A}\right)\right|}-\ln\left(\frac{1}{2A}\sqrt{z^2h^{-2}+\xi_1^2}\right)\right]$$

$$+\frac{Ah}{8\pi}\sum_{k=1}^{\infty}\left[\frac{1}{\sqrt{(k-\xi_2/A)^2+A^{-2}h^{-2}(z^2+h^2\xi_1^2)}}-\frac{1}{k}\right]$$

$$+\frac{Ah}{8\pi}\sum_{k=1}^{\infty}\left[\frac{1}{\sqrt{(k+\xi_2/A)^2+A^{-2}h^{-2}(z^2+h^2\xi_1^2)}}-\frac{1}{k}\right]$$

$$+\frac{Ah}{2\pi}\sum_{k=1}^{\infty}\sum_{m=1}^{\infty}\left[K_0(2\pi k\sqrt{A^{-2}h^{-2}z^2+(\xi_1/A+m)^2})\right.$$

$$\left.+K_0(2\pi k\sqrt{A^{-2}h^{-2}z^2+(\xi_1/A-m)^2})\right]\cos(\beta_k\xi_2). \quad (31.13)$$

It will be understood that the first terms in equations (31.10) correspond to the well-known singular solution obtained by Lord Kelvin for an unbounded isotropic medium loaded by a force P_1 at point $\xi_1 = \xi_2 = z = 0$; the remaining terms in (31.10) are regular functions at this point. It can be shown also that the expressions

$$\Phi_1(\xi_1, \xi_2, z) - \frac{A^2h^2}{8\pi}(z^2 + h^2\xi_2^2)(z^2 + h^2\xi_1^2 + h^2\xi_2^2)^{-3/2},$$

$$\Phi_2(\xi_1, \xi_2, z) + \frac{A^2h^2}{8\pi}h^2\xi_1\xi_2(z^2 + h^2\xi_1^2 + h^2\xi_2^2)^{-3/2},$$

$$\Phi_3(\xi_1, \xi_2, z) + \frac{A^2h^2}{8\pi}(z^2 + h^2\xi_1^2 + h^2\xi_2^2)^{-1/2}$$

are periodic in ξ_1 and ξ_2 with period A, which secures the periodicity of the displacements $U_1^{(1)}$, $U_2^{(1)}$ and $U_3^{(1)}$ in the same coordinates. We note, furthermore, that the following equations hold:

$$\Phi_3(\xi_1, \xi_2, z) = \Phi_3(\xi_2, \xi_1, z), \qquad \Phi_2(\xi_1, \xi_2, z) = \Phi_2(\xi_2, \xi_1, z),$$

and that the manner in which the functions $K_0(x)$ and $K_1(x)$ behave at large x secures good convergence for the double series in (31.11)–(31.13).

Thus far we have considered the case when the force P_1 directed along the asix ξ_1 is the only one acting on the body. The cases when either the force P_2 or P_3 is acting, respectively, along the ξ_2 or z axis, can be treated in a similar fashion. That is to say, we write

$$\begin{cases} U_1^{(2)} = \sum_{n=1}^{\infty}\sum_{m=1}^{\infty} U_{1nm}^{(2)}(z)\sin(\beta_n\xi_1)\sin(\beta_m\xi_2), \\ U_2^{(2)} = \sum_{n=0}^{\infty}\sum_{m=0}^{\infty} U_{2nm}^{(2)}(z)\cos(\beta_n\xi_1)\cos(\beta_m\xi_2), \\ U_3^{(2)} = \sum_{n=0}^{\infty}\sum_{m=1}^{\infty} U_{3nm}^{(2)}(z)\cos(\beta_n\xi_1)\sin(\beta_m\xi_2), \end{cases}$$

$$\begin{cases} U_1^{(3)} = \sum_{n=1}^{\infty}\sum_{m=0}^{\infty} U_{1nm}^{(3)}(z)\sin(\beta_n\xi_1)\cos(\beta_m\xi_2), \\ U_2^{(3)} = \sum_{n=0}^{\infty}\sum_{m=1}^{\infty} U_{2nm}^{(3)}(z)\cos(\beta_n\xi_1)\sin(\beta_m\xi_2), \\ U_3^{(3)} = \sum_{n=0}^{\infty}\sum_{m=0}^{\infty} U_{3nm}^{(3)}(z)\cos(\beta_n\xi_1)\cos(\beta_m\xi_2), \end{cases}$$

and obtain by the same kind of argument:

$$U_1^{(2)} = U_2^{(1)}\frac{P_2}{P_1},$$

$$U_2^{(2)} = \frac{P_2h^2}{16\pi G(1-\nu)}[(3-4\nu)(z^2 + h^2\xi_1^2 + h^2\xi_2^2)^{-1/2} + h^2\xi_2^2(z^2 + h^2\xi_1^2 + h^2\xi_2^2)^{-3/2}]$$

$$+\frac{P_2}{A^2G}\left[-z\eta(z)+2\Phi_3(\xi_2,\xi_1,z)+\frac{1}{2(1-v)}\Phi_1(\xi_2,\xi_1,z)\right]. \tag{31.14}$$

$$U_3^{(2)}=\frac{P_2h^2}{16\pi G(1-v)}zh\xi_2(z^2+h^2\xi_1^2+h^2\xi_2^2)^{-3/2}$$

$$-\frac{P_2}{2A^2G(1-v)}\frac{z}{h}\frac{\partial}{\partial\xi_2}\Phi_3(\xi_1,\xi_2,z),$$

$$U_1^{(3)}=U_3^{(1)}\frac{P_3}{P_1},\qquad U_2^{(3)}=U_3^{(2)}\frac{P_3}{P_1}, \tag{31.15}$$

$$U_3^{(3)}=\frac{P_3h^2}{16\pi G(1-v)}[(3-4v)(z^2+h^2\xi_1^2+h^2\xi_2^2)^{-1/2}+z^2(z^2+h^2\xi_1^2+h^2\xi_2^2)^{-3/2}]$$

$$+\frac{P_3}{A_2G}\left[-z\eta(z)+\frac{3-4v}{2(1-v)}\Phi_3(\xi_2,\xi_1,z)-\frac{Z}{2(1-v)}\frac{\partial}{\partial z}\Phi_3(\xi_2,\xi_1,z)\right],$$

where again the first terms correspond to the singular Kelvin solution and the remaining terms are regular functions securing the required periodicity of $U_i^{(2)}$ and $U_i^{(3)}$ with respect to ξ_1 and ξ_2.

Adding together (31.10), (31.14) and (31.15) we are now in a position to express the fundamental solution in the form

$$U_i(\boldsymbol{x})=[F_{ij}(\boldsymbol{x})+F_{ij}^*(\boldsymbol{x})]P_j, \tag{31.16}$$

where

$$\boldsymbol{x}=(x_1,x_2,x_3),\qquad x_1=h\xi_1,\qquad x_2=h\xi_2,\qquad x_3=z,$$

$$F_{ij}=\frac{1}{16\pi G(1-v)}\left[(3-4v)\frac{\delta_{ij}}{r}+\frac{x_ix_j}{r^3}\right],\quad r^2=x_ix_i, \tag{31.17}$$

and the functions $F_{ij}^*(\boldsymbol{x})$ in terms of the functions $\Phi_i(x_1/h,x_2/h,x_3)$, $i=1,2,3$, are given by

$$F_{11}^*=\frac{1}{A^2G}\left[-x_3\eta(x_3)+2\Phi_3\left(\frac{x_1}{h},\frac{x_2}{h},x_3\right)+\frac{1}{2(1-v)}\Phi_1\left(\frac{x_1}{h},\frac{x_2}{h},x_3\right)\right],$$

$$F_{12}^*=F_{21}^*=\frac{1}{2A^2G(1-v)}\Phi_2\left(\frac{x_1}{h},\frac{x_2}{h},x_3\right),$$

$$F_{13}^*=F_{31}^*=-\frac{x_3}{2A^2G(1-v)}\frac{\partial}{\partial x_1}\Phi_3\left(\frac{x_1}{h},\frac{x_2}{h},x_3\right),$$

$$F_{22}^*=\frac{1}{A^2G}\left[-x_3\eta(x_3)+2\Phi_3\left(\frac{x_1}{h},\frac{x_2}{h},x_3\right)+\frac{1}{2(1-v)}\Phi_1\left(\frac{x_2}{h},\frac{x_1}{h},x_3\right)\right], \tag{31.18}$$

$$F_{23}^*=F_{32}^*=-\frac{x_3}{2A^2G(1-v)}\frac{\partial}{\partial x_2}\Phi_3\left(\frac{x_1}{h},\frac{x_2}{h},x_3\right),$$

$$F_{33}^*=\frac{1}{A^2G}\left[-x_3\eta(x_3)+\frac{3-4v}{2(1-v)}\Phi_3\left(\frac{x_1}{h},\frac{x_2}{h},x_3\right)-\frac{x_3}{2(1-v)}\frac{\partial}{\partial x_3}\Phi_3\left(\frac{x_1}{h},\frac{x_2}{h},x_3\right)\right].$$

Combined with (31.11)–(31.13), equations (31.16)–(31.18) present the fundamental solution of the doubly periodic three-dimensional problem of the theory of elasticity.

32. SINGLY PERIODIC FUNDAMENTAL SOLUTION OF THE PLANE ELASTICITY

Associated with the problem we discuss in this section (i.e. a plane with periodicity in one direction only) are two-dimensional formulations of the problems (b11), (b22) and (b*11), (b*22), i.e. equations (19.16), (19.18) and (19.22), (19.24), all with $\lambda\mu = 11, 22$.

The fundamental solution periodic in ξ_2 with period A is defined in the plane case as a solution of the system

$$\frac{1}{h}\frac{\partial \tau_{i2}}{\partial \xi_2} + \frac{\partial \tau_{i3}}{\partial z} = -P_i\delta(\xi_2)\delta(z) \quad (i = 2, 3),$$

$$\tau_{22} = \frac{c_{11}}{h}\frac{\partial U_2}{\partial \xi_2} + c_{12}\frac{\partial U_3}{\partial z}, \qquad \tau_{33} = \frac{c_{12}}{h}\frac{\partial U_2}{\partial \xi_2} + c_{11}\frac{\partial U_3}{\partial z}, \tag{32.1}$$

$$\tau_{23} = \tau_{32} = c_{44}\left(\frac{1}{h}\frac{\partial U_3}{\partial \xi_2} + \frac{\partial U_2}{\partial z}\right),$$

which is valid in the region $|\xi_2| < A/2$, $|z| < \infty$, with the functions U_2 and U_3 periodic in ξ_2 with period A; P_2 and P_3 are the ξ_2- and z-components of the (concentrated) external force acting at the origin.

Without going into the details of the derivation, the desired solution of (32.1) is

$$\begin{aligned} U_2(\xi_2, z) = &-\frac{P_2}{8G(1-\nu)Ah}\left[4(1-\nu)z + \frac{3-4\nu}{2\pi}Ah\ln\left(2\left|\cosh\left(\frac{2\pi z}{Ah}\right) - \cos\left(\frac{2\pi\xi_2}{A}\right)\right|\right) + \frac{z\sinh\left(\frac{2\pi z}{Ah}\right)}{\cosh\left(\frac{2\pi z}{Ah}\right) - \cos\left(\frac{2\pi\xi_2}{A}\right)}\right] \\ &+ \frac{P_3}{8G(1-\nu)Ah}\frac{z\sin\left(\frac{2\pi\xi_2}{A}\right)}{\cosh\left(\frac{2\pi z}{Ah}\right) - \cos\left(\frac{2\pi\xi_2}{A}\right)}, \\ U_3(\xi_2, z) = &\frac{P_2}{8G(1-\nu)Ah}\frac{z\sin\left(\frac{2\pi\xi_2}{A}\right)}{\cosh\left(\frac{2\pi z}{Ah}\right) - \cos\left(\frac{2\pi\xi_2}{A}\right)} \end{aligned} \tag{32.2}$$

$$+\frac{P_3}{8G(1-v)Ah}\left[-2(1-2v)z-\frac{3-4v}{2\pi}Ah\ln\left(2\left|\cosh\left(\frac{2\pi z}{Ah}\right)\right.\right.\right.$$

$$\left.\left.\left.-\cos\left(\frac{2\pi\xi_2}{A}\right)\right|\right)+\frac{z\sinh\left(\frac{2\pi z}{Ah}\right)}{\cosh\left(\frac{2\pi z}{Ah}\right)-\cos\left(\frac{2\pi\xi_2}{A}\right)}\right].$$

It should be remarked that the two-dimensional fundamental solution (32.2) cannot be deduced from the three-dimensional solution (31.16)–(31.18) and is therefore of interest by itself. Noting that this solution is only unique up to constant terms, it can be shown (see, for example, Crouch and Starfield 1983) that in the limit as A tends to infinity, it reduces to the familiar Kelvin solution for the plane deformation of an unbounded isotropic medium loaded by a concentrated force P_2, P_3 at the origin.

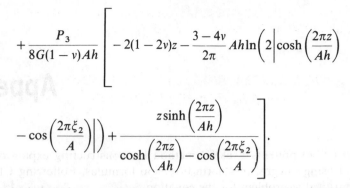

Appendix A

Ufland (1976) presents a useful method for constructing expansions on a composite interval using integral transform inversion formulas. Following Ufland we consider a boundary value problem for the equation

$$\frac{\partial^2 u}{\partial x^2} = \frac{\partial u}{\partial t} \tag{A.1}$$

on the interval $0 < x < \frac{1}{2}$ under the initial condition

$$u(x, 0) = f(x), \quad 0 < x < \tfrac{1}{2},$$

and the boundary conditions

$$\left.\frac{\partial u(x,t)}{\partial x}\right|_{x=0} = 0, \qquad \left.\frac{\partial u(x,t)}{\partial x}\right|_{x=\frac{1}{2}} = 0. \tag{A.2}$$

We also assume that both the functions $u(x, t)$ and its derivative, $\partial u(x, t)/\partial x$, undergo discontinuities at the inner point $x = \gamma < \frac{1}{2}$ of the interval $(0, \frac{1}{2})$, that is,

$$u(\gamma - 0, t) = \mu u(\gamma + 0, t), \tag{A.3}$$

$$\left.\frac{\partial u}{\partial x}\right|_{x=\gamma-0} = \nu \left.\frac{\partial u}{\partial x}\right|_{x=\gamma+0}, \tag{A.4}$$

where μ and ν are known numbers.

Taking the Laplace transform of (A.1) with respect to t and using the initial condition we find that

$$\frac{\partial^2 \bar{u}(x,p)}{\partial x^2} - p\bar{u}(x,p) = -f(x), \tag{A.5}$$

where

$$\bar{u}(x,p) = \int_0^\infty u(x,t)\mathrm{e}^{-pt}\,\mathrm{d}t.$$

It is required that the solution to (A.5) satisfy conditions (A.2) on $(0, \gamma)$ and $(\gamma, \frac{1}{2})$. We write

$$\bar{u}(x,p)|_{0<x<\gamma} = A_1(p)\cosh(\sqrt{p}x) - \frac{1}{\sqrt{p}}\int_x^\gamma f(\xi)\sinh(\sqrt{p}\xi)\cosh(\sqrt{p}x)\,\mathrm{d}\xi$$

$$-\frac{1}{\sqrt{p}}\int_0^x f(\xi)\cosh(\sqrt{p}\xi)\sinh(\sqrt{p}x)\,\mathrm{d}\xi, \tag{A.6}$$

$$\begin{aligned}\bar{u}(x,p)|_{\gamma<x<\frac{1}{2}} &= A_2(p)\cosh[\sqrt{p}(\tfrac{1}{2}-x)]\\ &-\frac{1}{\sqrt{p}}\int_\gamma^x f(\xi)\sinh[\sqrt{p}(\tfrac{1}{2}-\xi)]\cosh[\sqrt{p}(\tfrac{1}{2}-x)]\,\mathrm{d}\xi\\ &-\frac{1}{\sqrt{p}}\int_x^{1/2} f(\xi)\cosh[\sqrt{p}(\tfrac{1}{2}-\xi)]\sinh[\sqrt{p}(\tfrac{1}{2}-x)]\,\mathrm{d}\xi,\end{aligned} \tag{A.7}$$

and use the contact conditions (A.3) and (A.4) to obtain:

$$\begin{aligned}A_1(p) &= \frac{1}{\sqrt{p}\omega(p)}[\sinh(\sqrt{p}\gamma)[\sinh(\sqrt{p}(\tfrac{1}{2}-\gamma))\\ &+\delta\cosh(\sqrt{p}\gamma)\cosh(\sqrt{p}(\tfrac{1}{2}-\gamma))]\int_0^\gamma f(\xi)\cosh(\sqrt{p}\xi)\,\mathrm{d}\xi\\ &+\frac{\mu}{\sqrt{p}\omega(p)}\int_\gamma^{1/2} f(\xi)\cosh(\sqrt{p}(\tfrac{1}{2}-\xi))\,\mathrm{d}\xi,\end{aligned} \tag{A.8}$$

$$\begin{aligned}A_2(p) &= \frac{\delta}{\mu\sqrt{p}\omega(p)}\int_0^\gamma f(\xi)\cosh(\sqrt{p}\xi)\,\mathrm{d}\xi+\frac{1}{\sqrt{p}\omega(p)}[\cosh(\sqrt{p}(\tfrac{1}{2}-\gamma))\cosh(\sqrt{p}\gamma)\\ &+\delta\sinh(\sqrt{p}(\tfrac{1}{2}-\gamma))\sinh(\sqrt{p}\gamma)]\int_\gamma^{1/2} f(\xi)\cosh(\sqrt{p}(\tfrac{1}{2}-\xi))\,\mathrm{d}\xi,\end{aligned} \tag{A.9}$$

where

$$\delta=\frac{\mu}{\nu},\qquad \omega(p)=\sinh(\sqrt{p}(\tfrac{1}{2}-\gamma))\cosh(\sqrt{p}\gamma)+\delta\cosh(\sqrt{p}(\tfrac{1}{2}-\gamma))\sinh(\sqrt{p}\gamma). \tag{A.10}$$

The solution to (A.5) will then have the required discontinuities at $x=\gamma$ if we represent it in the form

$$\begin{aligned}\bar{u}(x,p)|_{0<x<\gamma} &= \mu\int_\gamma^{1/2} f(\xi)\frac{1}{\sqrt{p}\omega(p)}\cosh(\sqrt{p}(\tfrac{1}{2}-\xi))\cosh(\sqrt{p}x)\,\mathrm{d}\xi\\ &+\int_0^x f(\xi)\frac{\cosh(\sqrt{p}\xi)}{\sqrt{p}\omega(p)}[\sinh(\sqrt{p}(\tfrac{1}{2}-\gamma))\sinh(\sqrt{p}(\gamma-x))\\ &+\delta\cosh(\sqrt{p}(\tfrac{1}{2}-\gamma))\cosh(\sqrt{p}(\gamma-x))]\,\mathrm{d}\xi\\ &+\int_x^\gamma f(\xi)\frac{\cosh(\sqrt{p}x)}{\sqrt{p}\omega(p)}[\sinh(\sqrt{p}(\tfrac{1}{2}-\gamma))\sinh(\sqrt{p}(\gamma-\xi))\\ &+\delta\cosh(\sqrt{p}(\tfrac{1}{2}-\gamma))\cosh(\sqrt{p}(\gamma-\xi))]\,\mathrm{d}\xi,\end{aligned} \tag{A.11}$$

$$\bar{u}(x,p)|_{\gamma<x<\frac{1}{2}} = \frac{\delta}{\mu}\int_0^{\gamma} f(\xi)\frac{\cosh(\sqrt{p}\xi)\cosh(\sqrt{p}(\frac{1}{2}-x))}{\sqrt{p}\omega(p)}$$

$$+ \int_{\gamma}^{x} f(\xi)\frac{\cosh(\sqrt{p}(\frac{1}{2}-x))}{\sqrt{p}\omega(p)}[\cosh(\sqrt{p}\gamma)\cosh(\sqrt{p}(\xi-\gamma))$$

$$+ \delta\sinh(\sqrt{p}\gamma)\sinh(\sqrt{p}(\xi-\gamma))]\,\mathrm{d}\xi$$

$$+ \int_{x}^{1/2} f(\xi)\frac{\cosh(\sqrt{p}(\frac{1}{2}-\xi))}{\sqrt{p}\omega(p)}[\cosh(\sqrt{p}\gamma)\cosh(\sqrt{p}(x-\gamma))$$

$$+ \delta\sinh(\sqrt{p}\gamma)\sinh(\sqrt{p}(x-\gamma))]\,\mathrm{d}\xi. \tag{A.12}$$

From (A.11) and (A.12), reverting to originals and performing some manipulations, we find

$$u(x,t)|_{0<x<\gamma} = \frac{1}{(\frac{1}{2}-\gamma+\delta\gamma)}(\delta\int_0^{\gamma} f(\xi)\,\mathrm{d}\xi + \mu\int_{\gamma}^{1/2} f(\xi)\,\mathrm{d}\xi)$$

$$+ \sum_{n=1}^{\infty} \mathrm{e}^{-p_n^2 t}\cos(p_n x)\left[-2\frac{C_n^{(1)}}{C_n}\int_0^{\gamma} f(\xi)\cos(p_n\xi)\,\mathrm{d}\xi\right.$$

$$\left.+\frac{2\mu}{C_n}\int_{\gamma}^{1/2} f(\xi)\cos(p_n(\tfrac{1}{2}-\xi))\,\mathrm{d}\xi\right], \tag{A.13}$$

$$u(x,t)|_{\gamma<x<\frac{1}{2}} = \frac{1}{\mu(\frac{1}{2}-\gamma+\delta\gamma)}\left(\delta\int_0^{\gamma} f(\xi)\,\mathrm{d}\xi + \mu\int_{\gamma}^{1/2} f(\xi)\,\mathrm{d}\xi\right)$$

$$+ \sum_{n=1}^{\infty} \mathrm{e}^{-p_n^2 t}\cos(p_n x)\left(-\frac{C_n^{(2)}}{\mu}\left[-2\frac{C_n^{(1)}}{C_n}\int_0^{\gamma} f(\xi)\cos(p_n\xi)\,\mathrm{d}\xi\right.\right.$$

$$\left.\left.+\frac{2\mu}{C_n}\int_{\gamma}^{1/2} f(\xi)\cos(p_n(\tfrac{1}{2}-\xi)\,\mathrm{d}\xi\right)\right], \tag{A.14}$$

where p_n are the roots of the equation

$$\sin(p_n(\tfrac{1}{2}-\gamma))\cos(p_n\gamma) + \delta\cos(p_n(\tfrac{1}{2}-\gamma))\sin(p_n\gamma) = 0, \tag{A.15}$$

$$C_n^{(1)} = \sin(p_n(\tfrac{1}{2}-\gamma))\sin(p_n\gamma) - \delta\cos(p_n(\tfrac{1}{2}-\gamma))\cos(p_n\gamma), \tag{A.16}$$

$$C_n^{(2)} = \delta\sin(p_n(\tfrac{1}{2}-\gamma))\sin(p_n\gamma) - \cos(p_n(\tfrac{1}{2}-\gamma))\cos(p_n\gamma)$$

$$= -\frac{\cos(p_n\gamma)}{\cos(p_n(\frac{1}{2}-\gamma))}, \tag{A.17}$$

$$C_n = (\tfrac{1}{2}-\gamma+\delta\gamma)\cos[p_n(\tfrac{1}{2}-\gamma)]\cos(p_n\gamma)$$

$$-\left(\gamma+\frac{\delta}{2}-\delta\gamma\right)\sin[p_n(\tfrac{1}{2}-\gamma)]\sin(p_n\gamma). \tag{A.18}$$

The desired expansion is now found by setting $t = 0$ in (A.13) to give

$$f(x) = \tfrac{1}{2}A_0 \begin{cases} 1, & 0 < x < \gamma \\ \dfrac{1}{\mu}, & \gamma < x < \tfrac{1}{2} \end{cases} + \sum_{n=1}^{\infty} A_n \begin{cases} \cos(p_n x), & 0 < x < \gamma, \\ \dfrac{\cos(p_n\gamma)\cos(p_n(\tfrac{1}{2} - x))}{\mu\cos(p_n(\tfrac{1}{2} - \gamma))}, & \gamma < x < \tfrac{1}{2} \end{cases} \tag{A.19}$$

with

$$A_n = -\frac{2C_n^{(1)}}{C_n}\int_0^{\gamma} f(\xi)\cos(p_n\xi)\,\mathrm{d}\xi + \frac{2\mu}{C_n}\int_{\gamma}^{1/2} f(\xi)\cos[p_n(\tfrac{1}{2} - \xi)]\,\mathrm{d}\xi, \tag{A.20}$$

$$A_0 = \frac{2}{(\tfrac{1}{2} - \gamma + \delta\gamma)}\left(\delta\int_0^{\gamma} f(\xi)\,\mathrm{d}\xi + \mu\int_{\gamma}^{1/2} f(\xi)\,\mathrm{d}\xi\right). \tag{A.21}$$

It is readily seen that the expansion (A.19) defines a function with the required discontinuities at $x = \gamma$:

$$f(\gamma - 0) = \mu f(\gamma + 0), \qquad f'(\gamma - 0) = \nu f'(\gamma + 0). \tag{A.22}$$

If $\mu = \nu = 1$ and $\delta = 1$, it can be proved that (A.19) reduces to an ordinary Fourier series on $(0, \tfrac{1}{2})$:

$$f(x) = \tfrac{1}{2}A_0 + \sum_{n=1}^{\infty} A_n\cos(2\pi n x), \tag{A.23}$$

where

$$A_n = 4\int_0^{1/2} f(\xi)\cos(2\pi n\xi)\,\mathrm{d}\xi.$$

Appendix B

The method of integral transforms is known as a very effective and useful approach to the solution of boundary value problems in mechanics and mathematical physics. Therefore the generalization of this method for the applications to boundary value problems for a composite medium, i.e. for the case of rapidly oscillating coefficients of equations, can be considered as a rather useful and valuable result. A generalization of this kind in the case of Fourier integral transforms and series is considered here in accordance with the approach developed in Kalamkarov *et al.* (1991b). The generalized Fourier integral transforms and series can be applied to the analytical solution of heat conductivity and elasticity problems for laminated composite solids.

The following auxiliary problem is considered for the derivation of the new generalized integral transforms:

$$\frac{\partial}{\partial x}\left[A^{(\varepsilon)}(x)\frac{\partial u}{\partial x}\right]=B^{(\varepsilon)}(x)\frac{\partial u}{\partial t}, \tag{B.1}$$

$$u(x,0)=f(x). \tag{B.2}$$

Here $A^{(\varepsilon)}(x)=A(y)$ and $B^{(\varepsilon)}(x)=B(y)$ are rapidly oscillating functions, 1-periodic in $y=x/\varepsilon$, ε is a small parameter and $f(x)$ is a given piecewise-smooth and absolutely integrable function in the interval $(0,\infty)$.

Application of the Laplace transform to equation (B.1), taking account of the initial condition (B.2), yield

$$[A(y)\bar{u}']'-pB(y)\bar{u}=-B(y)f(x). \tag{B.3}$$

Here

$$\bar{u}=\int_0^\infty u(x,t)\mathrm{e}^{-pt}\,\mathrm{d}t,\quad \bar{u}'\equiv\frac{\partial\bar{u}}{\partial x}.$$

We denote by $\bar{u}_1$ and $\bar{u}_2$ two linearly independent solutions of the homogeneous equation

$$[A(y)\bar{u}']'-pB(y)\bar{u}=0. \tag{B.4}$$

The solution of the non-homogeneous equation (B.3) can be then represented in the form

$$\bar{u}(x,p)=\int_0^\infty B(\xi/\varepsilon)f(\xi)G(x,\xi,p)\,\mathrm{d}\xi, \tag{B.5}$$

where

$$G(x,\xi,p)=\begin{cases}-\dfrac{1}{A(\xi/\varepsilon)W}\bar{u}_1(\xi,p)\bar{u}_2(x,p), & \xi\leqslant x,\\[2ex] -\dfrac{1}{A(\xi/\varepsilon)W}\bar{u}_1(x,p)\bar{u}_2(\xi,p), & \xi\geqslant x.\end{cases}$$

$W=W(\bar{u}_1,\bar{u}_2)$ is the Wronskian of the functions $\bar{u}_1=\bar{u}_1(\xi,p)$ and $\bar{u}_2=\bar{u}_2(\xi,p)$.

Application of the inverse Laplace transform to (B.5) and substituting $t=0$ taking account of (B.2) yield

$$f(x)=\int_0^x B(\xi/\varepsilon)f(\xi)\hat{G}(x,\xi,0)\,\mathrm{d}\xi. \tag{B.6}$$

Here

$$\hat{G}(x,\xi,t)=\frac{1}{2\pi\mathrm{i}}\int_{\sigma-\mathrm{i}\infty}^{\sigma+\mathrm{i}\infty}G(x,\xi,p)\mathrm{e}^{pt}\,\mathrm{d}p. \tag{B.7}$$

The coefficients of equation (B.4) are periodic rapidly oscillating functions. It can be solved, therefore, by means of the asymptotic homogenization technique in the form of two-scale expansions (see Section 4):

$$\begin{aligned}\bar{u}_1=&\left[\sum_{k=0}^{\infty}\varepsilon^{2k}p^k\chi^{2k}N_{2k}(y)\right]\cosh(\sqrt{p}\chi x)\\ &+\left[\sum_{k=0}^{\infty}\varepsilon^{2k+1}p^{k+1/2}\chi^{2k+1}N_{2k+1}(y)\right]\sinh(\sqrt{p}\chi x),\end{aligned} \tag{B.8}$$

$$\bar{u}_2=\left[\sum_{k=0}^{\infty}(-1)^k\varepsilon^k\chi^k p^{k/2}N_k(y)\right]\mathrm{e}^{-\sqrt{p}\chi x}. \tag{B.9}$$

Here

$$\chi^2=\langle B\rangle\langle A^{-1}\rangle,\qquad \langle\varphi(y)\rangle=\int_0^1\varphi(y)\,\mathrm{d}y,$$

and 1-periodic functions $N_k(y)$ solve the following series of local problems:

$$\begin{aligned}&\frac{\mathrm{d}}{\mathrm{d}y}\left[A(y)\frac{\mathrm{d}N_k}{\mathrm{d}y}+A(y)N_{k-1}\right]+A\frac{\mathrm{d}N_{k-1}}{\mathrm{d}y}\\ &\quad+AN_{k-2}-\frac{B(y)}{\chi^2}N_{k-2}=0\quad(k=2,3,\ldots),\end{aligned} \tag{B.10}$$

$$N_0\equiv 1,\qquad N_1=-y+C_1\int_0^y A^{-1}(\xi)\,\mathrm{d}\xi+N_1^0,$$

where

$$C_1=1/\langle A^{-1}\rangle,\qquad N_1^0=\text{const}.$$

It can be shown that the Wronskian of the functions $\bar{u}_1$ and $\bar{u}_2$ (see (B.8) and (B.9)) can be represented in the following form:

$$W = A^{-1}(y)\left[-\chi\sqrt{p}C_1 + \chi \sum_{k=1}^{\infty} (\varepsilon\chi)^k p^{k+1/2} g_{2k} \right]. \tag{B.11}$$

Here

$$g_{2k} = \sum_{j=1}^{k} N_{2j-1} A\left(\frac{\mathrm{d}}{\mathrm{d}y} N_{2k-2j+2} + N_{2k-2j+1}\right)$$
$$- \sum_{j=0}^{k} N_{2k-2j} A\left(\frac{\mathrm{d}}{\mathrm{d}y} N_{2j+1} + N_{2j}\right). \tag{B.12}$$

It can be proved that the magnitudes g_{2k} are constant. The local functions $N_k(y)$ are determined from (B.10) up to the additive constants N_k^0. These constants can be determined from the recurrent series of algebraic equations

$$g_{2k} = 0 \quad (k = 1, 2, \ldots).$$

In particular, it is possible to assume

$$N_2^0 = N_4^0 = \cdots = 0$$

and determine $N_1^0, N_3^0, \cdots$ from the equations

$$g_2 = 0, \qquad g_4 = 0, \ldots .$$

The following simple formula for the Wronskian is then valid:

$$W = -A^{-1}(y)\chi\sqrt{p}C_1. \tag{B.13}$$

Let us now calculate the function $\hat{G}(x, \xi, t)$ (see (B.7)) taking account of (B.8), (B.9) and (B.13). The function $G(x, \xi, p)$ has a branching point $p = 0$ in the complex plane p and therefore

$$\hat{G}(x, \xi, t) = -\frac{1}{2\pi\mathrm{i}} \int_0^{\infty} G(x, \xi, \tau\mathrm{e}^{\mathrm{i}\pi})\,\mathrm{e}^{-\tau t}\,\mathrm{d}\tau + \frac{1}{2\pi\mathrm{i}} \int_0^{\infty} G(x, \xi, \tau\mathrm{e}^{-\mathrm{i}\pi})\,\mathrm{e}^{-\tau t}\,\mathrm{d}\tau.$$

Using formulae (B.8) and (B.9) we obtain:

$$\hat{G}(x, \xi, t) = \frac{2}{\pi\chi C_1} \int_0^{\infty} y_{\mathrm{c}}(x, \mu) y_{\mathrm{c}}(\xi, \mu)\,\mathrm{e}^{-\mu^2 t}\,\mathrm{d}\mu. \tag{B.14}$$

Here

$$y_{\mathrm{c}}(x, \mu) = \left[1 + \sum_{k=1}^{\infty} \varepsilon^k N_k(x/\varepsilon) \frac{\mathrm{d}^k}{\mathrm{d}x^k}\right] \cos(\chi\mu x). \tag{B.15}$$

Expressions (B.6) and (B.14) yield the following resulting integral transform:

$$f(x) = \left(\frac{2}{\pi}\right)^{1/2} \int_0^{\infty} F_{\mathrm{c}}(\mu) y_{\mathrm{c}}(x, \mu)\,\mathrm{d}\mu, \tag{B.16}$$

$$F_c(\mu) = \left(\frac{2\langle A^{-1}\rangle}{\pi\langle B\rangle}\right)^{1/2} \int_0^\infty B(\xi/\varepsilon) f(\xi)\, y_c(\xi,\mu)\, d\xi. \tag{B.17}$$

It is also proved that the function $y_c(x,\mu)$ (see (B.15)) solves the following equation:

$$\frac{\partial}{\partial x}\left[A(y)\frac{\partial}{\partial x} y_c(x,\mu)\right] + \mu^2 B(y) y_c(x,\mu) = 0. \tag{B.18}$$

The integral transform (B.16), (B.17) generalizes the classical Fourier cosine transform because in the case of constant coefficients $A(y) = B(y) = \text{const.}$ all the local functions $N_k(y)$, $(k = 1, 2, \dots)$, are equal to zero, and from (B.15) it follows that

$$y_c(x,\mu) = \cos(\mu x), \quad \chi = 1.$$

Let us now consider the application of the integral transform (B.16), (B.17) to the solution of the following boundary-value problem for a periodic laminated composite half-plane $x_2 \geqslant 0$:

$$\frac{\partial}{\partial x_1}\left[A^{(\varepsilon)}(x_1)\frac{\partial u}{\partial x_1}\right] + B^{(\varepsilon)}(x_1)\frac{\partial^2 u}{\partial x_2^2} = 0 \quad (x_2 > 0), \tag{B.19}$$

$$u(x_1, 0) = \varphi(x_1), \tag{B.20}$$

where $A^{(\varepsilon)}(x_1) = A(y)$ and $B^{(\varepsilon)}(x_1) = B(y)$ are 1-periodic functions in $y = x_1/\varepsilon$, $\varphi(x_1)$ is an even function and ε is a small parameter.

Application of the integral transform (B.16) yields

$$u = \left(\frac{2}{\pi}\right)^{1/2} \int_0^\infty V_c(x_2,\mu) y_c(x_1,\mu)\, d\mu. \tag{B.21}$$

From (B.19) and (B.20) taking account of (B.18) we obtain:

$$\partial^2 V_c/\partial x_2^2 - \mu^2 V_c = 0.$$

The bounded solution of this equation is

$$V_c(x_2,\mu) = C(\mu) e^{-\mu x_2}. \tag{B.22}$$

The boundary condition (B.20) and representation (B.21) yield

$$\left(\frac{2}{\pi}\right)^{1/2} \int_0^\infty C(\mu) y_c(x_1,\mu)\, d\mu = \varphi(x_1). \tag{B.23}$$

Application of the inverse transform (B.17) to (B.23) gives the explicit formula for the function $C(\mu)$:

$$C(\mu) = \left(\frac{2\langle A^{-1}\rangle}{\pi\langle B\rangle}\right)^{1/2} \int_0^\infty B(\xi/\varepsilon)\varphi(\xi) y_c(\xi,\mu)\, d\xi. \tag{B.24}$$

Formulae (B.21), (B.22) and (B.24) complete the analytical solution of the problem (B.19), (B.20).

The above technique can be used to derive integral transforms and series of various types. For example, the following generalized Fourier cosine series can be obtained:

$$f(x) = \frac{a_0}{2} + \sum_{n=1}^{\infty} a_n y_n^{(c)}(x), \quad 0 < x < l,$$

where

$$a_n = \frac{2}{l\langle B \rangle} \int_0^l f(\xi) B(\xi/\varepsilon) y_n^{(c)}(\xi)\,\mathrm{d}\xi \quad (n = 0, 1, 2, \dots),$$

$$y_n^{(c)}(x) = \left[1 + \sum_{k=1}^{\infty} \varepsilon^k N_k(y) \frac{\mathrm{d}^k}{\mathrm{d}x^k}\right] \cos\left(\frac{\pi n x}{l}\right).$$

This special function solves the equation

$$[A(y) y_n^{(c)\prime}]' = -B(y)\chi^{-2}(\pi n/l)^2 y_n^{(c)}$$

and satisfies the condition

$$y_n^{(c)}(0) = 1.$$

And 1-periodic local functions $N_k(y)$ are still determined from the above recurrent series of local problems (B.10).

The approach considered can also be used to derive the integral transforms and series in various cases of axial symmetry. For example, in Kalamkarov *et al.* (1991a) the generalized Fourier–Bessel series, generalized Hankel, Weber–Orr and Mellin integral transforms are derived and applied to some boundary value problems for axially symmetric multilayer composite solids.

References

Aleksandrov A. Ya. 1965, On the determination of reduced elastic properties of honeycomb fillers, in: *Raschety Elementov Aviats Konstr.* Mashinostroenie, Moscow, No. 4, pp. 59–70.

Aleksandrov A. Ya., Bryukker L. E., Kurshin L. M. and Prusakov A. P., 1960, *The Design of Three-Layered Panels* Oborongiz, Moscow.

Alekseev V. M., Tikhomirov V. M. and Fomin S. V., 1979, *Optimal Control* Nauka, Moscow.

Ambartsumyan S. A., 1974, *General Theory of Anisotropic Shells* Nauka, Moscow.

Ambartsumyan S. A., 1987, *Theory of Anisotropic Plates* Nauka, Moscow.

Andrianov I. V., Lesnichaya V. A. and Manevich L. I., 1985, *Homogenization Method in the Statics and Dynamics of Ribbed Shells* Nauka, Moscow.

Andrianov I. V. and Manevich L. I., 1983, Shell design using the homogenization method, *Uspekhi Mekh.*, **6**, 3–29.

Annin B. D., Kalamkarov A. L. and Kolpakov A. G., 1990, Analysis of local stresses in high-modulus fibre composites. *Proc. Int. Conf. on Localized Damage Computer-aided Assessment and Control*, Vol. 2 Comput. Mechanics Publ., Southampton, pp. 231–244.

Artola M. and Duvaut G., 1977, Homogénéisation d'une plaque reinforcée, *C. R. Acad. Sci., Sér. A*, **284**, 707–710.

Babuška I., 1976, Solution of interface problems by homogenization. Parts I, II, *SIAM*, J. Math. Anal., **7**, 603–645.

Babuška I., 1977, Solution of interface problems by homogenization. Part III, *SIAM, J. Math. Anal.* **8**, 923–931.

Bakhvalov N. S. and Panasenko G. P., 1984, *Homogenization in Periodic Media.* Mathematical Problems of the Mechanics of Composite Materials Nauka, Moscow.

Banerjee P. K. and Butterfield R., 1981, *Boundary Element Methods in Engineering Science* McGraw-Hill, New York.

Bateman H. and Erdélyi A., 1954, *Tables of Integral Transforms*, Vol. II McGraw-Hill, New York, Toronto, London.

Bensoussan A., Lions J.-L. and Papanicolaou G., 1978, *Asymptotic Analysis for Periodic Structures*, Vol. 5 North-Holland, Amsterdam.

Berdichevskii V. L., 1983, *Variational Principles in Continuum Mechanics* Nauka, Moscow.

Birger I. A., 1961, *Circular Plates and Shells of Rotation* Oborongiz, Moscow.

Bourgat J. F., 1979, Numerical experiments of the homogenization method for operators with periodic coefficients, in: *Computing Methods in Applied Sciences and Engineering*, 3rd. Int. Symp., 1977, Lect. Notes Math., No. 704 Springer, Berlin, pp. 330–356.

Caillerie D., 1981a, Equations de la diffusion stationnaire dans un domaine comportant une distribution périodique d'inclusions aplaties de grande conductivité, *C. R. Acad. Sci., Sér. 1*, **292**, No. 1, 115–118.

Caillerie D., 1981b, Homogénéisation des équations de la diffusion stationnaire dans les domaines cylindriques aplatis, *RAIRO Anal. Numér.*, **15**, 295–319.

Caillerie D., 1982, Plaques elastique minces à structure périodique de période et d'épaisseur comparables, *C. R. Acad. Sci., Sér. 2* **294**, 159–162.

Caillerie D., 1984, Thin elastic and periodic plates, *Math. Meth. in the Appl. Sci.*, **6**, 159–191.

Caillerie D., 1987, Nonhomogeneous plate theory and conduction in fibered composites, in: *Lect. Notes in Physics*, Vol. 272 Springer, Berlin, pp. 1–62.

Cherepanov G. P., 1983, *Fracture Mechanics of Composite Materials* Nauka, Moscow.

Christensen R. M., 1979, *Mechanics of Composite Materials* Wiley, New York.

Ciarlet Ph. G. and Rabier P., 1980, Les equations de von Kármán, in: *Lect. Notes in Math.*, Vol. 826 Springer, Berlin.

Cioranescu D. and Paulin J., 1979, Homogenization in open sets with holes, *J. Math. Anal. and Appl.*, **71**, 590–607.

Crouch S. L. and Starfield A. M., 1983, *Boundary Element Methods in Solid Mechanics* Allen & Unwin, London, Boston, Sydney.

Duvaut G., 1976, Analyse fontionnelle et méchanique des milieux continus. Application à l'étude des materiaux composites élastiques a structure périodique-homogénéisation, in: *Proc. Theor. and Appl. Mech. Prepr., 14th IUTAM Congr* (Delft 1976) Amsterdam, pp. 119–132.

Duvaut G., 1977, Comportement macroscopique d'une plaque perforée périodiquement, in: *Lect. Notes Math.*, No. 594, pp. 131–145.

Duvaut G. and Metellus A.-M., 1976, Homogénéisation d'une plaque mince en flexion de structure périodique et symétrique, *C. R. Acad. Sci.*, Sér. A., **283**, 947–950.

Endogur A. I., Weinberg M. V. and Yerusalimskii K. M., 1986, *Honeycomb Structures: Characterization and Design* Mashinostroenie, Moscow.

Ene H. I., 1983, On linear thermoelasticity of composite materials, *Int. J. Eng. Sci.*, **21**, 443–448.

Eshelby J. D., 1957, The determination of the elastic field of an ellipsoidal inclusion, and related problems, *Proc. Roy. Soc. Lond.*, **A241**, 376.

François D., Pineau A. And Zaoui A., 1991, *Comportement Mécanique des Matériaux*, Hermes, Paris.

Goldenblat I. I. and Kopnov V. A., 1968, *Strength and Plasticity Criteria in Structural Materials* Mashinostroenie, Moscow.

Gorbachev V. I., 1979, On the elastic equation of a cylindrical pipe of varying thickness under the action of surface loads and displacements, *Probl. Prochn.*, No. 5, 79–83.

Gorbachev V. I. and Pobedrya B. E., 1985, On some fracture criteria for composite materials, *Izv. Akad. Nauk Arm. SSR (Mekh.)*, **38**, 30–37.

Grigolyuk E. I. and Fil'shtinskii L. A., 1970, *Perforated Plates and Shells* Nauka, Moscow.

Grigolyuk E. I. and Kabanov V. V., 1978, *Stability of Shells* Nauka, Moscow.

Guz' A. N., 1986, *Three-Dimensional Stability of Deformable Solids* Vyshcha Shkola, Kiev.

Hashin Z., 1962, The elastic moduli of heterogeneous materials, *J. Appl. Mech.*, **29**, 143.

Hashin Z., 1983, Analysis of composite materials. A survey, *Trans. ASME: J. Appl. Mech.*, **50**, 481–505.

Hashin Z. and Rosen B. W., 1964, The elastic moduli of fiber-reinforced materials, *J. Appl. Mech.*, **31**, 223.

Hashin Z. and Shtrikman S., 1962, A variational approach to the theory of the effective magnetic permeability of multiphase materials, *J. Appl. Phys.*, **33**, 3125.

Jones R. M., 1975, *Mechanics of Composite Materials* McGraw-Hill, New York.

Kalamkarov A. L., 1986, Heat transfer from gear wheels with wavy or toothed surfaces, in: *Strength and Reliability in Chemical Equipment* Moscow Chemical Engineering Inst., Moscow, pp. 84–89.

Kalamkarov A. L., 1987a, On the determination of the effective properties of network plates and shells with periodic structure, *Izv. Akad. Nauk SSSR* (*Mekh. Tv. Tela*), No. 2, 181–185.

Kalamkarov A. L., 1987b, Thermal conductivity of a curvilinear nonhomogeneous anisotropic layer with periodic structure and wavy surfaces, *Inzh.-Fiz. Zh.*, **52**, 865–866.

Kalamkarov A. L., 1988a, Geometrical nonlinear problem of a thin layer of composite material with wavy surfaces and periodic structure, *Izv. Akad. Nauk SSSR* (*Mekh. Tv. Tela*), No. 5, 42–47.

Kalamkarov A. L., 1988b, On the design of multilayered strengthened shells: in: *Chemical Industry Equipment Design* Moscow Chemical Engineering Inst., Moscow, pp. 88–91.

Kalamkarov A. L., 1988c, Thermal conductivity of network plates and shells with regular structure, *Inzh.-Fiz. Zh.*, **54**, 510.

Kalamkarov A. L., 1989, Thermoelasticity problem for a structurally nonhomogeneous shells of regular structure, *Zh. Prikl. Mekh. i Tekhn. Fiz.*, No. 6, 150–157.

Kalamkarov A. L., Kudryavtsev B. A. and Bardzokas D. Ya., 1991a, New generalized integral transforms in axially symmetric boundary-value problems of composite mechanics, *Mekh. Komp. Mater.*, No. 6, 1005–1014.

Kalamkarov A. L., Kudryavtsev B. A. and Parton V. Z., 1987a, The asymptotic method of homogenization in the mechanics of composites with regular structure, *Itogi Nauki i Tekhn. VINITI.* (Mekh. Deform. Tv. Tela), **19**, 87–147.

Kalamkarov A. L., Kudryavtsev B. A. and Parton V. Z., 1987b, A curved composite material layer with wavy surfaces of periodic structure, *Prikl. Mat. i Mekh.*, **51**, 68–75.

Kalamkarov A. L., Kudryavtsev B. A. and Parton V. Z., 1987c, Thermoelasticity of a regularly nonhomogeneous layer with wavy surfaces, *Prikl. Mat. i Mekh.*, **51**, 1000–1008.

Kalamkarov A. L., Kudryavtsev B. A. and Parton V. Z., 1987d, Thermoelasticity of strengthened composite plates and shells of regular structure, in: *Proc. XIV Nat. Conf. on Plate and Shell* Theory, Vol. II Tbilisi Univ., Tbilisi, pp. 21–26.

Kalamkarov A. L., Kudryavtsev B. A. and Parton V. Z., 1989, Fundamental solution for a doubly periodic three-dimensional elasticity problem, *Izv. Akad. Nauk SSSR* (*Mekh. Tv. Tela*), No. 3, 44–50.

Kalamkarov A. L., Kudryavtsev B. A. and Parton V. Z., 1990, A new boundary-layer approach to the fracture mechanics of composites, *Prikl. Mat. i Mekh.*, **54**, 322–328.

Kalamkarov A. L., Kudryavtsev B. A. and Parton V. Z., 1991b, Generalization of Fourier integral transform in boundary-value problems for periodic nonhomogeneous medium, C. R. de *l'Acad. des Sci., Paris, Sér. I*, **312**, 309–313.

Karimov A. M., 1986, Free vibrations of an elastic composite layer of periodic structure, *Vectnik MGU* (*Mat. i Mekh.*), No. 3, 106–108.

Kesavan S., 1979a, Homogenization of elliptic eigenvalue problems, Part I, *Appl. Math. and Optim.*, **5**, 153–167.

Kesavan S., 1979b, Homogenization of elliptic eigenvalue problems. Part II, *Appl. Math. Optim.*, **5**, 197–217.

Kohn R. V. and Vogelius M., 1984, A new model for thin plates with rapidly varying thickness, *Int. J. Solids and Struct.*, **20**, 333–350.

Kohn R. V. and Vogelinus M., 1985, A new model for thin plates with rapidly varying thickness, II: A convergence proof, *Quart. J. Appl. Math.*, **43**, 1–22.

Kohn R. V. and Vogelius M., 1986, A new model for thin plates with rapidly varying thickness, III: Comparison of different scalings, *Quart. J. Appl. Math.*, **44**, 35–48.

Kolpakov A. G. and Rakin S. I., 1986, On the synthesis of one-dimensional composite materials with prescribed characteristics, *Zh. Prikl. Mekh. i Tekhn. Fiz.*, No. 6, 143–150.

Korolev V. I., 1971, *Elastoplastic Deformation in Shells* Mashinostroenie, Moscow.

Kovalenko A. D., 1970, *Fundamentals of Thermoelasticity* Naukova Dumka, Kiev.

Leont'ev N. V., 1984, Reduction problem for the composite material of a superconducting magnetic system, *Prikl. Probl. Prochn. i Plast.*, No. 27, 74–79.

Lewinski T., 1991, Effective models of composite periodic plates, *Int. J. Solids and Struct.*, **27**, 1155–1203.

Liebowitz H. (Editor), 1968, *Fracture, Vol. II. Mathematical Fundamentals* Academic Press, New York, London.

Lions J.-L., 1981, *Some Methods in the Mathematical Analysis of Systems and their Control* Gordon and Breach, New York.

Lions J.-L., 1985, Remarques sur les problèmes d'homogénéisation dans les milieux à structure périodique et sur quelques problèmes raides, in: *Les Méthods de l'Homogénéisation: Théorie et Applications en Physique* Eyrolles, Paris, pp. 129–228.

Lions J.-L., 1987, Homogenization and reinforced structures, in: *Proc. 2nd Int. Symp. Struct. Contr.*, Waterloo, July, 1985 Dordrecht, pp. 426–445.

Love A. E. H., 1944, *A Treatise on the Mathematical Theory of Elasticity* Dover, New York.

Lukovkin G. M., Volynskii A. L. and Bakeev N. F., 1983, Rubber dispersion mechanism for enchancing the impact strength of plastics, *Vysokomolek, Soed.* (A), **25**, 848–855.

Maksimov R. D., Plume E. Z. and Ponomarev V. M., 1983, Elastic properties of unidirectionally reinforced hybrid composites, *Mekh. Komp. Mater.*, No. 1, 13–19.

Malmeister A. K., Tamuzs V. P. and Teters G. A., 1980, *Resistance Ability of Polymeric and Composite Materials* Zinatne, Riga.

Mol'kov V. A. and Pobedrya B. E., 1985, Effective properties of a unidirectional fiber composite with a periodic structure, *Izv. Akad. Nauk SSSR* (*Mekh. Tv. Tela*), No. 2, 119–130.

Mol'kov V. A. and Pobedrya B. E., 1988, Effective elastic properties of a composite with an elastic contact between the fibers and the binder, *Izv. Akad. Nauk SSSR* (*Mekh. Tv. Tela*), No. 1, 111–117.

Mushtari Kh. M. and Galimov K. Z., 1957, *Nonlinear Theory of Elastic Shells* Tatknigoizdat, Kazan'.

Muskhelishvili N. I., 1963, *Some Basic Problems of the Mathematical Theory of Elasticity* Noordhoff, Groningen.

Nayfeh A. H., 1973, *Perturbation Methods* Wiley, New York, London, Sydney Toronto.

Nemat-Nasser S., Iwakuma T. and Hejazi M., 1982, On composites with periodic structure, *Mech. of Mater.*, **1**, 239–267.

Novozhilov V. V., 1948, *Fundamentals of Nonlinear Elasticity* Gostekhizdat, Moscow–Leningrad.

Novozhilov V. V., 1962, *Theory of Thin Shells* Sudpromgiz, Leningard.

Nowacki W., 1970, *Teoria sprężystości* Panstwowe Wydawnictwo Naukowe, Warszawa.

Nunan K. C. and Keller J. B., 1984, Effective elasticity tensor of a periodic composite, *J. Mech. and Phys. Solids*, **32**, 259–280.

Obraztsov I. F., Vasil'ev V. V. and Bunakov V. A., 1977, *Optimal Reinforcement of Composite Material Shells of Rotation* Mashinostroenie, Moscow.

Panasenko G. P., 1979, Higher order asymptotic analysis of the contact problems for periodic structures, *Mat. Zbornik*, **110**, 505–538.

Panasenko G. P. and Reztsov M. V., 1987, Homogenization of the three-dimensional elasticity problem for a nonhomogeneous plate, *Dokl. Akad. Nauk SSSR*, **294**, 1061–1065.

Parton V. Z. and Kalamkarov A. L., 1988a, Thermoelasticity of a regularly nonhomogeneous thin curved layer with rapidly varying thickness, *J. Thermal Stresses*, **11**, 405–420.

Parton V. Z. and Kalamkarov A. L., 1988b, On the stability equations for three-dimensional composite material solids, *Dokl. Akad. Nauk SSSR*, **300**, 308–311.

Parton V. Z. and Kalamkarov A. L., 1992, The general homogenized composite shell model and the applications, *Int. J. Solids and Struct.*, **29**, 1947–1955.

Parton V. Z., Kalamkarov A. L. and Koudriavtsev B. A., 1989, Métode d'homogénéisation en mécanique des solides déformables à microstructure régulière, *C. R. de l'Acad. des Sci., Paris, Sér. II*, **309**, 641–646

Parton V. Z., Kalamkarov A. L. and Kudryavtsev B. A., 1988, On the study of microstresses in the neighbourhood of a crack in a composite material, in: *Failure Analysis—Theory and Practice*, Vol. I, Proc. 7th European Conf. on Fracture Budapest, Hungary, 1988, pp. 427–432.

Paşa G. I., 1983, A convergence theorem for a periodic media with thermoelastical properties, *Int. J. Eng. Sci.*, **21**, 1313–1319.

Pobedrya B. E., 1984, *Mechanics of Composite Materials* Moscow State Univ., Moscow.

Pobedrya B. E., 1985, Zeroth approximation theory in the mechanics of a nonhomogeneous deformable solid, *Math. Methods and Physical and Mechanical Fields* (*Kiev*), No. 22, 34–40.

Pobedrya B. E. and Gorbachev V. I., 1977, On static problems in the elasticity of composite materials, *Vestnik MGU* (*Mat. i Mekh*), 101–110.

Pobedrya B. E. and Sheshenin S. V., 1979, On the influence matrix, *Vestnik, MGU* (*Mat. i. Mekh.*), 76–81.

Podstrigach Ya. S., Lomakin V. A. and Kolyano Yu. M., 1984, *Thermoelasticity of Nonhomogeneous Structures* Nauka, Moscow.

Podstrigach Ya. S. and Shvets R. N., 1978, *Thermoelasticity of Thin Shells* Naukova Dumka, Kiev.

Pontryagin L. S., Boltyanskii V. G., Gamkrelidze R. V. and Mishchenko E. F., 1976, *Mathematical Theory of Optimal Processes*, Nauka, Moscow.

Prudnikov A. P., Brychkov Yu. A. and Marichev O. I., 1983, *Integrals and Series. Special Functions* Nauka, Moscow.

Pshenichnov G. I., 1982, *Theory of Thin Elastic Network Plates and Shells* Nauka, Moscow.

Russel W. B. and Acrivos A., 1972, On the effective moduli of composite materials: slender rigid inclusions at dilute concentrations, *Z. Angw. Math. und Phys.*, **23**, 434.

Sanchez-Palencia E., 1980, *Non-Homogeneous Media and Vibration Theory*, Lect. Notes in Physics, Vol. 127 Springer, Berlin.

Sanchez-Palencia E., 1987, *Boundary Layers and Edge Effects in Composites*, Lect, Notes in Physics, Vol. 272 Springer, Berlin.

Sedov L. I., 1972, *A Course in Continuum Mechanics* Wolters-Noordhoff, Groningen.

Sendeckyj G. P., 1974, Elastic behavior of composites, in: *Composite Materials*, Vol. 2. *Mechanics of Composite Materials* Acadamic Press, New York.

Sheshenin S. V., 1980, Overall moduli of one composite, *Vestnik MGU* (*Mat. i Mekh.*), 79–83.

Tamuzs V. P. and Protasov V. D. (Editors), 1986, *Fracture of Composite Material Constructions* Zinatne, Riga.

Tartar L., 1978, Nonlinear constitutive relations and homogenization, in: *Proc. Int. Symp. Contemp. Develop. Cont. Mech. and Part. Differ. Equat.* (Rio de Janeiro, 1977) Amsterdam, pp. 472–484.

Ufland Ya. S., 1976, On some new integral transforms and their application to problems in mathematical physics, in: *Problems of Mathematical Physics*, Nauka, Moscow, pp. 93–105.

Vanin G. A., 1985, *Micromechanics of Composite Materials* Naukova Dumka, Kiev.

Vasil'ev V. V., 1988, *Mechanics of Composite Material Constructions* Mashinostroenie, Moscow.

Vasil'ev V. V. and Tarnopol'skii Yu. M., 1990, *Composite Materials. A Handbook* Mashinostroenie, Moscow.

Vinson J. R. and Sierakowski R. L., 1986, *The Behavior of Structures Composed of Composite Materials* Nijhoff Publ., Dordrecht.

Whittaker E. T. and Watson G. N., 1927, *A Course of Modern Analysis*, Part II Cambridge Univ. Press.

Zheludev I. S., 1968, *Physics of Crystalline Dielectrics*, Nauka, Moscow.

Index